Commission of the European Communities

toxicology

Opinion of the Scientific Committee on Cosmetology (11/86 — 10/90)

Commission Decision 78/45/EEC of 19 December 1977 concerning the institution of a Scientific Committee on Cosmetology (OJ L 13, 17.1.1978, p. 24)

Scientific Committee on Cosmetology

Directorate-General
Employment, Industrial Relations and Social Affairs

Directorate-General
Environment, Nuclear Safety and Civil Protection

1993

EUR 14208 EN

Published by the
COMMISSION OF THE EUROPEAN COMMUNITIES
Directorate-General XIII
Information Technologies and Industries, and Telecommunications
L-2920 Luxembourg

Cataloguing data can be found at the end of this publication

Luxembourg: Office for Official Publications of the European Communities, 1993

ISBN 92-826-5394-3

Printed in Belgium

CONTENTS

FOREWORD

The Scientific Committee on Cosmetology was set up by Commission Decision 78/45/EEC on 19 December 1977 (OJ N° L 13 of 17 January 1978, p. 24) in order to provide the Commission with informed opinions on any scientific and technical problems arising in connection with cosmetic products, and in particular on the substances used in their manufacture, on their composition and on the conditions for their use.

The members of the Committee are independent scientists highly qualified in the fields of medicine, toxicology, biology, chemistry or other similar disciplines.

The Committee is serviced by the Consumer Policy Service.

CLASSIFICATION OF INGREDIENTS

Ingredients are classified by the Scientific Committee on Cosmetology into four groups on the basis of assessment of the information provided in accordance with the guidelines.

GROUP A

Contains ingredients for which full data have been provided, and on the basis of which they do not present any health hazard, and which therefore may be used in cosmetic products for the purpose stated and at concentrations not exceeding the limits recommended.

GROUP B

Contains ingredients which, in the light of the data provided so far, do not appear to present a health hazard and which therefore may be used temporarily in cosmetic products for the purpose stated and at concentrations not exceeding the limits recommended ; additional information is needed for complete assessment.

GROUP C

Contains ingredients which have been submitted to the Scientific Committee on Cosmetology but which, in the light of the data provided, could not be evaluated, or for which the additional information requested under Group B has not been received within the prescribed time limit and, as a result cannot at present be recommended for use in cosmetic products.

GROUP D

Contains ingredients which, on the basis of the information provided, present a health hazard and which are therefore not recommended for use in cosmetic products.

Assessment have been made on the basis of the concentrations of ingredients requested by industry.

CONTENTS

32nd meeting of the SCC - November 6-7, 1986

REPORTS OF THE SCIENTIFIC COMMITTEE ON COSMETOLOGY

32d reunion

November 6 -7 1986

SUMMARY

ANTIOXIDANT AGENTS

		Class
Ascorbic acid and its salts	A01	A
Ascorbyl stearate and palmitate	A02	B
Isoascorbic acid and its salts	A04	D

HAIR DYES

2,4 diaminoanisole	A6	D
2,5 diaminoanisole	A34	D

SCIENTIFIC COMMITTEE ON COSMETOLOGY

List of participants

32nd meeting - 6/7 November 1986

Present

Mr DE GROOT

Mr GOULDING

Mr HILDEBRANDT

Mr LOPRIENO

Mr SCHOU

Mr O'MAHONY

Mr STUTTGEN

Mr AGACHE

Mrs DONY

Mrs KNAAP

Mrs ENJOLRAS

Apologies for absence

Mr MUSCARDIN

Mrs MACKIE

Commission

Mrs MASSE

Mr COLLIN)

Mr GONTIER) (DG XI/B/1)

REPORT FROM THE SCIENTIFIC COMMITTEE ON COSMETOLGY

CONCERNING CERTAIN ANTIOXIDANTS

THE COMMITTEE'S MANDATE

To give its opinion on the question whether the use of certain antioxidants in cosmetic products is admissible from the health point of view.

- A01 L. ASCORBIC ACID
- A02 L. ASCORBYL PALMITATE
  STEARATE
- A04 D. ISOASCORBIC ACID

at the maximal concentration of 2%

(1) L. A S C O R B I C A C I D XI/496/87

* Formula and synonyms

Colipa A01
CAS - RN 50-81-7
L. ascorbic acid (8 CI,9 CI)
3 - ceto - L - gulofuranolactone

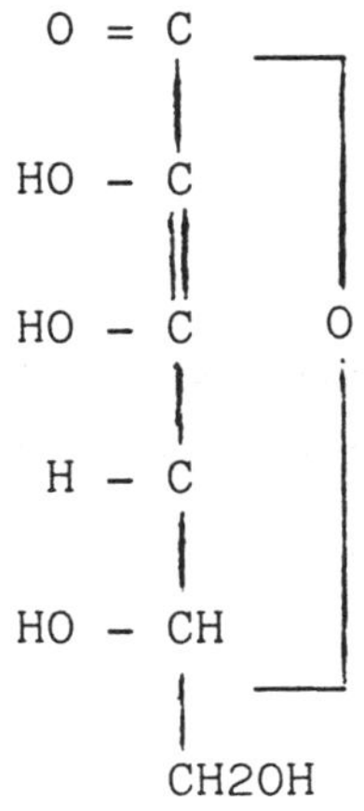

C6H802
MW : 176,12

* Characteristics

Soluble in water (10% W/V)
Slightly soluble in ethanol
Insoluble in oils and fats

* Use

Used in cosmetic products in concentrations not exceeding 2%. It has an antioxidant action by blocking the free radicals by UV radiations and prevents the formation of nitrosamines; its antinitrosating properties seem to correlate with the speed of oxidation indehydroascorbic acid.

* Recapitulation of the studies of toxicity

LD 50 varies from + 200 mg/kg to + 5 gr/kg depending on the animal tested and the route of administration.

Metabolism

Ascorbic acid is rapidly excreted and metabolized. In rats, after intraperitoneal injection of 1.5 to 5.9 mg of 14C-labelled ascorbic acid,19 to 29 % is converted to CO_2 and 0,4 % excreted asoxalate within 24 hours.

In man,oxalate formation and urinary excretion represent about half the quantities ingested,the rest being excreted as CO_2.

ORAL TOXICITY

In the short-term no effects were observed in rats by oral administration of 6.5 g/kg/bw for 10 weeks but 27.3 g/kg/bw/day caused 77 % mortality in four weeks.

In man, administered at a doe of 1 000 mg/day for three months it caused no toxic effects ; at a dose of 6 000 mg/day for 400 days nausea, vomiting, diarrhoea, headaches, fatigue, skin rashes and disturbed sleep were observedIn small doses it had a diuretic effect.

CUTANEOUS AND MUCOUS TOLERANCE

Ascorbic acid is well tolerated by the skin and mucous membranes and is not sensitizing.

In an eye irritation test (modified Draize) on rabbits, 0.1 ml of a 30 % aqueus solution instilled without rise-off produced slight to well defined redness, returning to normal in 72 hours. The cornea and iris were unaffected.

In a skin irritation test the application of 0.5 ml of a 30% aqueous solution with occlusion to the intact and abraded skin of a rabbit's shaved back produced barely perceptible erythema but no oedema.

In a Maurer optimization test on female guinea pigs, after intradermal injection with and without Freund's Complete Adjuvant, no evidence was observed of sensitization by intradermal injection of 0.1 ml of a 0,1 % physiological saline, or by occluded epicutaneous application of 0.05 ml/cm2 of a 10% dispersion in petroleum ether.

MUTAGENICITY

For bacteria ascorbic acid is not intrinsically mutagenic to *Salmonella typhimurium* (TA 100) if deionized water is used or if EDTA is added to tap water for the preparation of the incubation medium. But a mutagenic and cytotoxic effect is observed through generation of H2O2, notably in the presence of Cu++.
In an Ames test on strains TA 92, TA 1535, TA 100, TA 1537,

TA 94, TA 98 and TA 2637,ascorbic acid in solution in a phosphate buffer with the addition of an S9 fraction,did not produce a significant increase in the number of revertants, at the maximum non-cytotoxic dose of 5 mg/plate.

In vitro on mammal cells,ascorbic acid produces chromosome breaking and chromosome exchange in the ovary cells of Chinese hamsters,mutagenic doses being also toxic doses; the peroxide radicals are involved in the process.It produces a dose-dependent increase in sister-chromatid exchanges (SCES) in Chinese hamster ovary (CHO) cells and in human lymphocytes at concentrations from 10^{-4} to 10^{-2} M.

It reduces the rate of DNA synthesis in HeLa cells at the effective dose of 2.5 - 10^{-3} M ,showing no results typical of products producing DNA damage.

In a test on Chinese hamster fibroblasts,it did not produce any chromosome aberration,in a physiological solution at 0.3 mg/ml,but at non-toxic concentrations,with or without Cu^{2+},it produced a relatively high frequency of chromosome aberration in human fibroblasts.

A deactivating effect of the viruses depending on the presence of oxygen has been observed; the target molecule is DNA.

In vivo,a test via an intrahepatic intermediate host,after i.v. injection of S. typhimurium TA 100,in guinea pigs,did not indicate any genotoxic activity even at very high concentrations (5 000 mg/kg/bw/day).In contrast to the in vitro results,no sister-chromatid exchanges (SCEs) were observed in the spinal cord of the Chinese hamster for doses of 200 to 10 000 mg/kg/bw administered orally and intraperitoneally.

In Drosophila melanogaster,ascorbic acid induces lethal sex-linked recessive mutations.

The results of various systems of tests on bacteria and mammal cells show that ascorbic acid can have a comutagenic or anti-mutagenic effect.

CARCINOGENICITY

Ascorbic acid is not carcinogenic in long-term tests on animals. A two-year oral administration study on male and female rats with daily doses of 0, 1 000, 1 500 and 2 000 mg/kg/bw had no effect on haematology,urinary chemistry,blood enzyme activity or liver and kidney function,and caused no microscopic or macroscopic organ lesions

A bioassay of carcinogenicity by administration in the diet to male and female rats and mice of doses of25 000 and 50 000 ppm for 103 weeks indicated that ascorbic acid was neither toxic nor carcinogenic. Furthermore,a dose-related decline of certain spontaneous lesions in older female rats was observed and at the higher dose the number of male survivors treated (mice and rats) was greater thanthe number of survivors monitored.

It can inhibit or promote experimental carcinogenicity.In particular it reduces the incidence of skin tumours caused by chemical carcinomas from UV radiation.

TERATOGENICITY AND REPRODUCTION TOXICITY

Ascorbic acid is neither foetotoxic nor teratogenic.

In a reproduction study on three generations of guinea pigs (by administrationof 1.5 - 4.0 and 100 mg/kg/bw) an increase in the number of young was observed at 100 mg/kg/bw . The tissues of the newborn babies of mothers exposed to 1.5 and 4.0 mg/kg/bw were less satured than the mothers' tissues. There was no difference as regards the babies' size and viability.

There was no toxic effect on the mothers, no foetotoxic or teratogenic effects, nor adverse effects on post-partum development in a study involving oral administration of 250, 500 and 1 000 mg/kg/bw/day to mice,and 150 ,250 and 1 000 mg/kg/bw/day to rats,from day 6 to day 15 of gestation,and thereafter from days 0 to 21 after birth.

Similarly, there were no adverse effects on the mother or foetus, nor any increases in soft-tissue or skeletal malformations, after administration of 5.2 to 520 mg/kg of ascorbic acid, by oral intubation to pregnant CD-1 strain female albino mice from day 6-15 of gestation.

In a similar study on Wistar rats the incidence of complete closure of of the skull was greatest in the high-dose foetuses.

Ascorbic acid used in vitro on mouse embryo (limb bud) cell cultures has no measurable effect on the inhibition of proteoglycane synthesis, the extracellular matrix or the synthesis of cellular DNA.

C O N C L U S I O N

L(+) ascorbic acid is minimally toxic; it is well tolerated by the skin and mucous membranes and is not sensitizing. It is quickly metabolized and excreted. Its cutaneous absorption is not known. It has a mutagenic power in certain systems and is antimutagenic in other test systems; it may interact with DNA, inhibits cell transformation, causes mutations in the fruit fly and chromosome damage in mammal cells, but is inactive on bacteria and in vivo tests. It is not carcnogenic for animals, and can inhibit or promote experimental carcinogenesis.

It has no effect on reproduction, and is neither foetotoxic nor terato genic.

In cosmetics, it can prevent the formation of free radicals by peroxidation, and can inhibit nitrosamine formation.

In 1981 JEFCA changed the ADI, which had previously been set at 0-15 mg/kg/bw, to "not specified". The daily dose for human consumption recommended by the FDA is 45 mg/day for adults.

Classification A

References : "Unpublished Company data"

(2) L. ASCORBYL PALMITATE AND MONOSTEARATE XI/496/87

* Formula and synonyms

Colipa A02

CAS - RN 137 - 66 - 6

CAS - RN 25395 - 66 - 8

CAS - RN 51222 - 59 - 4

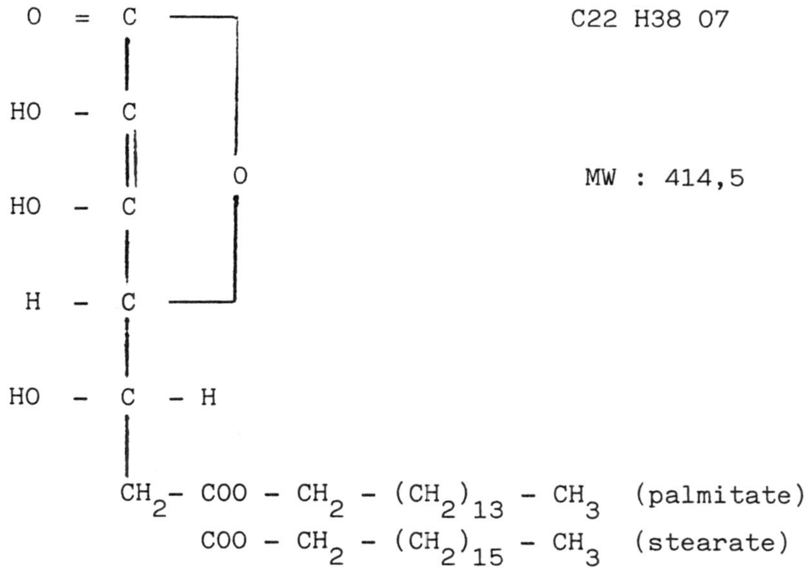

C22 H38 07

MW : 414,5

* Use

Ascorbyl palmitate and stearate are used in cosmetic products at concentrations not exceeding 2 %.

Antioxidant action : At level of 0.5 %, ascorbyl palmitate is a very efficacious antioxidant for vegetable oils with high level in onsaturated fat acids,used in cosmetic W/O emulsions.

Bleaching action : Esters would be able to reduce melanine In Japan it is used at levels of 0.1 to 4 % in cosmetic preparations to inhibit the formation of melanine.

* Recapitulation of the studies of toxicity

LD50 oral (mice) : 2 gr/kg/bw
(rats) : > 5 gr/kg/bw
(for a 33.3 % suspension of monopalmitate or a 15 % suspension of dipalmitate)

LD50 dermal (guinea pigs) : > 3 gr/kg/bw

ORAL TOXICITY

In a study on rats involving administration of 2 and 5 % ascorbyl palmitate in the diet for nine months, growth was significantly retarded at the 5 % dose and two rats of the tenhad bladder stones and hyperplasis of the bladder epithelium. One rat had an inflamed kidney. With the 2 % diet growth was slightly retarded but there was no differencein terms of mortality or histopathology.

No adverse effects were observed after administering in the diet 100, 200, 500, 1 000 and 3 000 mg/kg/bw of L-ascorbyl stearate.

In a two year study of rats fed on heat treated lard containing 2 or 5 % ascorbyl palmitate (equivalent to 424 and 1060 mg/kg/bw or 0.05 and 0.25 % of the total diet) growth rate decreased at the higher dose and two of eight animals had oxalate stones after nine months of treatment.

No effects were observed at the 2% dose (estimated NEL at 125 mg/kg/bw). Sine food-grade palmitic acid contains large quantities of stearic acid, it can be assumed that the ascorbyl palmitate used in this study contained stearic acid.

CUTANEOUS AND MUCOUS TOLERANCE

The monopalmitate and dipalmitate are well tolerated by the skin and mucous membranes.

In a skin irritation test (modified Draize),application of a 10 % aqueous solution of monopalmitate and full-strength dipalmitate with occlusion for 24 hours to the intact shaved skin of albino rabbits was non-irritating.

In an eye irritation test (modified Draize) on rabbits,instillation of 0.1 ml of a 10% aqeous solution of monopalmitate was non-irritating. The instillation of 0.1 ml of full-strength dipalmitate was minimally irritating.

After application of ^{14}C-labelled ascorbyl palmitate to the skin of guinea pigs affected by scurvy, ascorbic acid levels in the skin, liver, kidneys and blood were found to be 4 to 8 times higher than in the control.

C O N C L U S I O N

Ascorbyl palmitate is minimally toxic. It is well tolerated by the skin and mucous membranes. A study of which no details are available indicates that it does penetrate through the skin.The no-effect dose for rats is evaluated at 125 mg/kg/bw, the kidney and bladder being the target organs in the case of long-term administration.Mutagenicity and sensitization data are unavailable.

JEFCA has established an ADI of 0-125 mg/kg/bw for the stearate or the palmitate or the sum of both ascorbyl esters.

Classification : B Information required :
- details of skin absorption
- sensitization
- mutagenicity in at least two test systems (reversion on bacteria and chromosome aberration)
- information on possible use to inhibit melanine production and anti-nitrosating action

References :

Colipa : "Unpublished Company data"

RTECS

Dr Robinson's data

(3) D. ISOASCORBIC ACID

* Formula and synonyms

Colipa A04

CAS - RN 89-65-6

D - erythro - hex - 2 - enonic acid, γ lactone (8 CI,9 CI)

Sodium erythorbate (sodium salt)

CAS - RN 7378 - 23 - 6

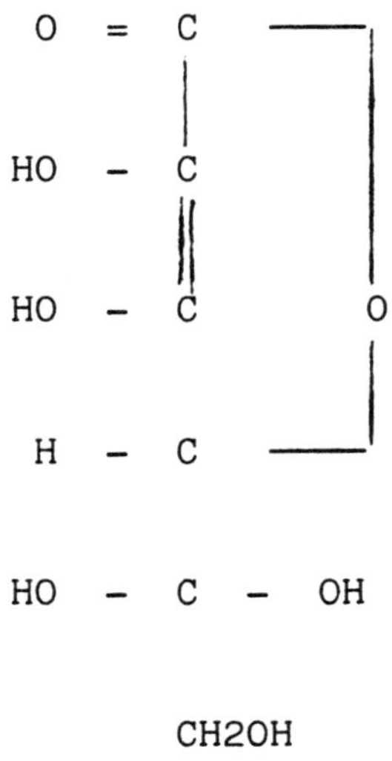

$C_6H_8O_6$

MW : 176,12

* Use

Isoascorbic acid is used in cosmetic products at concentrations not exceeding 2 %.
It is no longer permitted in foodstuffs.

* Recapitulation of the studies of toxicity

ORAL TOXICITY

Isoascorbic acid is absorbed via the gastro-intestinal tract by the same transport system as ascorbic acid and is competitive with it.

When given orally to guinea pigs simultaneously with ascorbic acid, it prevents ascorbate uptake by the guinea pig tissues at doses ± 200 mg/kg ,reducing ascorbate storage and bioavailability up to 50 % in some organs.

It is quickly excreted in the urine but its metabolites are unknown.

In mice fed a diet with 5 % erythorbic acid and 5 % ascorbic acid for two months,increasing dose levels to 10 % for five more months , erythorbic acid was found to replace 45 % of ascorbic acid in the liver and 28-39 % in the brain tissues. The blood ascorbate level was unchanged .

In studies on four humans, a daily intake of 50 mg isoascorbic acid for two weeks produced a deficiency of ascorbic acid and this was followed by 100 mg daily absorption for two weeks.All ascorbic acid concentrations continued to fall during the four-week period and urinary excretion increased considerably. 50-60 % of a 300 mg loading dose of ascorbic acid appeared in the urine. The fate of the retained portion is not known.

No adverse effects were observed when rats were fed 1% in the diet for 36 weeks.

Similarly, no effects are observed regarding growth rate,mortality or histopathology, in rats fed for two years on a 1 % diet, nor in a group of four beagle dogs receiving 1 g/day of isoascorbic acid orally for 240 days and another group receiving 5 g/day for 50 days and later 7.5 g/day for 190 days.

CUTANEOUS AND MUCOUS TOLERANCE

A single application of 2 g/kg of sodium isoscorbate to both the intact and abraded skin of rabbits produced no sign of toxicity during the fourteen -day observation period.

A single instillation of 100 mg of powder into the eyes of albino rabbits with rinse-off after five seconds produced no sign of irritation.Some redness reversible after 24-48 hours was observed in three of the twelve rabbits.

MUTAGENICITY

Erythorbic acid (99.6 % pure) exhibited a relatively weak mutagenic potential in the Ames test, at a dose of 50 mg/plate in respect of the T 100 strain with S_9 mix (222 his^+ revertants) and without activation (144 his^+ revertants).

It was negative in a chromosome aberration test carried out *in vitro* on a culture of Chinese hamster fibroblasts and in a dominant lethal test.

CARCINOGENICITY

Sodium erythorbate administered in doses of 1.25 and 2.5 % in rats' drinking water for four weeks did not show any carcinogenic potential.There was a slight reduction in certain spontaneous tumours in females treated at 2.5 %.

C O N C L U S I O N

Isoascorbic acid is well tolerated by the skin and mucous membranes.It is neither mutagenic nor carcinogenic.

Acute toxicity is not known and the toxicity data for short-term oral administration are not sufficient to establish a no-effect dose.

The sensitization and skin absorption data are missing.It competes with ascorbic acid (vitamin C) as regards bioavailability.

Classification : D This classification is justified by the competition with ascorbic acid and could be reviewed only if the Committee had data to show the absence of skin irritation

References :

- Colipa "Unpublished Company data"
- Dr Robinson's data
- Abe I
  Exp. Mol. pathol.; vol.41 , iss. 1,1984, p. 35-43
- Fukushima S. et al
  Cancer Letter (Shannon Irel.); vol. 23, iss. 1 ,1984,p.29-37

REPORT BY THE SCIENTIFIC COMMITTEE ON COSMETOLOGY

ON THE USE OF CERTAIN HAIR DYES

THE COMMITTEE'S MANDATE

To give its opinion on the question whether the use of certain hair-dyes is admissible from the health point of view.

- 2,4 diaminoanisole
- 2,5 diaminoanisole

(1) 2,4 D I A M I N O A N I S O L E A6

* Formula and synonyms

4 - methoxy - m - phenylenediamine CTFA
1 - methoxy - 2 , 4 - diaminobenzene
C.I. : 76050
CAS : 615 - 05 - 4

$C_7 H_{10} N_2O$ MW : 138,2

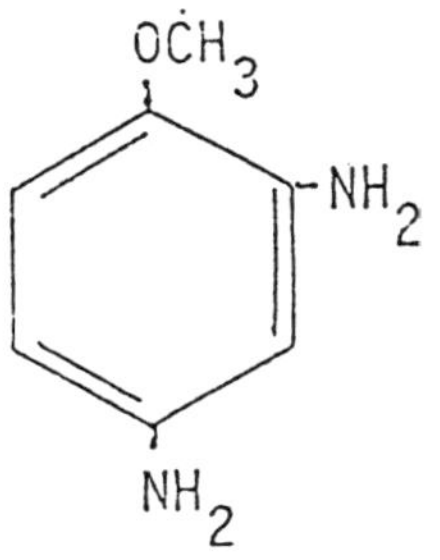

* Use

Oxidation hair dye
It is used as a coupler,giving rise to different colours with,f.i. p. phenylenediamine (blue),p. aminophenol (red)

No longer used by big companies ; used only by small companies.

Max.use : 2 % before mixing the solution.

* Recapitulation of the studies of toxicity

LD50 oral (rat) : 460 - 515 mg/kg/bw

i.p (rat) : 372 mg/kg

ORAL TOXICITY

+++++++++++++

- Subacute toxicity : RAT. 2,4 - DAA administered orally at the only one dose level of 23 mg/kg/day for 12 weeks to female rats showed a slight depression of the red and blood cell count and of growth rat.

  RAT Fischer 344 & Mouse B6C3F1. NCI bioassay.5 males and 5 females/ group of both species received in the diet these doses of 2,4-DAA sulphate : 0-0.75-0.125-0.209-0.348 and 0.580 % for 4 weeks followed by a two week observation period.The maximum tolerated doses were 0.5 % for rats and 0.24 % for mice.
- Short-term dermal toxicity : 2,4-DAA containing formulation (2.0 % and 4.0 % in water) tested on shaven intact skin of New Zealand rabbits by topical application produced no toxic effects at 3-7-13 weeks after treatment at the histopathological analyses.

CUTANEOUS AND MUCOUS TOLERANCE

+++++++++++++++++++++++++++++++++

DERMAL TOXICITY

++++++++++++++++

Dermal irritation : the compound applied to intact and abraded skin of rabbits as a 2,5 % (w/v) aqueous solution resulted "mildly irritating".

Eye irritation : the compound applied as a 2,5 % (w/v) aqueous solution on rabbits' eyes resulted non irritating.

Sensitization was tested in Guinea pigs treated with 3 % 2,4-DAA solution containing 2 % Natrosol,2 % Tween 80, 0.05 % Sodium sulphite and 10 % isopropanol(pH = 7) applied 6 days/week for 3 weeks.The results were negative. Another sensitization test on Guinea pig with 2,4-DAA (5-25 % in distilled water) by Magnusson & Kligman's method (reaction evaluated at 24, 48 and 72 h) showed delayed contact hypersensitivity in 1/10 and equivocal reaction in 2/10 animals at the first challenge application.

Dermal absorption. (14-C)-2,4-DAA (4 μg/cm2) of acetone solution applied for 24 h on abdomen of adult males and females rhesus monkey showed that 4.7 % of the dose penetrated in the skin. (14-C)-2, 4-DAA in a formulation applied for 20 min to scalp hair of the Rhesus monkey showed a penetration of ca. 0.02 % (in oxidative mixture) and 0.23 % (without hydrogen peroxide).

(14-C)-2,4-DAA (as ingredient of 3 different formulations) showed an absorption range of 0.26-1.1 % after cutaneous applications on rats evaluated at 24 h.

In hairless Wistar rats a topical application of radiolabelled 2,4-DAA alone or in a complete formulation did not show a difference in the amount penetrated through the skin.

Human studies : 2,4-DAA dihydrochloride (ring-labelled, 4 μg/cm2) applied on the ventral skin surface of the forearm to man for 24 h showed that 3.9 % of the dose penetrated in the skin.

2,4-DAA-14C applied to dry hair of 3 volunteers over a period of 5-8 min. and then left for additional 20 min. before rinsing off,was excreted in the urine at a value of 0,022 % with a t 1/2 of urinary excretion of 18 hrs.

MUTAGENICITY AND GENOTOXICITY

The studies have shown that 2,4-diaminoanisole is mutagenic : (1) on Salmonella (2) on S.pombe; (3) on mouse lymphoma cells; (4) on Chinese hamster V79 (in vitro); (5) on Salmonella/mouse or rat urinary assay (in vivo) ; on D.melanogaster (SLRL) for gene mutation induction ;(6) on Chinese hamster ovary (CHO) cells lines for

chromosome aberration (in vivo); (7) on S.cerevisiae for gene conversion and mitotic recombination; (8) on rat hepatocytes and HeLa cells for the induction of UDS; (9)on human fibroblast for DNA strand breaks and (10) DNA repair in E.coli ; (11) on bone marrow cells of mice (up to 23.3 mg/kg i.p.) for SCE (in vivo) ; and (12) on rat (0.66 nmol/kg i.p.) for DNA/damage alkaline elution assay.
Other mutagenicity studies have shown that the compound is negative on E.coli 343/113; on Salmonella/mouse (up to 3.5 mg) H.M.A. (in vivo); and in morphologically abnormal sperm test on mouse for gene mutation; that thecompound does not induce dominant lethals in rats Charles River treated for 10 weeks with doses of 10-20-40 mg/kg (i.p.) and for 8 weeks with a dose of 20 mg/kg (i.p.); and micronuclea in rats Sprague Dawley treated with 0.5%(w/v) suspension in tragacanth's gum with 0.05 % sodium sulphite (gastric intubation).

Biochemichal studies : 3-H-(ring labelled)-2,4-DAA incubated with purified rat liver nuclei and microsomes in the presence of NADPH led to the formation of radiolabelled product(s) bound covalently to nuclear protein. Pretreatment with ß-Naphthoflavone showed that PB-treatment increased binding (ca. 70 % for both frations).

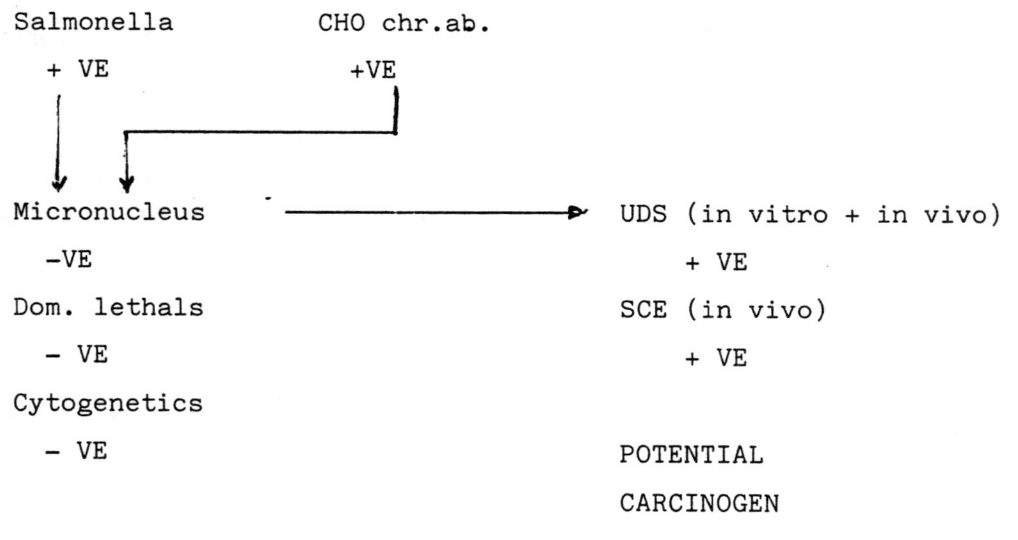

(according to J.ASHBY,1986)

CARCINOGENICITY

++++++++++++++

Long-term studies were carried out on mice and rats by a NCI bioassay. the compound fed in the diet at 0.12, 0.5 % for rats for 78 weeks followed by 29 weeks observation period and 0.12, 0.24 % for mice for 78 weeks followed by a 19 weeks observation period resulted carcinogenic in both sex and in both species producing follicolar-cell thyroid tumours and malignant tumours of the skin and its associated gland in both sex of rat and C-cell adenoma only in males and, in mice combined incidence of follicular-cell adenomas and carcinomas in female and only fullicular cell adenomas in males.

Various studies, some incomplete and inadequate for the evaluation, performed on rats and mice with formulations containing different level of 2.4-DAA (from 0.33 % to 4.0 %) showed negative results.

TERATOGENICITY

+++++++++++++

No embriotoxic or teratogenic effects were observed during gestation of Charles River rats treated topically at a dose of 2 mg/kg/bw with formulations containing 2.4-DAA (2 % and 4 % in water)on shaven skin on day 1-4-7-13-16-19. A retarded effect of ossification process (bones of the feet and the cervical and caudal vertebral centre) was observed in mice topically treated twice a week for four weeks before mating and until the 18th of gestation with a formulation containing 2 % of 2.4-DAA and mixed before use 1:1 with hydrogen peroxide equivalent to 0.5 mg/mouse.On the basis of 1 % dermal absorption,the dose producing the teratogenic effects is 5 µg/mouse (200 µg/kg).In a reproductive study on rats treated with 2 % 2.4-DAA containing formulation no negative effects were observed (this study was not available for the present analysis).

C O N C L U S I O N

IARC (1982) classified 2.4-DAA as a chemical with sufficient evidence of carcinogenic potential,The substance showing :

-Sufficient evidence of mutagenicity/genotoxicity in vitro and in some in vivo assays,

-Sufficient evidence of carcinogenicity in two species and in both sexes,

-Some evidence of teratogenicity.

Therefore, the Committee may not allow the use of 2.4-DAA.

Classification : D

Bibliography :

- N. LOPRIENO & G. BONCRISTIANI : A review on toxicology of 2.4 - Diaminoanisole. Pisa, Italy, sept. 1986.
- S C C , 2nd series EUR 8634 (op 1980)

(2) 2,5 D I A M I N O A N I S O L E A34

* Formula and synonyms :

2 - methoxy - p - phenylenediamine

1 - methoxy - 2.5 - diaminobenzene

C.I :

CAS : 5307 - 02 - 8

$C_7H_{10}N_2O$ MW : 138

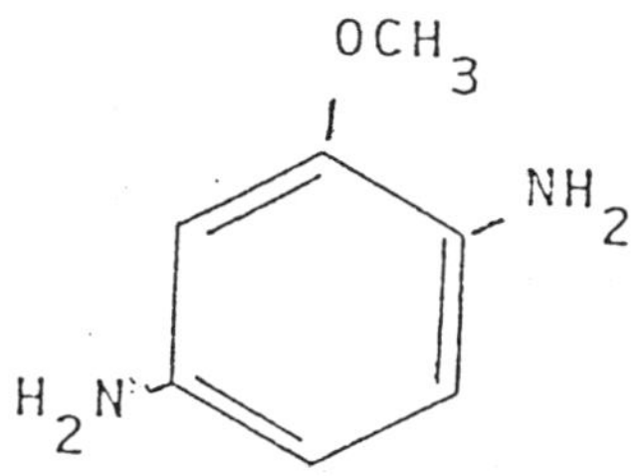

* Use

- Banned in Italy
- It is used in oxidative and developer hair dyes with couplers, such as m-phenylenediamine, m-aminophneol, m-dihydroxybenzenes (the resulting reactions produce different indo-amines of different colour).

Production and use : 300 kg.

* Recapitulation of the studies of toxicity

LD50 oral (rat) : 60 mg/kg
(mouse) : 34 mg/kg
i p (rat) : 28 mg/kg

O R A L T O X I C I T Y

Short-term oral toxicity : the compound administered as sulfate, 0-1-5-25 mg/kg/day X a minimum of 91 consecutive days by oral gavage to rats showed that 1 mg/kg/day represents the dose with "no-effect level."

A similar study in a dog (0-1-3-9 mg/kg/day/7 day/wks X 91 consecutive days) gave "no-effect level" at 1 mg/kg/day.

CUTANEOUS AND MUCOUS TOLERANCE

DERMAL TOXICITY

Dermal irritation studies were carried out on rabbit using two different methods (a & b). The results showed that the compound has resulted "slightly irritating" (dose : 0.5 ml or 0.5 g pasta, IIP = 0.08) in method (a) and "mildly irritating" (dose not reported,IIP = 1.1) in method (b).

Eye irritating : The compound (100 mg as sulfate) applied on rabbits' eye gave in all animals a positive reaction with lesions to conjunctiva (redness : diffuse, crimson red,individual vessels not easily discernible ; chemosis : obvious swelling with partial eversion of lid in two animals swelling with lids about half closed).The grade of ocular reaction was evaluated at 1-2-3-4-7-17 and 21 days.

In another study on rabbit reading at 1 hour, (1-2-3-4-7 days)the compound (= Imexine OAN 0.5 % in water) did not result irritating.

Sensitization was tested in Guinea pig treated with 1 % in paraffin oil of test compound as sulfate using the method of Magnusson and Kligman (reaction evaluated at 24-48-72 h). 3/10 animals showed a temporary skin-sensitization (very slight oedema) and 1/10 a permanent skin sensitization (slight oedema).

Dermal absorption. (14-C)-2,5-DAA.2HC1 (aqueous solution) applied epicutaneously without hydrogen peroxide on Him : OFA-Sprague Dawley rats at 2 mg (free base)cm2 showed that 1.02 % of the amount applied was absorbed. (14-C)-2,5- DAA.2HCI in commercial preparation (1:1 with 6 % H2O2) after epicutaneous applications on 9 cm2 of clipped skin areas and 2 mg (free base)/cm2 on unclipped skin)showed an absorption of 0.28 % (clipped skin) and 0.07 % (unclipped).

Short-term dermal toxicity: 2.5-DAA containing formulation (6.0 in water) tested on shaven intact skin of New Zealand rabbits by topical application produced no toxic effects at 3-7-13 weeks after treatment at the histopathological analyses.

Human studies : Skin absorption of a hair dye product with test compound (1.9 %,mixed 1 : 1 with H202) applied (27-32 min.) on washed hair of 5 female volunteers was less than 0.102 % of the amount applied, under hypothesis that the distribution of the dye is equal in the whole body. Patch-test performed in human with sensitization to aromatic amino compound (paragroup allergy) revealed a sensibilization for the testcompound similar to p-phenylenediamine.

MUTAGENICITY AND GENOTOXICITY

++++++++++++++++++++++++++++++++++

The compound has produced positive results in Salmonella in the presence of different metabolic activation systems. The compound resulted positive in vitro for genotoxicity on primary rat hepatocytes for the induction of UDS and in human skin fibroblast for the induction of single strand breaks : in both studies purity was not reported. Other mutagenicity studies have shown that the compound is negative for gene mutation on Bacteriophage TD4/E.coli test system, in a spot test on Neurospora crassa, in one study on Salmonella typhimurium and in mouse lymphoma cell lines in the absence of metabolic activation system. The compound did not induce chromosome aberration in vivo by micronucleus test on bone marrow cells on rat (5 mg/kg in 1 % MC by gastric intubation in 2 equal doses separated of 24 hours) and in two different dominant lethal studies on rat : (a) 20 mg/kg i.p. for 8 weeks and (b) until to 25 mg/kg/day for 10 weeks (30 mg/kg/day for first 2 weeks). Induction of mitotic gene conversion on Saccharomyces cererisiae using different experimental procedures was negative.

Salmonella   CHO chr.ab.

+ VE   ?

↓

Micronucleus ⟶ UDS in vitro

-VE   + VE

Dom. lethals   DNA breaks in vitro

-VE   + VE

CARCINOGEN ?

CARCINOGENICITY

++++++++++++++

Long-term study was carried out with a formulation containing test compound (6% in water, as sulfate) by dermal application on mouse (0.05 ml/w. for ca. 2 years) : no biologically significant difference was observed between treated and controls groups.

TERATOGENICITY

+++++++++++++

No biologically significant difference in the mean number of corpora lutea, implantation site and live fetuses, and sex ratio and no significant changes for soft-tissue or skeletal anomalies were seen on rat treated topically at a dose of 2 ml/kg b.w. with a formulation containing 2.5-diaminoanisole (6 % in water)on shaven skin on day 1-4-7-10-13-16-19 of gestation. Study inadequate (treatments every 3 days : only one dose tested.

C O N C L U S I O N

The compound has very high acute toxicity compared to many other hair dyes and a NEL for sistemic toxicity of 1 mg/kg/day.
Due to the inadequacy of long-term carcinogenicity study available, the evidence of carcinogenicity could be assessed.
The teratogenicity study is inadequate.
Short-term studies on mutagenicity/genotoxicity indicate that the compound reacts with DNA *in vitro*.
The request made by the SCC in 1980 was not followed by any more detailed study.
The human absorption has been calculated to be around 25-30 µg/kg,a value close to the NEL in the animal toxicity.

Classification : D

Bibliography : - N. LOPRIENO & G. BONCRISTIANI : A review on toxicology of 2.5-Diaminoanisole. Pisa,Italy sept.1986
- SEC 2d series EUR 8634 (op. 1980)

CONTENTS

33rd meeting of the SCC - February 2, 1987

33rd MEETING OF SCC

20 FEBRUARY 1987

Members of the Committee

Professor A.P. DE GROOT

Doctor R. GOULDING

Professor A. HILDEBRANDT

Professor N. LOPRIENO

Professor J. SCHOU

Professor D.P. O'MAHONY

Professor G. STUETTGEN

Professor P. AGACHE

Professor L. MUSCARDIN

Professor J. DONY

Doctor O. ENJOLRAS

Doctor A. KNAAP was excused

REPORTS OF THE SCIENTIFIC COMMITTEE ON COSMETOLOGY

Opinions expressed on February 2 1987

33rd reunion

SUMMARY

ANTIOXIDANT AGENTS

BHA	A05	EEC N° 320	C
BHT	A06	321	D
DAHQ	A011		C
DBHQ	A012		C

PRESERVATIVE AGENTS

Suttocide A	P84	C
Decominol	P87	C

ANTI-DANDRUFF AGENT

ONADINE MSD	P50	EEC N° 41	D

Complementary opinions on hydroxy-8-quinoleine

REPORT BY THE SCIENTIFIC COMMITTEE ON COSMETOLOGY

ON THE

USE OF CERTAIN ANTIOXIDANT AGENTS IN COSMETIC PRODUCTS

(opinion expressed on 20 February 1987)

THE COMMITTEE'S MANDATE

To give its opinion on the question whether the use in cosmetiç products of certain antioxidant agents is admissable from the health point of view.

- BHA
- BHT

} in a maximal use concentration of 1%

- TAHQ
- TBHA

} in a concentration of 0,2%

(1) B H A : BUTYLATED HYDROXYANISOLE

-Formula and synonyms

EEC n° 320

Colipa A05

C11 H16 02

MW : 180,2

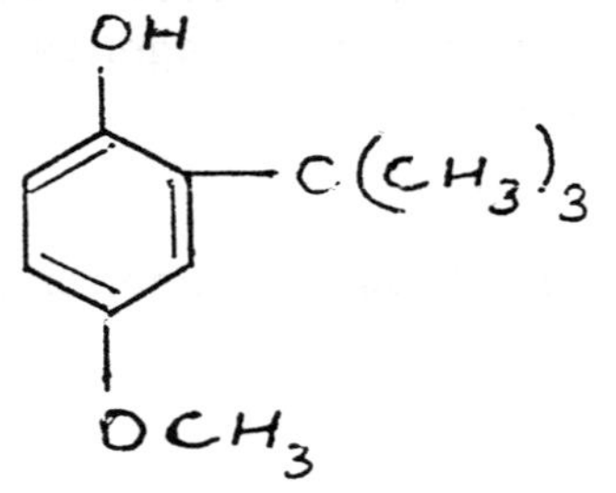

3-BHA : approx 90%

3 - tert-butyl-p-hydroxyanisole

phenol,2 -(1,1 dimethylethyl) - 4 -methoxy

CAS - RN 121 - 00 - 6

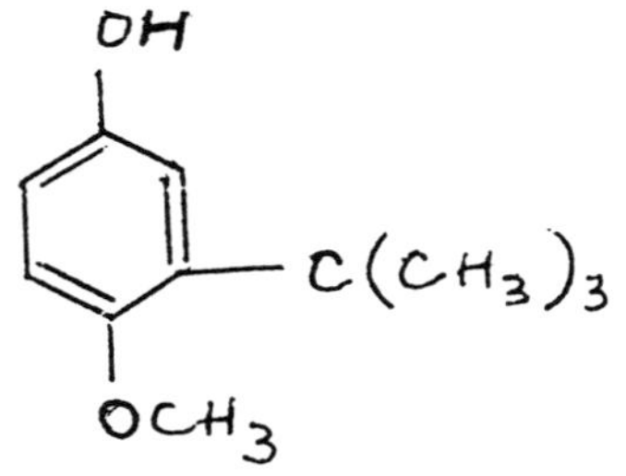

2 BHA : approx 8% MW 180.2

2 - tert-butyl-p-hydroxyanisole

phenol,3 -(1,1 dimethylethyl) - 4 - methoxy

The commercial product is a mixture of two isomers (3-BHA and 2-BHA) with the 3-isomer being the major component

- Physical characteristics

White or slightly yellow,waxy solid ; odour aromatic

Insoluble in water

Slightly soluble in fats and oils

Soluble in alcohol,propylene glycol,chloroform and ether

- Use

As an antioxidant in cosmetics at a concentration not exceeding 1 %
(eye make-up,lipstick,skin-care prepar. and perfumes
BHA protects the lipids from peroxydation and acts synergistically with BHT and propyl gallate and is often used in combination with these a.o. for greater efficacy

- Cosmetic quality

Purity : $\geq$ 98,5 of a mixture of 3-BHA (approx. 90 %) and 2-BHA (approx. 8 %)

Impurities :	4-hydroxyanisole	$\leq$ 0,5 %
	1-t-butyl-2,5-dimethoxybenzene	$\leq$ 0,5 %
	2,5-di-t-hydroxyanisole	$\leq$ 0,2 %
	hydroquinone dimethyl ether	$\leq$ 0,1 %
	sulphated ash	$\leq$ 0,01 %
	lead	$\leq$ 20 ppm

- Other qualities
  - It offers moderate protection to the skin from UV radiation
  - It inhibits the formation of nitrosamines in an acid pH
  - It shows an antimicrobial activity towards gram + bacteria and fungal organisms.

- Recapitulation of the studies of toxicity

LD50 : oral (mice) = 1,10 - 2 gr/ kg bw
(rats) = 2,10 > 5 gr/ kg bw
(comes from studies before 1960)

LD50 : dermal (rabbit) = 2,10 > 5 g r/ kg bw
(formulation with 0,1 % BHD)

ORAL TOXICITY

A daily gavage at a dose of 1 gr. in olive oil for four to seven days is lethal to rabbits. Adrenal changes and an increase in Na^+ and K^+ is observed.

In rats 0,5 gr/kg bw by gavage for 6 days has a transient depressive action on organic acid transport.

Studies on animals and humans show that BHA is absorbed via the gastro-intestinal tract and metabolized.

In humans, BHA is conjugated with glucuronic acid in the liver.

Urinary excretion products consist mainly of glucuronides with small quantities of sulphates and free BHA. No dealkylation or hydroxylation products are detected. The time required to excrete the dose administrated varies from 23 to 50 hours.

Although solubility in lipids favours storage in tissues, studies on pigs, rats and dogs and a pharmacokinetic study on rabbits show that the quantities stored are limited because of the metabolism and the rapid excretion.

A recent kinetic study on humans indicated the formation of tertbutyl hydroquinone by metabolic conversion. This product has generally been identified as a decomposition product resulting from UV exposure.

In a large number of short- and long-term studies by oral administration to rats, rabbits, guinea pigs, monkeys, dogs and chickens, the most frequently noted effect was an increase in liver weight and changes in the endoplasmic reticulum; these changes were not generally accompanied by persistent hepatoxicity. The lowest effect level is 25 mg/kg/bw/day for mice and 50 mg/kg/bw/day for monkeys.

In a recent series of studies there was a proliferation of the epithelium of the forestomach, which can be linked to the forestomach tumours induced by BHA in rats.

Three studies on rats involving administration in the diet at con-concentrations up to 2% for 9 and 27 days, 1, 2 and 4 weeks and 13 weeks showed that the effects on the forestomach - proliferation of the epithelium, increased mitotic rate, hyperplasia, hyperkeratosis and inflammatory changes - increased with the duration of exposure; after the treatment was stopped the lesions regressed slowly but not completely, and the degree of reversibility was inversely proportional to the prior duration of exposure. The no-effect dose is evaluated at 0,25 mg/kg/bw/day.

These observations are confirmed in other species having a forestomach such as the more susceptible hamster,in which papillomas appear after 24 weeks on a 1% BHA diet.
Animal species having no forestomach : daily oral intubation of 1 000 mg/kg/bw of BHA has no effect on the stomach or oesophagus of the guinea pig ; administration of 0,25 , 10,5 and 1% in the diet of beagle dogs for six monthss has no effect on the stomach,oesophagus, duodenum or liver.Histopathological and histometric examinations showed no changes in the mucous membranes of the stomach,oesophagus or duodenum.The liver weight increased at each dose but caused no histopathological changes.It produces a greater proportion of oesophaglal basal layer cells in mitosis in monkeys after 85 days at a dose of 250 mg/kg/bw/day ; these changes are not accompanied by histological changes and the dose of 125 mg/kg/bw/day has no effect.

CUTANEOUS AND MUCOUS TOLERANCE

DERMAL ROUTE TOXICITY

In special short-term studies on dermal effects,application of BHA (up to 20 %) in lanolin to the ears of guinea pigs for six weeks caused ultrastructural morphological lesions of the epidermis,with destruction of the superficial connective tissue and collagen break-up, BHA causes slight to moderate irritation of the skin of albino rabbits when applied to the shaved,intact or abraded skin with occlusion in preparations containing 0,1 to 0,2 %.
It causes slight to very slight irritation to the eyes of albino rabbits after single or repeated instillation of formulations containing 0,1 and 0,2 %.

In humans

In a number of tests to assess irritation potential,sensitization and photosensitization,

- performed on 10 to 353 human subjects.
- with cosmetic preparations containing concentrations of 0,01 to 0,2 % BHA,
- by simple application with occlusion,
- by repeated application with occlusion followed by a challenge application according to various protocols including modified Draize,Philips and lauryl sulphate maximization,

- by repeated applications with and without occlusion followed by a challenge application and exposure to UV radiation at 360 nm or 280 to 370 nm, according to various protocols including Draize-Shelansky,Schwarz-Peck and Shelansky-Shelansky.

BHA is generally found to be minimally to slightly irritant,non-sensitizing and non-photosensitizing.

Nevertheless,individual cases of contact dermatitis after applications of formulations containing 0,005 and 0,052 % BHA and a variety of allergic reactions following the ingestion of products containing BHA are reported in the literature.

The North American Contact Dermatitis Group estimates a 2 % incidence of skin sensitization among 548 subjects exposed to 2% BHA. In 3/112 patients treated for contact dermatis responding positively to a 2 % BHA patch test,the biopsy and control patch tests revealed a lymphocyte allergy,the patients having been sensitized after repeated local application.

A recent study over a two-year period showed that 7/1 096 patients suffering from facial eczema were allergic to BHA (1% patch test)and five also reacted to TBHQ (1% patch test).

MUTAGENICITY AND CARCINOGENICITY

BHA has exhibited neither mutagenic properties nor genotoxic effects in a number of in vitro and in vivo test systems.

On bacteria,it was negative in two tests on Salmonella typhurium (TA 98, 100 and TA 535, 1537 and 1538) and on S. cerevisiae (D 4) with or without activation.

On mammal cells,it produced no chromosome aberration on cultures of human embryo lung cells exposed for 24 hours at a concentration of 2, 20 and 200 ug/ml in isopropyl alcohol,nor on Chinese hamster fibroblasts at a concentration of 10-4 M in ethanol.

Similarly, in vivo, after oral administration to rats at doses of 15,150 and 1 500 mg/kg, it does not cause significant aberrations of the chromosomes of the spinal cord and it was not mutagenic in a lethal dominant test. No sex-linked recessive lethal increases were observed in D. melanogaster fed on BHA.

After administration at doses of 15, 150 and 1 500 mg/kg to mice, no mutations were induced in an intermediate-host test, on S. typh. Conversely, in a similar test on S. cerivisiae, it increased mitotic recombinations at all doses.

BHA can have an antimutagenic effect as shown by the results obtained in various test systems on bacteria and mammal cells.

Short-term studies on mice using subcutaneous and intraperitoneal injection show no evidence of carcinogenicity, nor do long-term studies by topical application of 0,1 to 10 mg BHA in acetone once a week for 323 to 509 days, or 1 mg in acetone twice a week for 30 weeks.

Long-term oral studies on rats and dogs at doses up to 0,5 % in the diet were negative, but it should be noted that none of these studies is acceptable in terms of standard protocols.

A recent study in Japan, in which BHA was administrated in concentrations of 0,5 and 2,0 % in the diet showed that BHA was carcinogenic for rats' stomachs ; it induces a significant increase in the incidence of papillomas and carcinomas of squamous cells in both sexes. The incidence of hyperplasia of the forestomach associated with neoplasic changes is dose-related.

BHA can inhibit or promote experimental carcinogenesis. It inhibits the carcinogenic power of benzopyrene and 7,12-dimethylbenzanthracine on mouse skin.

TERATOGENICITY AND EMBRYOTOXICITY

No embryotoxic or teratogenic effects were observed in a series of teratogenicity studies on mice,rats and rabbits at doses of 60 to 1 000 mg/kg/bw/day.

However,mice exposed to BHA at 5 000 ppm via their mothers' diets during gestation and lactation and then directly up to the age of six weeks showed behavioural changes:reduced sleep,learning speed and orientation reflexes.

In another study on mice,by administration in the diet of 0,5%/kg during gestation,the 30-day-old mice exhibited reduced exploratory reflexes, body weight and cerebral cholinesterase.

In a postnatal study,on rats fed with 1 250,2 500 or 5 000 ppm BHA in the diet for two weeks before mating, and during gestation and lactation,an increase relative to the mortality dose was observed up to the age of six weeks,but not at the age of three months.Administration in the diet of BHA at 50,200 and 400 mg/kg/bw to pigs (Danish landrace SPF) from mating to the 110th day of gestation had no effect on reproduction,nor any teratogenic effects.

However,a reduction in weight gain was observed at the highest dose and a dose-related increase in liver and thyroid weight; histological changes appearing in the thyroid particularly at the highest dose could indicate a reduction in thyroid activity.

CONCLUSION

JEFCA has set an ADI of 0,5 mg/kg/bw of BHA,BHT or the sum of both.

Owing to the ADI value,the absence of information on skin absorption and sensitization problems which may be linked to concentration,the Committee cannot accept use at 1% in cosmetics and wonders about the actual concentration in use.

Classification : C Information required : - skin absorption
- actual concentration in use

References

- Report of the Joint FAO/WHO Expert Committee on Food Additives. WHO Tech Rept Series No (1983)
- Cosmetic Ingredient Review Report : BHA (CTFA) (1983)
- COLIPA Data CSC/294/83-1,2
  Acute Dermal T
  MB Research Laboratories , Pennsylvania USA
  Project number MB 77-1719 July 25 1977
  Sensitization on Human
  - William L. Epstein MD Project n° WLE 10 May 16 , 1977
  - A.M. Kligman, J. Invest. Dermatol. 47 (5) , 393 - 409 1966
- Roed-Petersen J et al Br J Dermatol 94 233-41 (1976)
- White I R et al Contact Dermatitis 11 265-7 (1984)
- Allen J R et al Fd Cosmet Toxicol 10 769-99 (1972)
- Ford S M Fd Cosm Toxicol 18 5020 ; 21-6 (1980)
- Ishidate M et al Fd Chem Toxicol 22(8), 623-36 (1984)
- SCF Data III/1244/84
- Altman H J et al (1984a)
  Naunyn Schmiedebergs Arch Pharmacol Suppl. 325 Abst 110 (Summary as presented in SCF Data III/1244/84
- Nera EA et al Toxicology 32 197-213 (1984)
- Iverson F et al Toxicology 35, 1-11 (1985)
  (submitted as COLIPA data CSC/461/85)
- Ito N et al Gann 74 459-61 (1983)
- Hayoshi Y (1984) Statement as given at Toxicology Forum Meeting, Washington.
- Iverson F et al An 85-day study of butylated hydroxyanisole in the cynomolgus monkey.(Internal Report,Toxicology Research Division,Bureau of Chemical Safety,Food Directorate,Canada - submitted as part of SCF Data III/1244/84)
- Witschi H R et al. Fundamental Appl Toxico 4,795-801 (1984)
- Dr Robinson's data slips
  Journal of the American College of Toxicology vol3 N°5 1984
  N.L.M'S National Interactive Retrieval Service Medlars II

(2) B. H. T. BUTYLATED HYDROXYTOLUENE

* Formula and synonyms

- Colipa AO 6
- EEC n°32
- CAS-RN 128-37-0

- Phenol,2,6-bis (1,4-dimethylethyl)-4-methyl (9 C1)
  BHT

- $C_{15}H_{24}O$

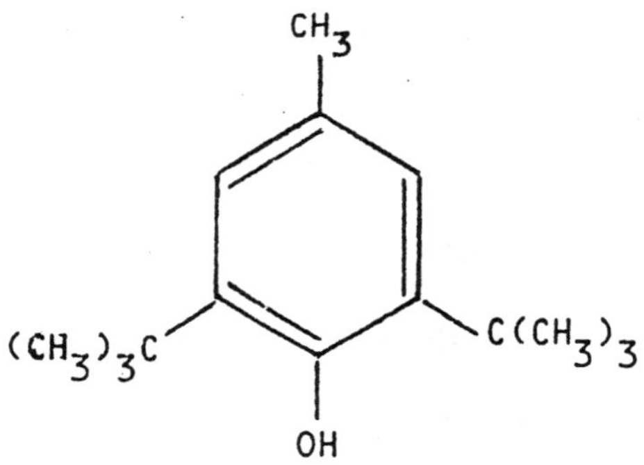

- MW : 220,34

* Characteristics

White,crystalline or flaked solid; odour characteristic
Insoluble in water
Soluble in fats,alcohol,ether and other organic solvents
In a study of the influence of pH and light on the stability of antioxidants,BHT was stable regardless of light or pH except at high pH ($\geqslant$ 9).

* Use in cosmetics

As an antioxidant at a concentration not exceeding 1% (often 0,025 - 0,3 %)

* Other qualities

- it inhibits the formation of nitrosamines (in a concentration of 0,05 - 0,2 %) in creams and shampoos that are preserved with bronopol (0,01 - 0?05).

- it offers protection to the skin from UV radiations in a (0,5 % in diet or topical 1 %) phototoxicity study on hairless mice (LO,WB,BLACK,HS 1973)
- it inactivates certain viruses (Herpes simplex) because of its solubility in fats and its action on the cell membrane
- it reduces the ulcerations caused by adriamycine (in rats).

* Recapitulation of the studies of toxicity

LD_{50}	mouse	(oral)	836-2000 mg/kg/bw
	rat	(oral)	1700-2450 mg/kg/bw
	cat	(oral)	940-2100 mg/kg/bw
	rabbit	(oral)	2100-3200 mg/kg/bw
	guinea pig	(oral)	10700 mg/kg/bw
		(intraperitoneal)	190 2270 mg/kg/bw

ORAL TOXICITY

— BHT given orally is rapidly absorbed from the intestine.The highest tissue levels are found in adipose tissue but accumulation in fat does not seem to occur with repeated administration in rats or mice, and the half-life in liver and fat is 7-10 days in the rat.Samples of human subcutaneous adipose tissue taken from UK adults,consuming an estimated 1 mg of BHT daily,contained 0.01-0.49 ppm; samples from USA adults, consuming an estimated 2 mg of BHT daily,contained 0.34-3.19 ppm.

The principal metabolism is metabolic oxidation in liver and lungs. By metabolic conversion in human,the studies indicated the formation of 5-carboxy - 7 -(1-carboxy - 1 -methylethyl) 3,3 -dimethyl - 2 - hydroxy 2,3 -dihydrofuran (glucuronide) as principal product.

With one oral dose 40 mg/kg/bw BHT C^{14} (in humans) 50 % of radio-activity is excreted in urine between 24 hours.We have a slow excretion (25 %) for 10 days.

Lbng-term and chronic oral toxicity

Systemic effects

- Effect on the liver

Studies (on rats) showed a relation between BHT exposure and enlargement of the liver. It is a rapide and revertible action.
More recent studies on infant and juvenile monkeys have reported that daily doses of 50 or 500 mg BHT/kg/bw for 4 weeks had no effect on urine or serum chemistry or on haematology.Histological examination of the major organs however did show an effect on the liver. BHT treated animals showed enlargement of hepatocytes and their nuclei with moderateproliferation of endoplasmic reticulum and prominent lipid droplets. In the high-dose group there was fragmentation of the nucleolus in some hepatocytesJuveniles also showed enzyme changes.

In rats with a 60 day-diet at the level of 250 mg/kg and in rats and mice with a 6 day-diet at 0,5 % ,high levels of P-450 cyt were observed.
BHT in the diet of SD rats resulted in a marked decrease in the NADPH-cytochrome P-450 reductase activity of isolated liver microsomal preparations. This effect was not observed when BHT was added <u>in vitro</u> to liver microsomes. An increase in GSH-8 transferase was observed in liver of rats fed 0.4 % BHT in the diet. Dietary BHT is also reported to have an effect on the carboxylation process in the conversion of rat liver microsomal protein to prothrombin. It has also been suggested that phenolic antioxidants such as BHT may protect from carcinogens by lowering cyclic guanosine monophosphate levels.
BHT intreferes with the normal lipid metabolism in the liver. According to the treated animal kind (rat or monkey) increasing of serum cholesterol level or reduction of total cholesterol levels are found . That could show a difference in metabolism betweem rodents and monkeys.

- Haematology

In haematology studies BHT is reported to have caused extensive internal and external haemorrhaging,in association with hypoprothrombinaemia leading to death in some,but not all strains of laboratory rats,mice and guinea pigs. The minimum dietary level of BHT to cause such effects was 170 ppm; with continuous feeding,deaths occurred within 40 days at dietary levels of 6900 ppm and above. These effects were not seen in hamsters,dogs,rabbits or quail. Further recent study into the mechanism of BHT induced haemorrhage has indicated that a number of factors may be involved,including reduction in vitamin K and plasma zinc levels,reduced activity of clotting factors,alterations in platelet function and vascular permeability.

- Effect on the thyroid

Increases in numbers of thyroid follicles, thyroid weight and thyroid iodine uptake have been observed in rats given 500 ppm BHT in the diet for up to 90 days, but this was not accompanied by any change in thyroid hormone levels in the blood and long-term studies on rats maintained on diets containing up to 10.000 ppm (1%) have not shown any adverse effects on the thyroid.

- Effect on the lungs

3 or 5 days after a i.p. 0.004 to 2?5 gr/kg injection in the mice, enlargement of the lungs and disorganization of the cell components are found. These changes are linked with a lung weight and a DNA synthesis increasing similar effects are observed with an oral administration.
BHT is metabolized in the lungs and recent studies could show that quinones of similar metabolic products proceeding from microsomal oxidation of toxic phenols have a part in the adverse effects on the lungs.
Adverse effects on adrenal glands,kidneys,heart cells,small intestine contractions and the growth rate are also observed.

CUTANEOUS AND MUCOUS TOLERANCE

DERMAL ROUTE TOXICITY

The possible skin lightening properties of BHT have been investigated by patch testing and normal usage on black-skinned volunteers.A single application of 0.5 or 3 % BHT in a cream base was applied to the back (intact skin) of 50 black factory workers,with occlusion for 48 hours.Slight erythema and flaking of the skin was observed in 9/50 cases.When 3% BHT in the same cream was applied twice daily over 7 weeks two volunteers suffered adverse reactions - one mild erythema and folliculitis in the test area after 5 weeks and the other erythema and skin flaking after 6 weeks. In no case was there any evidence of skin lightening.

In a maximisation test on human volunteers , 1.0 ml of BHT 2% in pet was applied for 5 alternate days for 48 hour periods, with occlusion to the forearm (pretested with sodium lauryl sulphate(SLS) for 24 hours before initial patch). After 14 days a challenge dose of 0.4 ml test material,at a concentration of 2% pet,was applied to the right side of the back,for 48 hours,with occlusion,preceded by 30 mm application of 5% aqueous SLS to the left side. Relatively little irritation was produced by SLS,and no significant reaction at the non SLS treated sites in the 28 subjects tested.

In control patch testing of 112 patients with BHT 2% pet and BHA 2% pet ,two patients had positive reactions to both and one to BHA only and one to BHT only.A patch test positive to BHT 2% pet was suspected to have caused eczema of the hands for 6 months and of the feet and lips for 3 months in a 24 year old female. When the patient was almost free of symptoms 5% alcoholic solutions of BHA and BHT were swabbed,each on one finger. After 15 minutes both of the treated areas itched, and shortly after an urticarial reaction developed. The symptoms subsided in about 1 hour. The results indicated that BHT and/or BHA produce allergic and not irritant reactions.

In a modified Draize test on 5 guinea pigs each animal was given a total of 10 intracutaneous injections over a 2 week period.The first injection was 0.04 mg BHT in 0.05 ml 10% aqueous ethanol, the second to tenth injections 0.08 mg BHT in 0.1 ml 10% alcohol and a final injection of 0.04 mg BHT. It is reported that signs of local irritation and necrosis were observed. All animals gained weight and appeared healthy. When one animal was given a sensitizing dose of 0.08 mg BHT in 0.1 ml 10% aq alcohol followed by a challenge dose of 0.16 mg in 0.2 ml 10% aqueous alcohol two weeks later,no signs of anaphylaxis were observed.

In an acute dermal study , a dose of 5g/kg to 10 rabbits, produced slight redness in 6 animals, moderate redness in 4 , slight oedema in 3 , and moderate oedema in 3. Recovery data are unavailable.

In a dermal toxicity study,a semi permanent hair dye formulation used as is and containing 0.25% BHT was applied to the abraded skin of 12 rabbits (six male,six female) twice weekly for 13 weeks. The dose used was 1 mg/kg of the 1:1 oxidation mixture. The rabbits were kept in a restrained position for one hour following each application,then shampooed,rinsed,dried and returned to their cages. Body weight gain, relative organ weights,clinical chemistry and results of urinalysis of test animals were comparable with controls. No dye discolouration of the urine was seen at any time during the test. No gross or microscopic lesions were observed.

MUTAGENICITY

The majority of mutagenicity tests for BHT have proved negative. Testsfor bacterial mutation, both in plate assays (concentrations of up to 1000 g/plate), with and without activation,and in a host mediated assay in the mouse (oral doses 30 mg/kg, 0.9 g/kg or 1.4 g/kg) have all been negative. Tests for induction of sex-linked recessive lethal mutations, X-chromosome loss, and translocations in Drosophila also proved negative. However conflicting results have been reported for gene mutation and chromosome damage in mammalian cells in *vitro* and in dominant lethal assays.

When human embryonic lung (WI-38) cells were analysed at anaphase following exposure to 2.5,25 or 250 µg/ml of BHT a sharp increase in the number of abnormal anaphase figures was observed,particularly evident at the highest dose level tested. Data are unavailable and the significance of the assay is unclear. Chromosome aberrations in metaphase human lymphocyte cultures exposed to 5-40 µg/ml BHT for the final 24 hours of a 72 hours culture have also been reported. The abnormalities described as "fragmented" or "prophase-like" were seen at all dose levels and no clear dose response was obtained.

A more recent study has reported negative results using hamster cells.

All four in vivo studies in bone marrow were reported to be negative. Administration of BHT in the diet for 21 months had no effect on chromosome aberrations in regenerating mouse liver.

A dominant lethal effect has been detected in the first mating week in groups of rats fed 1333 ppm and 4000 ppm BHT in the diet for 10 weeks. Other dominant studies are available which showed negative results. (It is understood that the positive result might be regarded as a questionable response.All currently available dominant lethal studies are to be assessed by the relevant expert Committee within the UK.)

CARCINOGENICITY

According to the studies,contradictory effects are observed in the incidence of lung tumours.

In carcinogenicity studies carried out to date,administration of BHT at range 1000-7500 ppm in the diet to mice for 16 months or more was associated with a dose-related increased incidence of lung tumours, slightly increased incidence of stomach tumours and decrease in reticulum cell sarcomas, compared with controls. The incidence of hepatic cysts was also increased but liver tumours were not seen in either treated or control mice. In two recently conducted studies on rats maintained on diets containing in one case 2500 or 10000 ppm and the other 3000 or 6000 ppm BHT for 104 weeks, no evidence of carcinogenicity was found. However a recent 2-generation long-term carcinogenicity study in Denmark has shown a significant enhanced development of hepatocellular adenomas and carcinomas in rats of both sexes dosed with BHT at concentrations of 0,25,100 or 500 mg/kg/bw.
A dose related response was observed in male F_1-rats.At weaning treated F_1-rats had exhibited a dose related reduced body weight which persisted throughout the study. This was most pronounced in male F_1-rats.

The study does not discriminatebetween an initiating or promoting property of BHT. In addition it differed from other studies in that exposure to BHT commenced in utero thus extending the period of BHT exposure to 141-144 weeks. It has been reported that BHT can cross the placental barrier and be excreted in rat milk. Additionally none of the tumours found in the Danish study were seen before 111 weeks, the termination point in the other two studies. In the Danish study it should also be noted that survival of the controls was markedly poorer (33%) than that of the top-dose group (83%). It has also been suggested that the lowered body weight of rats dosed with BHT should not be regarded as a "toxic" effect but rather an adaptive response due to possible hyperactivity of the thyroid (see earlier report).

Single i.p. injections of 400-500 mg/kg BHT induced acute lung toxicity in the mouse.Testing of structural analogues to BHT and metabolic studies suggest enhancement of lung tumour yield by chemical carcinogens in susceptible species of mice.

TERATOGENICITY

Effects on reproduction and teratogenic properties are not clear. Recent studies show that BHT goes through the placenta and is found in the milk of rats.
Reproduction studies in a variety of species including non-human primates indicated a no-effect-level of 1000 ppm BHT in the diet or 50 mg/kg/bw. Effects reported in these studies included depression of growth rate of parents and offspring in a two-generation study on rats fed 3000 ppm BHT in the diet, but not at 300 or 1000 ppm, intrauterine deaths associated with maternal toxicity in rabbits given 3-320 mg/kg/day by gavage during foetal development,and developmental delays,reduced pup numbers and pup weight 12 days after birth in mice fed 5000 ppm BHT in the diet but not at 1000 ppm.

Rhesus monkeys maintained on a diet containing BHA and BHT,providing an intake equivalent to 50 mg/kg/bw/dayfor each antioxidant were bred for 1 year and continued on the diet for a second year. Clinical and behavioural observations on infants up to 2 years of age showed no abnormal findings.

In a teratology study on rats, a dose of 2 mg/kg of a semi-permanent hair dye formulation used as it was applied topically to the abraded skin of groups of 20 mated CR CD female rats on days 1, 4, 7, 10, 13, 16 and 19 of gestation. Except for changes in the colour of the skin and hair at the site of dye application no signs of toxicity were seen throughout the study. Body weight gains were similar in test and control animals. No significant differences in the mean number of corpora lutea,implantation sites and live foetuses were noted.

CONCLUSION

In spite of a lot of studies ,obtained results cannot clearly estimate the usual risk of BHT.In consideration of data showing that BHT could have a teratogenic effect on behaviour and growth, the current ADI of 0,5 mg/kg/bw needs supplementary data to be admitted again. BHT is more toxic than BHA and a usual concentration of 1% in cosmetic products could be cause of a too high systemic exposure level.

Classification : D

Bibliography

- ALLEN J R et al Arch. dermatol 109 , 216-7 (1974)
- BENTLEY-PHILLIPS, B., BAYLES, M.A.H.
  "Butylated Hydroxytoluene as a Skin Lightener".
  Arch. Dermatol. 1974, 109, 216-7.
- BLACK, H.S.
  "Effects of Dietary Antioxidants on Actinic Tumor Induction".
  Res. Comm. in Chem. Path. and Pharmacology 1974, 7 (4),783-6.
- BLACK, H.S., CHAN, J.T., BROWN, G.E.
  "Effects of dietary constituents on Ultraviolet Light-mediated Carcinogenesis".
  Cancer Res. 1978, 38 , 1384-7
- BURNETT C. et al
  J. Toxicol. Environ. Hlth 1 ,1027 - 40 (1976)
- CHAN, J.T., CHAN E.Y., BLACK, H.S., WYBORNY, L.E.
  "Effects of lipid soluble antioxidants on cytotoxicity induced by photochemical products of cholesterol".
  Experientia 1980 , 36 , 439-40.
- CIR Safety data test CTFA Company test code 1438 1979-1980.
- CIR Safety data test CTFA Product type 7B/13120 - 01
  Company test code E 35 /167 Date of test 2/22/79

- Colipa data CSC/295/83 - 2
295/82
- DEICHMANN, W.B., CLEMPER, J.J., RAKOCZY,R., BIANCHINE, J.
"Toxicity of Ditertiarybutylmethyl phenol".
Arch. Ind. Hlth. 1955, 11 , 93-101.
- DERIOS, G., CHAN, J.T., BLACK, H.S., RUDOLPH, A.H., KNOX, J.M.
"Systemic protection by anioxidants against UVL-induced erythema".
J. Invest. Derm. 1978 , 70, 123-5.
- Dr. ROBINSON data
Environmental Research vol. 29, p 1-29 (1982)
- GILLES, D., LUCAS, J.B.
NIOSH Report PB-249 387, 1975.
- Hill Top Research Miamiville Ohio
International Bio-Research Inc, Woodson Tenent Laboratories
November 76 (76-757-72)
- HIROSE M. et al Fd Cosm. Toxicol 19 , 147-51 (1981)
- Joint FAO/WHO Expert Committee on Food Additives (1983)
WHO Techn. Rept Series n° 696
- LO, W.B., BLACK, H.S.
"Inhibition of carcinogen formation in skin irridiated with ultraviolet light".
Nature 1973, 246, 489-91.
- MACDONALD, M.E.
Inveresk Research International Report N° 467, 1976.
- MAISACH, H.I., GELLIN, G., RING, M.
"Is the antioxidant BHT a depigmenting agent in man".
Contact Dermatitis 1975, 1 , 295-6.
- MAJORS, P.A., RUBENKOENIG, H.L., WELSH, A.L.
Unpublished report by Hill Top Research Institute,Inc.,1959.
- MALLETTE,F.S., VON HAMM, E.
"Studies on the toxicity and skin effects of compounds used rubber and plastics indutries".
- MATSUO M. et al Fd. Chem. Toxicol. 22, 345-54 (1984)
- MENEGHINI, C.L., RANTUCCIO, F., LOMUTO,M.
"Additives,vehicles and active drugs of topical medicaments as causes of delayed-type allergic dermatitis".
Dermatologica 1971, 143 , 137-47.
- MONOGRAPH : BHT prepared by International Life Science Institute (ILSI) Washington October (1984).

- MIZUTANI T. et al. Toxicol. appl. Pharmacol 62, 273-81 (1982).
- NCI Study (1979) NCI - CG - TR - 150
- NERA E.A. et al Toxicology 32 , 197-213 (1984)
- NLM'S National Interactive Retrieval Service Medlars II
- OLSEN P. et al Inst. Toxicol. Denmark (1984)
  Data as presented to SCF ref III.1254/84 + (Annex)
- RACINE Ph. Int J. Cosmet Sci, 3, 125-37 (1981)
- ROED-PATERSEN, J., HJORTH, N.
  "Contact Dermatis from antioxidants"
  Br. J. Derm. 1976, 24 , 233-41.
- SHELBY M.D. et al. Personal Communication (1984) (part of NTP Columbia University).
- Stamford Research Institute (1972-1977) Internal Report.
- TAKAHASHI O et al
  Fd. Chem. Toxicol. 22 , 97-103 (1984)
- VOLLUM, D.I.,
  "Hypomelanosis from an antioxidant in polyethylene film".
  Arch. Derm. 1971 , 104 , 70-2.

(3) D. A. H. Q. DITERTIARY AMYL HYDROQUINONE

* Formula and synonyms

DAHQ = 2,5 - ditertiary amyl hydroquinone is the aromatic phenol that conforms to the formula

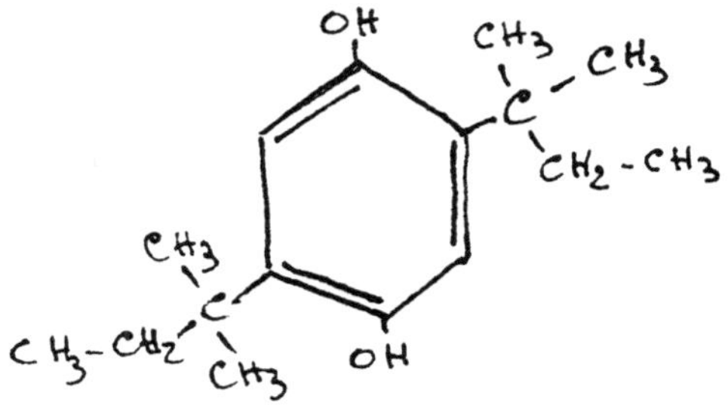

Other names : - 1,4 - Benzenediol 2,5 bis (1,1 - Dimethylpropyl)
- 2,5 - bis (1,1 Dimethylpropyl) 1,4 -Benzenediol
- 2,5 Di - t - Amylhydroquinone
- 2,5 Di - t - Pentylhydroquinone
- Diamylhydroquinone (DAH) CTFA
- Santovar A (Monsanto) (Trade name)
- Colipa n° A011

* Physical characteristics

A fine free-flowing buff-coloured powder
Odourless
Soluble in ethanol, acetone, ether
Insoluble in water

* Use

a) As an antioxidant in cosmetic foundation creams, cuticle softener, cream rouges, at a level not exceeding 0,2 %.
b) As an antioxidant for uncured rubber and used in adhesive films for their adhesive tape application.

* Recapitulation of the studies of toxicity

ORAL TOXICITY

- Acute : LD50 (rats) = 2 gr/kg/bw

- Long term-study

In a long term-study 115 male and 60 female rats were fed diets containing 0, 0.0125, 0.025, 0.05, 0.1 and 0.2% DAH. Body weights,food intake and general appearance were observed weekly for about 200 days then half the males of each group were autopsied and all females receiving a commercial laboratory diet. The remainder were autopsied at 500 days. Body weights of animals both at 200 and at 500 days were similar to those of controls.Weightsof organs for animals autopsied after 200 days did not change with increase in concentration. The gross appearance of these tissues and those from animals autopsied after 500 days did not differ significantly from that of control animals. Microscopic examination of tissue sections revealed no lesions which could be attributed to the test materials.

CUTANEOUS AND MUCOUS TOLERANCE

In an eye irritation test, 0.1 ml undiluted test material (ie 0.2%) in a make up foundation applied to the eyes of six albino rabbits with no rinse off caused slight conjunctivitis in all animals. Full recovery occurred in 48 hours. The cornea and iris membranes were unaffected.

In a skin irritation test, a total of 0.5 g of undiluted test material in a make up foundation,was applied to the intact skin of rabbits four times daily for seven days,without occlusion. No evidence of irritation was found.

When 0.2 ml of make up foundation containing 0.2% 2.5 ditertiary amyl hydroquinone was applied to the intact skin of 12 individuals, with occlusion,once daily for 21 days,no irritation occurred.

When a suspension of DAH made up by dissolving 0.1% of crystals in 5 ml of ethanol, adding 0.1 ml of Tween 80 and diluting to 100 ml with 0.9% NaCl solution was applied to the skin of guinea pigs,evaporation of the alcohol left a visible ,crystalline deposit which usually was not irritating.When dissolved in cottonseed oil, an erythema appeared in 24 hours, disappearing in the next 24 hours. Repeated applications of the oil solutions increased hyperemia, led to papular eruption followed by scab-formation over entire area. When treatment was stopped, lesions gradually disappeared, leaving a normal-looking skin after about a month. Skin sensitization was not found.

CARCINOGENICITY

Tertiary amyl hydroquinone (20-50 µg/ml) is reported to have given significant protection against photodecomposition of benzo(a) pyrene 0.1-0.2 µg/ml irradiated with longwave UV light.

TERATOGENICITY

In a study of the effect of antioxidants on the incidence of foetal resorption in rats as compared with normal animals held on normal diets, 2500 mg of 2,5-di-tert-amyl hydroquinone, given as a single dose on day one of gestation, gave a low percentage of resorptions (4.5% from 10 litters with 128 normal foetuses).

MISCELLANEOUS STUDIES

In a study of the effect of lipid antioxidants on normal rat erythrocytes, 2,5 ditertiary amyl hydroquinone was non-haemolytic.

2,5-Ditertiary-amyl hydroquinone was found in extracts of several human tissues, including neurological tumours, adult brain and foetal brain. On further investigation it was also found in adhesive tapes used in the hospital laboratory to label glassware, but not in quantities such that would be responsible for contamination of the tissues. One possible source was thought to be an antioxidant preparation used in rubber manufacture.

CONCLUSION

Are required :
- Short-term toxicity test to establish a NEL (no effect level)
- Information on dermal absorption
  potential sensitization
  oral short-term test mutagenicity

(gene on mutation on bacteria and chromosomal aberration in mammalian cells)

Industry should be asked whether there is a need for this substance.

Classification : C

References :

- COLIPA data CSC/296/82
- Wilson R.H. et al Proc. Soc.Exp.Biol. Med. 104 29-31 (1960)
- Middleditch B.S. et al Biomed.Mass Spectrom 7 (3) 105-8 (1980)
- Telford I.R. et al Am. J . Anat. 110 29-36 (1962)
- Kumar P.S. et al Int.J.Vit.Nutr.Res. 49 (4),352-8 (1980)

(4) D. B. H. Q. DITERTIARY BUTYL HYDROQUINONE

* Formula and synonyms

DBHQ = 2,5 - Ditertiary butyl hydroquinone is the aromatic compound that conforms to the formula :

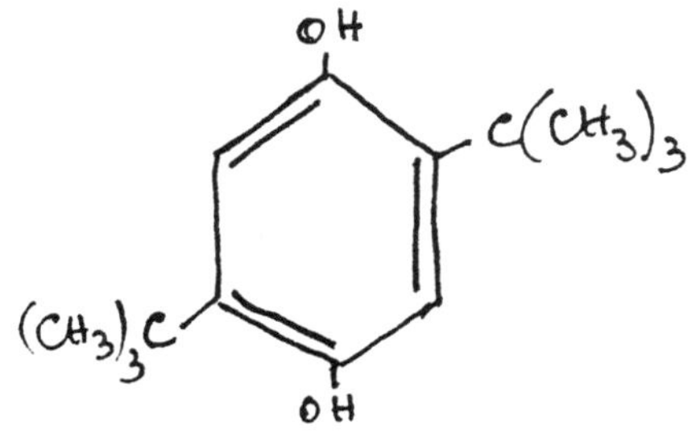

Other names : - dibutylhydroquinone
- DBH
- Colipa A012

* Physical characteristics

Soluble in organic solvents,oils and alkaline solutions
Insoluble in water

* Use

As an antioxidant in decorative cosmetics (lipsticks,eyeshadows, coversticks) at a concentration not exceeding 0,2 %.

* Recapitulation of the studies of toxicity

ORAL TOXICITY

Acute toxicity : LD50 (chicken) : 10 g/kg

A 20% solution in ethanol given by gavage to one rat at doses of 500 and 1000 mg/kg caused mild depression. A 30% solution in cottonseed oil given to five rats at 1500 mg dose level produced slight diarrhoea. In each case the effect was attributed to the solvent.

When DBH was given to chickens in the diet at level of 2 gr or more/ kg toxic symptoms were noted.

Short- and long-term studies

Chicken : continued administration of 0,25% in the feed (time not stated) led to nervous symptoms,increased mortality, decreased growth reduced egg production and hatchability.

Oral administration of DBH at the dose of 200 mg/kg to dogs for 2 months and to rats for an undisclosed time produced no evidence of toxic effects.

In a long-term oral study on 115 male and 60 female rats DBH was added with thorough mixing to two diets to give the following concentrations 0, 0.0125, 0.025, 0.05, 0.1 and 0.2%. Body weights, food intake and general appearance were observed weekly for about 200 days. There was no difference between treated and control animals At the end of this time half the male animals and all the females receiving a commercially available laboratory diet were autopsied. Gross appearance of tissues and organ weights were recorded.
No adverse effects were noted. The gross appearance of tissues and those autopsied after 500 days did not differ significantly from that of control animals. Histology studies revealed no lesions which could be attributed to the test material.

CUTANEOUS AND MUCOUS TOLERANCE

In skin irritation tests on rabbits and guinea pigs ,0.1 ml of a 10% solution of DBH in ethanol or cottonseed oil was placed on the skin five times daily, for two weeks. Evaporation of alcohol left a visible,crystalline deposit which usually was not irritating.When dissolved in cottonseed oil erythema appeared within 24 hours. Repeated applications increased hyperemia,led to papular eruption followed by scab-formation over the entire area. When treatment was stopped,the lesions gradually disappeared.Skin appeared normal after 28 days.
It is not clear from the report how many animals were used,the site of application,whether the skin was intact or abraded, with occlusion or not.
In a sensitization test on guinea pigs,a suspension of DBH was made by dissolving 0.1 g of crystals in 5 ml of ethanol, adding 0.1 ml of Tween 80 and diluting to 100 ml with 0.9 NaCl solution.

No evidence of sensitization was observed. Again information is lacking on number of animals,induction route,exposure time and number of applications.

Two reports are available of allergy to DBH when used as an anti-oxidant in cream eyeshadow .Patch testing at 1% in methyl ethyl ketone was positive.The concentration used in the cream was given as 0.52 %.

MISCELLANEOUS INFORMATIONS

Intraperitoneal injection of 200 mg/kg DBH into rats five times weekly for 4 weeks is said to have caused no adverse effects.

In a study on possible depigmentation effect, 0.4% DBH was fed to guinea pigs in the diet for one week, then reduced to 0.2% for a further two weeks. No bleaching of hair occurred. When DBH was given five times daily for 2½ months to a black kitten a partial browning of hair occurred which remained irreversible after the treatment was stopped. During the final six weeks, the daily dose was 100 mg.

CONCLUSION

Are required :
- information on dermal absorption
- adequate data on cutaneous irritation and potential sensitization
- short-term oral toxicity to establish a no effect level (NEL)
- mutagenicity test (gene on mutation on bacteria and chromosomal aberration in mammalian cells.)

Classification : C

References

- Wilson R.H. and Poley G.W. Proc. Soc. Exp. Biol.,Med. 104 29-31 (1960)
- COLIPA Data CSC/356/83.
- Calnan C.D. Contact Dermatitis Newsletter N°13. January (1973)
- Calnan C.D. Contact Dermatitis Newsletter N°14. August (1973).

REPORT BY THE SCIENTIFIC COMMITTEE ON COSMETOLOGY

ON THE USE OF

SODIUM HYDROXYMETHYLAMINOACETATE AS A PRESERVATIVE AGENT

(opinion expressed on 20-2-87)

* The committee's mandate

To give its opinion on the use of sodium hydroxymethylaminoacetate as a preservative agent in a maximal concentration of 0.5% in cosmetic products.

* Formula and synonyms

$C_3H_6NO_3Na$
MW 127,1

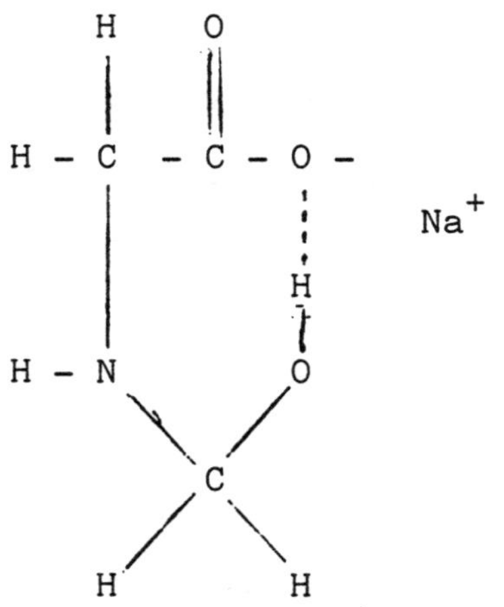

Other names: - sodium hydroxymethyl glycinate
- Suttocide A
- Colipa n° P84

* Characteristics

The compound is strongly alkaline
Highly soluble in water
Soluble in methanol,propylene glycol and glycerin
Insoluble in most organic solvents

* Use

A recently introduced preservative for use in cosmetics at concentrations of 0,05% to 0,5%.

* Recapitulation of the studies of toxicity

ORAL TOXICITY

Acute : LD50 (rat) 1,067 gr/kg/bw and 1,410 gr/kg

In a sub-chronic oral toxicity study, 4 groups of 10 rats/sex received by gavage 0 (control),10, 40 or 160 mg/kg/bw/day as a 2% aqueous solution for 90 days. There were no clinical signs of toxicity or changes in body weight gain, food intake, haematology,clinical chemistry or urine examinations. Gross or microscopic examinations did not reveal any treatment-related changes. It appears, therefore, that the dose-levels were too low to establish the target organ or the type of systemic toxicity induced by this compound. Otherwise, the study was well conducted and reported.

No signs of toxicity established in an oral 90-study at the highest dose level of 160 mg/kg/bw/day appears to indicate a low systemic toxicity. An effective dose and a target organ have therefore not been established.

CUTANEOUS AND MUCOUS TOLERANCE

Dermal LD50 (rabbit) > 2 gr/kg/bw

The undiluted material applied dermally under occlusion caused burned skin probably as a result of the alkaline properties.

Skin irritation tests in rabbits, showed a 5% aqueous solution to be moderately irritating, while a 0.5%solution revealed only slight, transient irritation.

In a repeated dermal application test, guinea pigs received 0.5 ml aqueous dilutions of 50, 7.5, 0.75 and 0.38% under occlusion on days 1, 3 and 6 of a one week period. No signs of oedema or irritation were observed.

Eye irritation tests in rabbits conducted with 100 mg undiluted powder showed moderate irritation when the eye remained unwashed, and mild irritation when the eye was washed after treatment.

A 5% aqueous solution was mildly irritating if not washed out,and not irritating if washing was applied.

Sensitization was examined in guinea pigs, by the Landsteiner test, the maximization test and the Buehler test.In the Landsteiner test,

0.1 ml 0.1% solution in saline was injected intradermally ten times, once every other day. After a 2 weeks rest period, the intradermal challenge injection of 0.05 ml 0.1% solution did not reveal sensitizing potency.
In the maximization test, the induction treatment consisted of 6 intradermal injections of 0.1 ml 5% solution, followed 8 days later, by topical application of 0.3 g moistened powder. On day 22, a topical challenge treatment with a 50 % aqueous dilution produced a positive reaction in 7 out of 10 animals. When the challenge was repeated with 5% and 0.5%, 7 days later, 4/10 respectively 2/10 animals reacted positively. These results indicate sensitizing properties of the test substance.
In the Buehler test, 0.5 ml 0.5% aqueous solution was applied topically 10 times during 3 weeks. After 2 weeks rest, the challenge with the 0.5% solution was negative in each of the ten test animals.

No information on irritation or sensitization in humans was provided. The substance showed slight skin irritation on rabbit skin and sensitizing properties in guinea pigs even at use concentration.

MUTAGENICITY

An Ames test with up to 0.5 mg/plate in 5 strains of S.typhimurium, with and without metabolic activation did not indicate mutagenic properties.

CONCLUSION

The available information shows many gaps which are considered unacceptable, especially for a new substance.
Therefore,evaluation is not possible until the information is available as required in the guidelines.

Are required :
- mucous irritation test
- skin sensitization test in humans
- systemic toxicity (NEL and target organ)
- mutagenicity test (chromosomal aberration in mammalian cells)

Note : The committee noted that the concentration needed for preservation of cosmetics is probably considerably less than 0.5 %.

Classification : C

REFERENCES - LABORATORY REPORTS

Colipa Subm I 9/86

Acute Oral and Dermal Toxicity

- Griffiths, J.T. and Parent, R.A. FDRL Report N° 6185a, June 12 1979. Approximate acute oral toxicity (LD50) in rats.
- Frances, S. and Ferrer, L. Wells Lab Report H-9304 December 20, 1979. Acute oral toxicity in rats.
- Griffiths, J.T. and Parent, R.A. FDRL Report N° 6185a, April 4, 1979. Acute dermal toxicity in rabbits.

Primary Skin Irritation

- Griffiths, J.T. and Parent, R.A. FDRL Report N° 6261a-1, July 5 1979. Primary skin irritation test with rabbits (as received).
- Griffiths, J.T. and Parent, R.A. FDRL Report N° 6261a-2, July 5 1979. Primary skin irritation test with rabbits (5% aqueous).
- Griffiths, J.T. and Parent, R.A. FDRL Report N° 6550A, May 21 1980. Primary skin irritation test with rabbits (0.5% aqueous).
- de Azevedo, J.P. and Rupp, F. Wells Lab Report N° H-8713, October 1, 1979. Report on primary skin irritation (as received).
- de Azevedo, J.P. and Rupp, F. Wells Lab Report N° H-8713A, October 1, 1979. Report on primary skin irritation (5.0% aqueous).
- Levenstein, I. Leberco Labs Report 04515, April 4, 1980. Primary skin irritation in rabbits (full strength).
- Levenstein, I. Leberco Labs Report 04515, April 4, 1980. Primary skin irritation in rabbits (full strength).
- Reagan, E.L. and Becci, P.J. FDRL Study N° 8453A, June 20, 1985. Dermal range finding study of Suttocide A in Guinea pigs.

Eye irritation

- Griffiths, J.T. and Koschier, F.J. FDRL Report 6261a-1, August 1, 1979. Eye irritation test in rabbits when administered as received.

- Griffiths, J.T. and Koschier, F.J. FDRL Report 6261a-2, August 1, 1979. Eye irritation test in rabbits when administered as 5% aqueous solution.

- de Azevedo, J.P. and Frances, S. Wells Lab Report H-8712, October 4, 1979. Report on eye irritation testing in two rabbits.

Contact Sensitization

- Levenstein, I. Leberco Lab Report 10864, December 22, 1980. Guinea pig sensitization study.

- Reagan, E.L. and Becci, P.J. FDRL study N° 8158, December 5, 1984. Dermal sensitization study : maximization test (GPMT) of Suttocide A in guinea pigs.

- Reagan, E.L. and Becci, P.J. FDRL study N° 8453A, June 20, 1985. Dermal Sensitization study : maximization test (GPMT) of Suttocide A in guinea pigs.

Subchronic Oral Toxicity

- Johnson, W.D. et al. FDRL Study N° 7824. May 15, 1984. 90-Day Oral (gavage) toxicity study in rats with Suttocide A.

Mutagenicity

- Haworth, S.R. et al. Microbiological Associates Report Study N° T2114.501, September 15, 1983. Salmonella/Mammalianmicrosome plate incorporation mutagenicity assay (ames test).

REFERENCES - PUBLISHED LITERATURE

- Knudsen, L.F. and Curtis, J.M. (1947) J.Am.Stat.Assoc. 42:282.

- Litchfield, V.T. and Wilcoxen, F. (1949) J.Pharmacol. Exper. Therap. 96:99.

- Draize, J.H. et al. (1944) J. Pharm. Exper.Therap. 82:377.

- J. Soc. Cosmetic Chemists (1962) 13 : 281.

- Landsteiner,K. and Jacobs,J. (1935) J. Exp. med. 61:643.

- Magnusson,B. and Kligman,A.M. (1969) J. Invest.Dermatol. 52 : 268.

- Buehler,E.V. & Griffith,F. (1975) In : Animal Models in Dermatology H.I.Maibach(ed) pp56.

REPORT BY THE SCIENTIFIC COMMITTEE ON COSMETOLOGY

ON THE USE OF

3- DECYLOXY- 2-HYDROXY-1-AMINOPROPANE HYDROCHLORIDE AS A PRESERVATIVE

(opinion expressed on 20 February 87)

* The committee's mandate

To give its opinion on the use of 3-decyloxy-2-hydroxy-1-aminopropane as a preservative agent in cosmetics in a maximal concentration of 0,5 %.

* Formula and synonyms

$C_{13}H_{29}NO_2Cl$ MW : 267,5

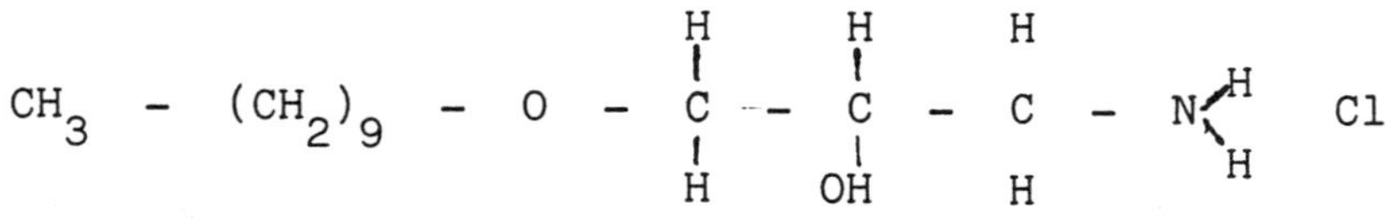

Other names : - Decominol
- Ster 4
- PVA 44
- Colipa P 87

* Characteristics

Soluble in water,ethanol,chloroform
Insoluble in hexane

* Use

Proposed for use as a preservative in all types of cosmetics up to 0,5%

* Recapitulation of the studies of toxicity

ORAL TOXICITY

Acute : LD50 (mouse and rat) : 1,2 gr/kg/bw
Damage to gastric mucosa found.

Subacute toxicity

Dog . Oral. Two male beagle dogs were given doses of a.i. in capsules The initial dose was 2.5 mg/kg/bw/ day for 3 days ; the dose was then increased for the next 3 days,and so on.The doses used were : 2.5,10, 40 , 160 and 640 mg/kg/bw/day, over 15 days.Some temporary drowsiness was noted after dosing at the 40 mg/kg level and above, and with some of the doses of 160 mg/kg/bw and all of the doses of 640 mg/kg/bw vomiting occurred. Blood was present in the faeces at the top dose. Some changes in clinical chemistry and haematology do not seem to be significant. At autopsy, ulceration or bleeding points were found in the gastric mucosa. Slight renal changes may have been drug-related.

Subchronic toxicity

Dog. Oral. A 13-week study was carried out in 8 m and 8 f beagles. The a. i. was administered by capsule at doses of 0, 0.2, 1.0 and 10 mgKg/bw/day for 91 consecutive days. No changes in behaviour were noted. The laboratory and autopsy findings were not markedly abnormal; at the top dose there was a fall in the relative weight of the thyroid gland.

Rat. Oral. A 13-week study in 60 m and 60 f SPF rats was carried out by administering 0, 10, 40 and 160 mg/kg/bw by gavage as a 0.5 % aqueous solution. Behavioural changes were seen early in the experiment at dose levels of 40 mg/kg and above ; at the top-doselevel effects seen were piloerection, decrease in spontaneous activity, and neglect of grooming. Top dose males lost weight after 50 days. Clinical chemistry studies showed uraemia at 40 mg/kg (females only) and 140 mg/kg (both sexes). Tests for occult blood in the faeces were negative. At autopsy,dose related gastric gastric damage was seen macroscopically. An increase in relative liver weight,together with evidence of mesenchymal inflammation was seen at the top dose level.

Thus the NEL in this study was below 10 mg/kg/bw/day,but effects seen at this level were limited to local irritant effects on the gastric mucosa, presumably due to the direct irritant effect of a 0.5 % solution.

CUTANEOUS AND MUCOUS TOLERANCE

Mucous membrane irritation

Rabbit. A 1% solution of a.i. in water was used in a Draize test, with and without rinsing. Irritation was found in all cases, more marked in the eyes not subjected to rinsing. Complete recovery had occurred in a week.

Skin irritation

Rabbit. A 1% aqueous solution with occlusion was mildly irritating. Repeated daily applications of 2% aqueous solution for one month, without occlusion, gave slight irritation in the first week,but none thereafter.

Phototoxicity

Guinea pig. Twenty male animals were tested. A solution of 0.5 % of a.i. in ethanol/water (1:1) was chosen as the maximum dose which was non-irritant. The material was allowed to remain in contact with the skin for 15 minutes before exposure to ultra-violet light.Suitable controls were used. The test was negative.
In the phototoxicity test, 1% of a.i. in ethanol caused some erythema. but this solution diluted with water (1:1) was non-irritant.

Photo-allergenicity

Guinea pig. The animals used in the preceding test were used again. Three sensitizing exposures were carried out, each in the same way as for the phototoxicity tests (supra), at intervals of 3 days. After a 16-day rest, a challenge application and exposure to ultra-violet were carried out. Suitable controls were used. The test was negative.

Sensitization

Guinea pig. A Magnusson-Kligman maximization test in 10 animals was negative when 0.1 ml of 1% solution was used for intradermal induction and a 0.5% solution for the challenge.

MUTAGENICITY

An Ames test was negative up to 50 µg/plate ; 100 µg was strongly inhibitory to the organisms. A gene mutation test in CHO-cells (HGPRT-test) at concentrations up to 0.015 mg/ml was negative. Higher concentrations were not examined because of toxicity to the cells (ref. 18). A micronucleus test in female SD rats treated with one oral dose of 1000 mg/kg was negative.

SYSTEMIC TOXICITY

LD50	intravenously	(mouse)	60 mg/kg/bw
LD50	subcutaneously	(mouse)	2000 mg/kg/bw
		(rat)	4000 mg/kg/bw

Rat. Intravenous. Groups of 10 males received 0, 1.75, 3.5 and 7 mg/kg/bw/day for 15 days into the tail vein. The top dose animals developed necrotic changes in the tail, and thrombosis and fibrosis in the vein, and in them the intended course could not be completed. Abnormalities noted were piloerection and haematuria at all doses ; the haemoglobin and haematocrit values fell markedly in the top dose group. Autopsy showed relatively slight changes. The NEL in this experiment seems to have been $<$1.75 mg/kg/bw/day.

CONCLUSION

The tests show we have a skin and mucous irritating substance (even at the concentration of 1 %). It can also cause hepatic and renal damages in caseof systemic administration.
NEL is probably lower than 10 mg/kg/bw/day (oral administration in rats). Damages of gastric mucosa (dogs and rats) are found.
The oral absorption of the compound appears to be very poor.

The Committee requires information : - on percutaneous absorption
- on results of tests of chromosomal aberration and mutagenicity in mammalian cells.

Classification : C

REPORT BY THE SCIENTIFIC COMMITTEE ON COSMETOLOGY

ON THE USE OF

OMADINE MDS IN SHAMPOOS

* The Committee's mandate

To give its opinion on the use of dithio-2,2'-bispyridine - dioxide 1,1' addition product with magnesium sulphate trihydrate in anti-danduff shampoos at the maximal concentration of 1 %.

* Formula and synonyms

EEC n°41

Colipa P 50

Chemical name : Dithio - 2,2' - bispyridine - dioxide 1,1',with magnesium sulphate trihydrate.

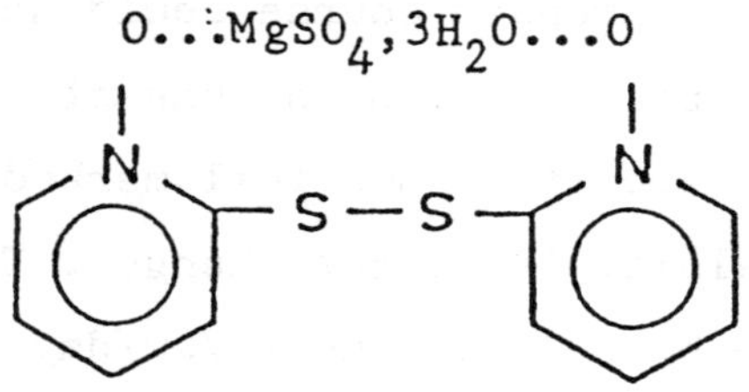

MW : 426,3

Other names : - OMADINE$^{(R)}$ MDS Trade name
- Pyrithion - disulphide magnesium sulphate
- [Bis(2- pyridyl - 1 - OXO) disulphide] magnesium sulphate trihydrate.

* Characteristics

OMADINE MDS is more soluble in water and ethanol than pyrithione itself.

* Use

Used as an anti-danduff agent in shampoos up to 1 %

* Recapitulation of the studies of toxicity

LD50 oral (rats) : 1,1 - 1,35 gr/kg
dermal (rabbits) : 8 gr/kg

ORAL TOXICITY

Acute : In rats single oral doses of 0,125 gr/kg/bw or above were associated with decreased activity, decreased respiration and prostation.
A single oral or intraperitoneal dose of 20 mg/kg in mice did not induce abnormalities in behaviour or growth rate and no microscopic changes were detected in tests or epididymides after 7-21 weeks.

A sub-chronic (90-day) oral study in rats was conducted with 1, 3, 10 and 30 mg/kg/bw/day. Signs of intoxication (abdominal hypotension, kyphosis, and hind leg ataxia) occurred in all treatment groups. No microscopic changes were found in the pancreas but a report on the microscopy of other organs was not available.

In a 6-month oral study in monkeys 5, 15 or 50 mg/kg (4/sex/dose) was administered daily by gavage. Decreased body weight and food intake occurred with 15 and 50 mg/kg. Mortality (isolated cases) occurred at all dose levels, but this was not dose-related and the significance is unclear.

An oral dose administered to rats is excreted mainly in the urine. The compound disappears from the blood in a rapid early phase (c. one hour) followed by a slow phase (20-100 hours depending on the species). The main urinary metabolite is the S-glucuronide of 2-mercapto-pyridine-N-oxide The predominant and persistent metabolite in blood is 2-methylsulfonylpyridine.

CUTANEOUS AND MUCOUS TOLERANCE

DERMAL TOXICITY

A skin irritation test in rabbits with 0.5 g (undiluted substance) produced distinct dermal changes. Aqueous solutions up to 5% did not produce skin irritation in humans.

No clear signs of phototoxicity were observed in guinea pigs treated topically with 0.1 ml of up to 5% suspensions in corn oil, followed by UV-A irrediation.

Distinct eye irritation occurred in rabbits with 0.1 ml of a 5% aqueous suspension (score 11/110). Severe eye irritation was induced in monkeys by applying 100 mg of the diluted powder, but 0.1 ml of a 1% aqueous solution was negative.

A sensitization test in guinea pigs using 0.1 ml of a 0.2% aqueous solution in Freunds complete adjuvant injected into the hind foot pad and scapula regions for induction and 0.05 ml of 0.1 and 0.5% aqueous solutions applied topically as the challenge dose produced negative results. Negative results were also obtained when a 10% aqueous solution was applied topically as both the induction and challenge treatment in a Buehler type test.

No photosensitizing properties were observed upon applying 0.1 ml of 1.5% suspension in corn oil and UV-A irradiation.

Dermal absorption has been shown to occur in animals and man.

- In man, 3-15% of a dermal dose was recovered in the urine after application of 4, 12 or 40 $\mu g/cm^2$ in methanol,without occlusion.

- In an in vitro study with abdominal skin of humans, exposure for 1 000 minutes (without rinsing), resulted in recovery of 14% in the horny layer, 0.6% in the epidermis, 1.0% in the corium and 1.0% in the penetration chamber.

- In an in vitro study with human scalp skin exposed for 10 minutes to an ointment containing 1% omadine, the amounts recovered were 1.76% in horny layer, 3.32% in epidermis, 1.32% in corium and 0.07% in penetration chamber.

- When rats were treated with a hair dressing formulation containing 0.1 omadine and applied to the back skin for 4 hours 4.45% of the applied dose was excreted in the urine and 0.31% in the faeces after 24 hours ; at the same time while the treated skin contained 2.55% of the applied dose.

- Dermal absorption in rats, treated with 0.15 ml of shampoos containing 1.5 , I.0 or 0.5% omadine was 0.42-0.66%,calculated from the total amounts detected in urine,faeces and carcass after 48 hours.

The proportion remaining in the treated skin was 4.15-4.76%. Assuming that 50% of the amount in the skin gradually enters the body, the total absorption would be 2.5-3.7% of the amount applied.

MUTAGENICITY

Equivocal results were obtained in a Salmonella assay for gene mutation. Some indication of mutagenic potential was obtained in one study, with an increase in revertants being seen at the highest concentration employed. However this was not confirmed in a repeat experiment. One in vitro test with Chinese hamster lung cells indicated that the compound had the potential to produce chromosome damage but the test was negative when repeated in a modified form. An increase in gene mutation was observed in Chinese hamster ovary cells, but the effects were not dose-related. No dominant lethal properties were observed in tests with mice and rats. The compound did not induce unscheduled DNA synthesis in hepatocytes of rats when tested at concentrations of 0.03 up to 30 ug/ml. This finding indicates absence of DNA damage.

CARCINOGENICITY

In a dermal carcinogenity study, mice were treated with 2 or 20 mg/kg three times/week for 18 months. The top-dose group showed increased mortality in both sexes, liver hypertrophy in males, a relatively high incidence of hepatocellular adenoma and carcinoma, and an increased incidence of malignant lymphoma. The authors considered the toxicological significance of the increased tumour incidences as unclear. Daily treatment of the skin of hairless mice with a 1% solution in 4.8% shampoo base for 20 minutes during 14 days did not affect proliferation of basal cells in the epidermis.

TERATOGENICITY AND REPRODUCTION

In a reproduction study, oral administration of 1.0, 3.0 or 7.5 mg/kg/bw/day to female rats from day 14 before mating through day 21 of the lactation period was associated with high mortality of females in the top-dose group (14/20) and several other signs of intoxication.In the mid-dose group the dams showed decreased numbers of implantations and of viable pups and increased pre-implantation loss. The authors considered 1.0 mg/kg a no-effect level.

In a peri and post-natal study pregnant female rats were treated orally with 1, 3 or 7.5 mg/kg/bw/day from day 15 of gestation up to day 20 of lactation. Dams of the high-dose group showed mortality and other signs of intoxication. In this group there was a decreased number of viable pups and decreased survival and growth rate of pups during lactation.

Teratogenicity has been examined in several studies.

- In one oral rat study doses of 3, 10, 30 or 100 mg/kg were employed ; maternal toxicity occurred at the two high-dose levels. In addition to these two levels therewas an increased incidence of a major skeletal abnormality consisting of bifurcated ribs.

- In a dermal rat study with dose levels of 1, 10 and 30 mg/kg skin changes occurred in the dams of all groups. Signs of embryotoxicity and growth depression of the dams were seen only in the high-dose group.

- In two dermal studies in swine (with dose levels of 30, I00 and 300 mg/kg in one study and of 10 and 30 mg/kg in the other) signs of maternal toxicity and skeletal anomalies occurred at all levels. The number of sows in each group (4-6) was too small to allow any definite conclusion.

- An oral teratogenicity study in rabbits showed loss of body weight with 5 mg/kg. With 3 and 5 mg/kg post-implantation losses were too high to allow any conclusion concerning teratogenic potential No evidence of teratogenicity was found with 1 mg/kg.

Dermal applications at 5 mg/kg/bw/day were reported to produce foetotoxicity in rabbits (Colipa note of February 1984,no details given) while 30 mg/kg/bw/day dermally produced embryotoxicity in rats and 10 mg/kg/bw/day produced maternal toxicity in swine.The proposed application rate for man is about 2 mg/kg/bw/ application.
The margins between application rates and toxic levels are very small.

CONCLUSION

When evaluating the extensive information on this substance, it appears that toxicity occurs at low levels of exposure and that the no-effect level may well be lower than 1 mg/kg/bw/day.

There is evidence of mutagenic and teratogenic properties.

Dermal absorption is likely to be considerably higher than 0.07 % in humans and 0.03 % in rats.

The toxicity of the substance , the dermal exposure and the potential dermal absorption occur with a too tight margin of safety to permit the use of this substance in cosmetics (even limited to anti-dandruff shampoos).

Classification : D

Information : COLIPA - Submission I : May 11, 1982
- Submission II : February 15, 1984
- Submission III : November, 1984
- Submission IV : June, 1985
- Submission V : September, 1985
- Submission VI : December, 1985
- Submission VII : November, 1986

1. Consumer Product Testing 1979
   Study n° 78437
2. Food and Drug Research Labs 1971 Lab n° 0497
3. " " " " 1971 Report n° 8549
4. Consumer Product Testing 1978 study n° 78107-5
5. Huntingdon Research Center 1973 Project n° 731-340
   Study listed on p. 10 was repeated due to excess amount of Omadine MDS on fur post exposure which was ingested during preening
6. Huntingdon Research Center 1973 Project n° 733-370
7. Consumer Product Testing 1978 study n° 78107-1
8. Food & Drug Research Labs 1971 report n° 0734
9. " " " " " " " 0821
10. Industrial Bio-Test 1973 report n° 601-03209
11. International Research & Development Corporation 1982
    report n° 397-039
12. Food & Drug Research Labs 1974 report n° 4074-1274
13. " " " " 1972 " " 8877

14. Consumer Product Testing 1979 report n° 78422

15. International Research and Development Corporation 1982 report n° 397-028

16. " " " " 1982 report n° 397-027

17. Industrial Bio-Test 1974 report n° 651-64101 B

18. " " " 1975 " " 651-04101 D

19. " " " 1976 " " 623-08160

20. International Research & Development Corporation 1981 report n° 397-029

21. Biodynamics 1981 project n° 80-2499

22. " 1977 " " 76-1631

23. International Research & Development Corporation 1980 study n° 397-023

24. Industrial Bio-Test 1976 report n° 622-08049

25. Bioassay systems 1980 report n° 10068

26. " " 1981 " " 10171

27. " " 1980 " " 10250

28. Industrial Bio-Test 1974 report n° 622-05693

29. " " " 1975 " " 622-06511

30. SRI International 1981 report n° LSC-4675 fiscal report - August 21 1981

31. " " 1981 " " LSC-4675 interview report - July 1981

32. Biometric Testing 1976 report n° A-1099

33. Pharmacologist 1974 16,252

34. Reference : Howard Maibach, M.D. 1977 Olin reports, also Wedig, J., Feldman, R., and Maibach, H. 1977. Percutaneous penetration of the magnesium sulfate adduct of dipyrithione in man.
Toxicol. Appl. Pharmacol. 41:1-6

35. Industrial Bio-Test 1975 report n° 632-06372

36. " " " 1975 report n° 632-06541

37. Bioassay systems 1983 report n° 11450

38. A.D. Little, Inc. 1983 report n° C-88661

39. The dermal absorption of Omadine MDS in rats HRC 25.7.84

45. Rougier et al (Subm V)
The measurment of stratem corneum reservoir
A predictive method for in vivo percutaneous absorption studies : influence of appl. time.
J. Invest. dermatol. 84,66 1985

46. Maibach & Wolfran
Percutaneous penetration of hair dyes, Journal Soc. Cosm. Chem. 32,223 1981

47. Report by Pr. Stottgen 3.12.85 (Subm VI)

48. Comparative absorption of pyrithiones (Subm VII)
D. Howes and J. Black
Toxicology 5 (1975) 209-220

- data sjeet Nat. Inst. Publ. Health, the Netherlands 1983
- toxicol. appl. pharmacol. 43 (1978) 373-379

COMPLEMENTARY OPINION ON HYDROXY-8-QUINOLEINE

(expressed on 20 February 1987)

(docs CSC/575/86 - XI/709/85)

Considering the non-effect dose : 75 mg/kgbw in the rat
150 mg/kgbw in the mouse

and the limitation of the use,the Committee can accept the use of hydroxy-8-quinoleine in a concentration of 0.03 % as a stabilizing agent of H_2O_2 in the non-rinsed capillary products.

CONTENTS

34th meeting of SCC - June 6, 1987

35th meeting of SCC - October 13, 1987

REPORTS BY THE SCIENTIFIC

COMMITTEE ON COSMETOLOGY

34th REUNION - JUNE, 30th 1987

LIST OF PARTICIPANTS

Present

Mr. AGACHE

Mr. DE GROOT

Mrs. DONY

Mrs. ENJOLRAS

Mr. GOULDING

Mrs. KNAAP

Mr. LOPRIENO

Mr. MUSCARDIN

Mr. O'MAHONY

Mr. SHOU

Mr. STUTTGEN

Apologies for absence

Mr. HILDEBRANDT

Commission

Mrs. MASSE

Mr. COLLIN (expert)

Mr. GONTIER (DG XI/B/1)

SUMMARY

Opinions expressed on June 30th, 1987 concerning the use of :

- PRESERVATIVE AGENTS

1. Chlorhexidine (EEC n° 31)
2. Alkyl Trimethyl Ammonium Bromure (EEC n° 55, P72)
3. Germal II (P79)

- HAIR DYES

2 NPPD (B25)

REPORT BY THE SCIENTIFIC COMMITTEE ON COSMETOLOGY ON THE USE OF 1,4 - DIAMINO - 2 - NITROBENZENE

. THE COMMITEE'S MANDATE

To give its opinion on the question whether the use of 2 NPPD as direct hair dye at the maximal concentration of 1% is admissible from the health point of view.

. FORMULA AND SYNONYMES

CI 76070

CAS Reg. n° 5307 - 14 - 2

COLIPA B25

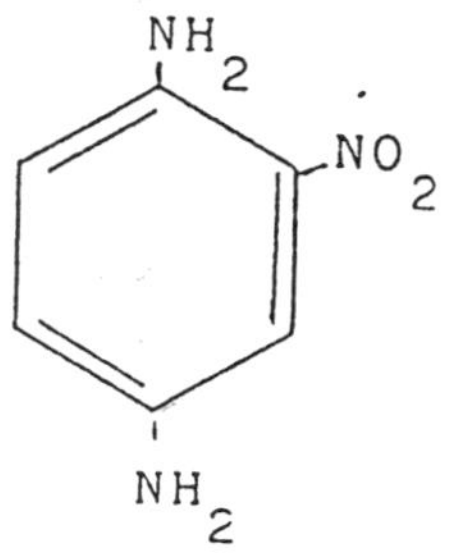

$C_6 H_7 N_3 O_2$

1,4 - DIAMINO - 2 NITROBENZENE

2 - NITRO - 1,4 - PHENYLENEDIAMINE
2 - NITRO - 1,4 - BENZENEDIAMINE
2 - NITRO - 4 - AMINOANILINE
2 - NITRO - 1,4 - DIAMINOBENZENE
2 - NITRO - p - PHENYLENEDIAMINE (2 NPPD)
o - NITRO - p - PHENYLENEDIAMINE
DIAMINONITROBENZENE
m - NITRO - p - PHENYLENEDIAMINE
o - NITRO - p - PHENYL ENEDIAMINE

. USE

As direct hair dye up to 1%.
It is used as a direct dye or in combination with oxidant dyes.
It produces brown and red shades on the hair.
Production and use : 2500 Kg.

. PRECEDING CONCLUSIONS
SEC, SECOND SERIES, EUR 8634 (1983) P 7.

"In view of the positive carcinogenity findings in animals, at the doses used, the SCC recommends that its use might be discontinued. Nevertheless, this decision could be modified because of the product's low percutaneous resorption and because the carcinogenicity tests by the dermal route were conducted on a mixture of the substance and not with 2 NPPD alone".

17-10-87 : Banned in Italy and Denmark
Recommended for banning in F.R.G.

. RECAPITULATION OF THE STUDIES OF TOXICITY

LD50 oral (rat) : 1800 - 3080 mg/kg
i.p. (rat) : 348 mg/kg

- Metabolism. 2 - NPPD 14 - C radiolabelled showed at 4 days after dosing to rats by differents route (oral, i.p. or s.c.) that ca. 4% of the dose was retained in the body, mainly in the intestinal tract and lesser in the liver and kidney. The metabolic pattern in urine showed 6 radioactive components for rat and 7 for mouse.

ORAL TOXICITY

Subacute toxicity : RAT Fischer 344 & Mouse B6C3F1. NCI bioassay.
5 males and 5 females/group received in the diet for 4 weeks period, followed by 2 weeks observation period, these doses of 2 - NPPD : 0-315-680-1465-3155-6800 ppm for rats and 0-810-1180-1740-2550-3750-5550-8080-11830 ppm for mice. The maximum tolerated dose was 1100 ppm for males and 2200 ppm for females rats and 4400 ppm for mice.

CUTANEOUS AND MUCOUS TOLERANCE

DERMAL TOXICITY

- Dermal irritation : the compound applied as a 2.5% (W/V) preparation (0.5% aqueous gum tragacanth containing 0.05% anhydrous sodium sulphite) resulted non irritant.
- Eye irritation : the compound applied as 2.5% (W/V) suspension (0.5% aqueous gum tragacanth containing 0.05% anhydrous sodium sulphite) on rabbit's eyes resulted "mildly irritating".
- Sensitization : was tested in Guinea pigs treated with 3% 2-NPPD solution containing 2% Natrosol, 2% Tween 80, 0.05% Sodium sulphite and 10% isopropanol (pH = 7) applied 6 days/week for 3 weeks, sensitization evaluated 2 weeks later. The results showed a weak reaction (4/20 animals have an allergic effect).

- Dermal absorption : (14-C) 2-NPPD 14-C (86µg) in ethanol solution (40 µl) showed a high penetration to the clipped skin of rats (males : 11.7 µg/cm2, females 24.6 µg/ml) and mice (31-36 µg/cm2). 6.3 µg/cm2 (4%) of 3-H labelled compound (157 µg/cm2 as a 0.8% (W/V) penetrated in the skin of rats at 48 hours after application of a 50% solution of a semi permanent hair colorant shampoo base for 20 min. before rinsing off, and a similar study 4.6 µg/cm2 (4.8%) of 14-C labelled compound (96 µg/cm2, 0.5%) penetrated in the rat skin after 48 hours after treatment with 50% aqueous shampoo solution for 20 min. and 6.4% after 30 min. (14-C) -2NPPD (4 µg/cm2 on clipped sin for 24 hours) showed a skin penetration of 29.9% (percent of the dose applied) in monkey (ventral surface of forearm) and 17.7% in swine (back site). Dermal absorption in rhesus monkey through the scalp was 0.55% of the applied dose.

- Short term dermal toxicity : 2-NPPD containing formulation (1.1% in water) tested on shaven intact or abraded skin of New Zealand rabbits by topical application produced no toxic effects at 3-7-13 weeks after treatment at the histopathological analyses.

- Human : The cutaneous resorption study of 2-NPPD (20 mM/1 in isopropanol) in vivo on man show that ca. 49% (24 h), ca. 67% (48 h) and 65% (72 h) of the compound penetrates in the skin. In the normal application 0.75% of absorption by scalp has been observed in a period of 30 days.

MUTAGENICITY

Studies have shown that 2-nitro-p-phenylenediamine is mutagenic :

(1) on Salmonella typhimurium in the absence and in the presence of S9mix ± hydrogen peroxide ;

(2) on mouse lymphoma cells for gene mutation induction ;

(3) on two chinese hamster cells lines for chromosome aberration (in vitro) ;

(4) on rat hepatocytes and HeLa cells for the induction of UDS, and DNA repair in E.coli.

The mutagenicity studies have shown that the compound does not induce dominant lethals in rats Charles River treated for 8 weeks with a dose of 20 mg/k (i.p.) and for 10 weeks until to 40 mg/kg (i.p.), micronuclea in rats Sprague Dawley treated with 2 g/kg (gastric intubation), SCE in bonve marrow of chinese hamsters treated i.p. (up to 300 mg/kg) or by gavage (up to 500 mg/kg), mitotic gene conversion in the yeats. The 2-NPPD induced morphological transformation in C3H/10T½CL8 and reduced lymphocite transformation.

CARCINOGENICITY

Long term studies were carried out on mice and rats by a NCI bioassay, the compound fed in the diet at 550-1100 ppm for males and 1100-2200 for females rats for 78 weeks (27 weeks observation period) and, 2200 and 4400 ppm for mice for 78 weeks (12-13 weeks observation period) showed that 2-NPPD resulted positive in the females mice producing a significative increase of hepatocellular neoplasms (adenomas) and no conclusive evidence for the carcinogenicity in males mice and rats were obtained. After the reevaluation of the hystopatology the presence of adenoma was confirmed in one case and it was excluded in a second case (parenchyma cell proliferation). A formulation containing 2-NPPD at 0.015% level tested in A and DBAf mice by repeated topical application in aqueous acetone solution showed lymphoid tumours in both strains, but only in DBAf the incidence was statistically significant. Other studies carried out on mouse and rat (formulation 7401 containing 1.1% of 2-NPPD), and dog (Dye/Base composite containing 0.24% : only summary conclusions, data not available for evaluation) resulted negative.

TERATOGENICITY

2-NPPD (suspensions in sterile distilled water : 0-32-64-128-160-192-224-256 mg/kg/day) administered once a day 1% b.w. (10 ml/kg) by subcutaneous injection on days 6-15 of gestation of female mice (CD-1) showed teratogenic effects and maternal toxicity in a same range doses (160 mg/kg/day and above). The compound showed a significant reduction in average weight gain by the dams during pregnancy (32 and 128-256 mg/kg) and in average fetal weights (128-256 mg/kg/day); it produced a significant increase in the average percent malformed fetuses : principally cleft palate and, also significant fused ribs, white foci in the left ventricle of the fetal heart (160 and above mg/kg) and bilateral open eye (224 mg/kg). The dose without teratogenic effects was 128 mg/kg/day. A retarded effect of ossification process (bones of the feet and the cervical and caudal vertebral centre) was observed in mice topically treated twice a week for four weeks before mating and until the' 18th day of gestation with a formulation containing 1% of 2-NPPD equivalent to 0.5 mg/mouse. No teratogenicity effects were observed with formulations containing the compound on rat (1.1%, 1% or 0.24%) and rabbit (0.24%). In two reproductive studies on rat treated with 0.24% or 1.1% 2-NPPD containing formulation no negative effects were observed. (One study was not available for the present analysis).

CONCLUSION

No other long term carcinogenicity studies have been produced last evaluation of this compound by SCC in 1980. The compound has been found in a main study carcinogenic for females mice only, but the results have been questioned on the basis of other pathological evaluation analysis. No carcinogenic effect was observed in male mice and male and female rats. The compound has been found mutagenic and genotoxic in vitro assay (bacterial and mammalian cells) for different genetic end points. In vivo mutagenicity studies (dominant lethal, micronuclea and SCE in bone marrow cells) have produced negative results. The low scalp penetration in human (0.75% after 30 days) and the low animal toxicity observed may suggest the absence of health risk by this compound.

The evidence for carcinogenicity is inadequate (no conclusive evidence). The evidence for mutagenicity is positive in vitro and negative in vivo. More appropriate genotixicity studies, namely DNA damage in mouse liver, DNA binding in liver cells, may provide evidence more adequate for final evaluation of the toxicological potential of this compound.

The SCC in his plenary meeting of June 30th, 1987 asked for the following information :

1. a UDS test in vivo on hepatic mouse cells to obtained detailes of the reaction potential to DNA ;
2. a short-term study by oral administration to determine the NEL.

CLASSIFICATION : C

BIBLIOGRAPHY

N. LOPRIENO & G. BONCRISTIANI : A review on toxicology of 1,4 - Diamino - 2 - nitrobenzene.
Pisa, Italy, Sept. 1986.

REPORTS OF THE SCIENTIFIC COMMITTEE ON COSMETOLOGY CONCERNING THE USE OF CERTAIN PRESERVATIVE AGENTS IN COSMETIC PRODUCTS

THE COMMITTEE'S MANDATE

To give its opinion on the question whether the use of :

- chlorhexidine at the use level of 0.5 %
- alkyl trimethyl ammonium bromide at the maximal use of 0.1 %
- diazolidinylurea at the maximal use level of 0.5 %

as preservative agents in cosmetic products is admissible from the health point of view.

(1) CHLORHEXIDINE

* Formula and synonyms

EEC n°31

Colipa P 24

1,6-di-(N-p-chlorophenyl-N'-diguanido)-n-hexane

(chlorhexidine)

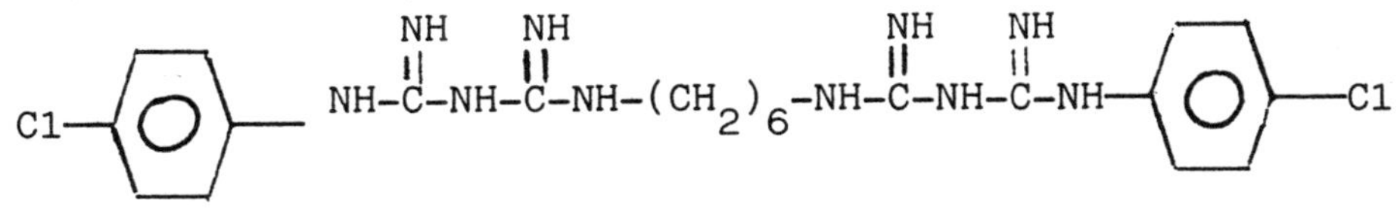

$C_{22}N_{10}Cl_2H_{30}$

* Characteristics

Insoluble in organic solvents and fats

Poorly soluble in water

* Use

As a preservative agent at the use level of 0.5 % active ingredient.

* Recapitulation of the studies of toxicity

The IV LD50 = 25 mg/kg Values are comparable for acetate and gluconate.

ORAL TOXICITY

Acute : The oral LD50 (rats and mice) is relatively high : 2gr/kgbw

A sub-chronic (90-day) oral study in rats with levels up to 0.3 % chlorhexidine gluconate in drinking water revealed infiltration of lymphocytes and vacuolation in several organs, and giant cells in testicles and lymphnodes.Similar changes were seen in a rat study with the acetate .The no-effect level was 50 mg/kgbw.

A chronic oral rat study with the gluconate in the drinking water revealed histiocytes in mesenteric lymphnodes at all dose levels (5, 25 and 50 mg/kg b.w.).

SKIN AND MUCOUS TOLERANCE

DERMAL ROUTE TOXICITY

Eye injury was found by E.M. examination of rabbit eyes treated with aqueous solutions of the gluconate at concentrations of 0.1% and higher.

Despite extensive use in man, sensitization has not been a problem.

A sub-chronic (90-day) dermal study in monkeys, bathed daily with a skin cleanser containing twice the normal level of chlorhexidine gluconate and subsequently rinsed with water, was negative.
After 8 weeks of treatment no

Dermal application of ^{14}C-labelled chlor hexidine as a 5% aqueous dilution or as a 4% handwash formula to human forearm skin, without occlusion, for 3 hours, and measuring labelled material in urine and faeces for up to 10 days, revealed only 0.009% of the dose applied in the faeces, while the urine did not contain measurable levels ($<$0.5 mg/l urine).
Repeated scrubbing the forearm skin of healthy humans with 5 ml of a 4% preparation or 3 minutes, 5 times daily, 5 days a week, for 3 weeks, did not result in detectable levels in the blood.
Hospital staff washing hands usually 5 times each day with 5 ml of a 5% digluconate preparation for one minute, followed by rinsing with water and then repeating the washing procedure for a further 2 minutes ,failed to show detectable blood levels ($<$ 0.01 mg/ml) when examined after 3 weeks exposure (Case et al. Chemotherapy 3 (1976 367 - 374); Colipa subm. II,ref.2).
Babies were bathed daily by applying a small quantity (amount not specified) 4% chlorhexidine detergent-based preparation to the head (excluding the face) and washing off with water. The preparation was then applied to the rest of the body. Next, the baby was immersed in water and dried. Blood samples taken at 1 and 4 hours after bathing showed low levels of chlorhexidine (less than 1 µg/ml) in only 15 out

of the 34 babies examined (Cowen et al. Arch. Dis. Children 54 (1979) 379-383 ; Colipa subm. II,ref.4).

In another study, 51 babies were sponged from the neck down with 30 ml of a 1:10 aqueous dilution of a 4% chlorhexidine gluconate solution. Blood samples taken after one hour showed no detectable levels of the test compound (i.e. 0.1 µg.ml) (O'Neill et al. Current Therap Res. 31 (1982) 485-489 ; Colipa subm. II,ref 5).

MUTAGENICITY

Mutagenic properties of chlorhexidine were established in an Ames test. A positive effect was found also in a DNA-repair test with E. coli.

A chromosome aberration test in Chinese hamster ovary cells in vitro with 1, 10 and 100 µg/ml,conducted with and without metabolic activation, was negative.

In a micronucleus test in mice, two intravenous applications of up to 2 x 30 mg/kg (the maximum tolerated dose) did not increase the number of polychromatic erythrocytes containing a micronucleus.

CONCLUSION

The potential dermal exposure of humans to chlorhexidin is close to the lowest level which induced lymphatic histiocytosis in a chronic rat study. Nevertheless the safety margin is probably sufficient, because percutaneous absorption in humans is only minimal.

Mutagenicity tests in microbial systems showed positive results,but a chromosome aberration test and a micronucleus test were negative.

Classification : A

The minimal percutaneous absorption authorizes the classification A though the NEL was below 5 mg/kgbw in the rat.

Information

- Data sheet National Institute of Public Health - The Netherlands
- Toxicol. appl. Pharmacol. 52 (1980) 255-261
- Colipa dossier : - Submission I, January 24, 1984
  - Submission II, February, 1987
- BLAISE and TOXLINE
- Cowen et al. Arch. Dis. Children 54 (1979) 379-383
- O'Neill et al. Current Therap. Res.31 (1982) 485-489
- Case et al. Chemiotherapy 3 (1976) 367-374

SUBM I - January 1984

1. L'OREAL - Laboratories of "Recherche Fondamentale"
Department of Toxicology
25 July 1983

SUBM II - February 1987

2. D.E. Case et al 1976
"Chlorhexidine : Attempts to Detect Percutaneous Absorption in Man" Chemotherapy, Vol. 3, 1976, p. 367-374

3. D.E. Case 1980
"Chlorhexidine : Attempts to Detect Percutaneous Absorption in Man" Int. Congr. Symp. Ser. - R. Soc. Med. 1979 (publied : 1980) Series n° 23, p. 39-43

4. J. Cowen et al 1979
"Absorption of Chlorhexidine from the Intact Skin of Newborn Infants"
Archives of Disease in Childhood, 54, 1979, p. 379-383

5. J. Oneill et al 1982
"Percutaneous Absorption Potential of Chlorhexidine in Neonates"
Current Therapeutic Research, Vol. 31, n° 3,
March 1982 p. 485-489

6. Noël De Graeve 1985
"Evaluation of Clastogenic Potential of Chlorhexidine Digluconate Using Chinese Hamster Ovary-Cells"
U.T.G., Genetic Toxicology Testing Unit - Sart Tilman (Liège) - Belgique - Report 05.07.1985

(2) <u>C. T. A. B.</u>

* <u>Formula</u>

Alkyl (C_{12} - C_{22}) trimehtylammoniumbromide/chloride

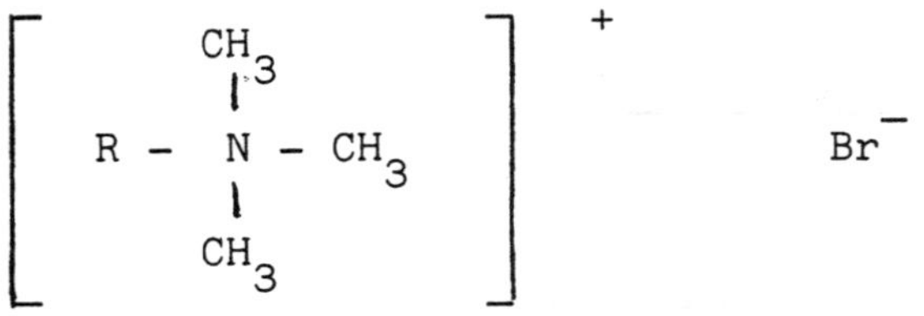

R = alkyl MW 364,48

* <u>Characteristics</u>

Soluble in water and alcohols

Insoluble in ether

* <u>Use</u>

As a preservative agent in cosmetics at the use level up to0.1 %.

* <u>Recapitulation of the studies of toxicity</u>

<u>PARENTERIC TOXICITY</u>

<u>LD50</u> intraperitoneal (rat) : 56 mg/kg bw
(mouse) : 39,8 mg/kg bw
(rabbit) : 125 mg/kg bw
subcutaneous (mouse) : 75-80 mg/kg bw

Single intravenous administration of 4 mg/kg to dogs caused a fall in blood pressure, while 12 mg/kg i.v. in monkeys produced traces of haemoglobin in urine without severe effects.

<u>ORAL TOXICITY</u>

Acute : <u>LD50</u> (rats) : 1 gr/kg bw

In a sub-acute oral study, groups of 6 mice fed a diet containing 0.1 % or more showed growth depression ; with 0.2 % they died in 35-63 days, while with 0.4 % the survival period was only 9-12 days. Mice fed 0.5 % in the diet died between 3-12 days,showing destruction of mucosal epithelial cells and haemorrhages in the intestinal tract; with 0.05 % marked hypertrophy and hyperplasia of the duodenal and jejunal mucosa occurred after 3 months, but with 0.02 % no changes were found after 6 months.

In a one-year oral study, rats received 0, 10, 20 or 45 mg/kg bw in drinking water. The top-dose group showed reduced weight gain and food conversion, wetting and brown discolouration of the fur over the anterior ventral region and increased caecal weight. Caecal enlargement also occurred with 20 mg/kg.No changes were seen with 10 mg/kg.

Upon oral administration of the ^{14}C-labelled compound to rats, 80 % of the dose was still present in the gastrointestinal tract after 8 hours, and 20 % was excreted in the bile during the first 12 hours. These results indicate slow intestinal absorption. Only small amounts were found in the organs, which disappeared largely within 4 days. Excretion of radioactivity in the faeces and urine amounted to 92 % and 1 % respectively within 3 days. Examination of bile and urine by TLC indicated that some metabolism occurs in the rat. No radioactivity was found in expired CO_2.

SKIN AND MUCOUS TOLERANCE

DERMAL TOXICITY

A primary skin irritation study in rabbits with a 1 % solution did not produce any changes. With 5 % there were mild effects (erythema, oedema, scab formation) lasting less than 48 h.; with 25 % there was mild tomoderate erythema and oedema, while with 50 % thse changes were moderate to severe. Dermal application of 1 % to humans produced reactions in 2.8 % of the treated subjects.

14 cases of hypersensitivity to this substance have been reported when used in treatment of burns. In another study, skin sensitivity was confirmed by patch testing in 46 patients.
Eye irritation occurred in the rabbit eye with concentrations of 0.1% and above.With 0.1 % the changes disappeared in 24 hr.

The percutaneous absorption of the ^{14}C-labelled substance in the rat was 3.15 % from a 3 % aqueous solution which was not rinsed, 0.093 % from a 0.5 % concentration in a hair rinse formulation (rinsed after 5 min.) and 0.59 % from a 1 % aqueous solution (rinsed after 15 min.)

MUTAGENICITY

Mutagenic properties were examined in an Ames test with S.typhimurium strains TA 1535, TA 1537 and TA 1538,both with and without metabolic activation and using up to 500 µg cetyl trimethyl ammonium chloride/ plate. The results did not show any significant increase in the number of revertant colonies.

Another Ames test with 5 strains of S. typhimurium with and without metabolic activation and using up to 625 µg/plate,was likewise negative.

A test with E. coli by the Mohn method in vitro, with up to µg test compound/ml did not reveal any mutagenic potential.

Negative results were also obtained in a morphological transformation assay with cells of 13 days old Syrian hamster embryos,and up to 300 µg test substance/ml.

TERATOGENICITY

A teratogenicity study in mice, treated once intraperitoneally with 10.5 or 35.0 mg/kg bw on day 8, 10, 12 or 14 of gestation, revealed an increased incidence of cleft palate,and skeletal defects in skull and sternum in both treatment groups, and embryolethality in the high-dose group.

In a teratogenicity study in rats, groups of 20 animals were treated daily by gavage with a substance designated Bromal Lot no.16593 at levels of 8, 25 or 50 mg/kg from day 5 through day 14 of pregnancy. The top dose group showed signs of toxicity both in the dams and in the foetuses, and an increased incidence of unossified bones and of several visceral abnormalities. The authors considered all defects congenital in origin, not indicative of teratogenicity. No increased incidence of abnormalities occurred in the intermediate dose group.

CONCLUSION

This substance has shown considerable systemic toxicity including indications of teratogenicity.
However, absorption through the intestinal wall and through the skin is low.
A chromosomal aberration test on mammalian cells is needed.

Classification : B

(3) DIAZOLIDINYLUREA

* Formula and synonyms

- N - (hydroxy methyl) - N - (1,3- dihydroxymethyl - 2,5 - dioxo - 4 - imidazolidinyl (- N' - hydroxymethyl) urea.

Germall II

CEE n° 17

Colipa P 79

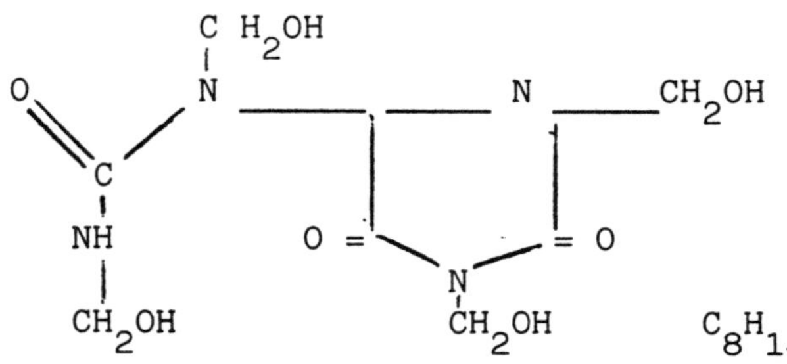

$C_8H_{14}N_4O_7$

MW 278.23

* Characteristics

Soluble in water

Insoluble in most organic solvents

* Use

Use level as a preservative up to 0.5 %

* Recapitulation of the studies of toxicity

ORAL TOXICITY

Acute : LD50 (mouse) : 3700 mg/kg bw

LD50 (rat) : 2570 mg/kg

A 7-day oral toxicity test was carried out in which the a.i. was given by gavage to groups of 5 rats/sex in doses of 0, 100, 200, 400, 800 and 1600 mg/kgbw/day. There were slight transient behavioural changes at 200 m/kg and more marked and prolonged ones at 400 mg/kg and above.Necropsy showed gastric changesin some animals.

A 29-day oral toxicity test in rats was carried out , in which the a.i. was given by gavage to groups of 5 animals/sex in doses of 0, 100, 300 and 900 mg/kg bw/day. All animals survived. There were transient behavioural changes at 100 mg/kg, and more severe and long-lasting changes at higher doses. There was a reduction in gain of body weight, and a decrease in food intake, in top dose males. The top-dose rats showed gross and microscopic evidence of gastric lesions, and the weight of the adrenals was increased. Similar changes, though less marked, occurred with 300 mg/kg. The no-toxic effect level was stated to be greater than 100 mg/kg bw/day.

In a 90-day oral rat study , the test substance was fed in the diet at levels providing 0, 10, 25 or 100 mg/kg bw. There were no treatment-related changes in growth rate, haematology and clinical chemistry or in the results of the pathological examinations. Therefore,100 mg/kg was considered to be a no-toxic effect level.

The supplementary information received since the previous evaluation confirms an oral no-effect level of 100 mg/kg bw or slight lower.

CUTANEOUS AND MUCOUS TOLERANCE

The dermal LD50 (rabbit) when applied undiluted as a powder is >2000 mg/kg.
Skin irritation tests in rabbits with 0.5 ml of 1.0 % and 5.0% solutions in water were negative.

An eye irritation tst in rabbits with 0.1 ml of 1.0 and 5.0 % aqueous solutions did not induce any changes.
In a sensitization test in guinea pigs by the Landsteiner-Draize method, the induction treatment consisted of 10 intracutaneous injections of 0.1 ml 0.1 % in saline solution.After a 2 weeks rest period, 0.05 ml of a 0.1% solution was given intracutaneously as a challenge. The test was considered negative.

In a maximization test in guinea pigs, induction was performed with the undiluted substance applied topically, and with a 5% solution applied intradermally. The induction areas were pretreated with sodium lauryl sulphate. The response to the challenge treatment,conducted 2 weeks later, suggested the test substance to be a weak sensitizer.

Phototoxicity tests in groups of 9-11 human subjects with 0.2 g subscreen cream or sunscreen lotion containing 0.5% diazolidinyl urea,did not induce signs of phototoxicity,when applied 10 times on a closed patch for 24 hours, followed by UV- or UV-A-irradiation at a distance of 10 cm during 15 minutes. A similar phototoxicity test in 11 subjects with one application of 0.2 g sunscreen lotion containing 0.5% and irridiated with UV-A was likewise negative.

A sensitization study in 110 women was conducted with a sunscreen formulation containing 0.25% of the preservative applied with 10 successive patches under semi-occlusion. The challenge patch with the same formulation,after a 2 week rest, resulted in only one dubious positive reaction. No cross-sensitization was observed in 71 patients (who had shown sensitization to various preservatives and under other chemicals) upon patch testing with 0.5 and 1.0% Germall II in water petrolatum.

In several photosensitization tests, groups of 28-30 human subjects were treated with 0.2 g of 0.5% in a sunscreen cream ,or in a sunscreen lotion 10 times, each time for 24 hours and each treatment followed by UV-, UV-A- or UV-B- irradiation for 15 minutes.After a 10-18 days rest period, the challenge treatment with an occluded patch for 24 hours followed by UV-, UV-A- or UV-B- irradiation did not reveal photoallergic properties, although one slight transient reaction was observed.
A skin cleanser with 0.3% was examined in a prophetic patch test on 104 subjects and in a repeat insult patch test on 52 subjects.
No positive reactions were observed (summary report only).

MUTAGENICITY

The Ames test with dose levelsup to 800 µl/plate repeatedly showed relatively high numbers of revertants. The results were inconclusive. A micronucleus test in the mouse with oral dose levels of 1200, 2000 and 2800 mg/kg was negative.

CONCLUSION

The confirmed NEL of 100 mg/kg bw justifies the continued approval of the substance.

Classification A for this relatively new preservative that finds increasing use in cosmetics may be considered if results

- of an oral teratogenicity study,
- and of a gene mutation in mammalian cells become available.

Classification : B

SCIENTIFIC COMMITTEE ON COSMETOLOGY

List of participants

35th meeting -13 October 1987

Present

Mr. DE GROOT
Mrs. DONY
Mrs. ENJOLRAS
Mr. FIELDER
Mr. GOULDING
Mr. LOPRIENO
Mr. MUSCARDIN
Mr. O'MAHONY
Mr. SCHOU
Mr. STUETTGEM

Apologies for absence

Mr. AGACHE
Mr. GRANADOS JARQUE
Mr. HILDEBRANDT
Mrs. KNAAP
Mr. RAMOS MORGADO

Commission

Mrs. MASSE
Mr. COLLIN (expert)
Mr. GONTIER (DG XI/C/2)

SUMMARY

Opinions expressed on October 13th, 1987 concerning the use of :

- PRESERVATIVE AGENTS

1. HEXETIDINE	(EEC n° 15)	(P20)
2. BENZYL FORMAL	(EEC n° 16)	(P21)
3. CHLORACETAMIDE	(EEC n° 22)	
4. PHENOXY PROPANOL	(EEC n° 56)	(P54)
5. SUTTOCIDE A		(P84)

- COLOURING AGENTS

1. CI	13065	
2. CI	42535	
3. CI	42555	
4. CI	44045	
5. CI	73900	(46500)
6. CI	61554	
7. ACID RED	195	

- HAIR DYES

4 NOPD (B24)

REPORT OF THE SCIENTIFIC COMMITTEE ON COSMETOLOGY CONCERNING THE USE OF CERTAIN PRESERVATIVE AGENTS IN COSMETIC PRODUCTS

(opinion expressed on 13 October 1987)

THE COMMITTEE'S MANDATE

To give its opinion on the question whether the use of certain preservative agents :

- Hexetidine
- Benzylformal
- 2-Chloroacetamide

- Phenoxypropanol
- Suttocide A

in their usual level is admissible from the health point of view.

(1) HEXETIDINE

* Formula and synonyms

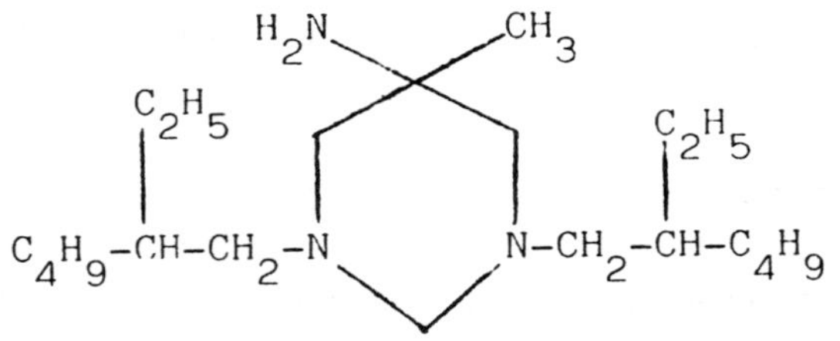

$C_{21}H_{45}N_3$

MW 339.60

CAS Reg.no. 141-94-6

5-amino-1,3-di-(2-ethylhexyl) - hexahydro-5-methylpyrimidine

Other names : Hexetidine
Sterisol
EEC n° 15
Colipa P 20

* Characteristics

A viscous oil with a density of 0,87 g/ml

Practically insoluble in water

Easily soluble in ethanol, acetone and other organic solvents

* Use

Used up to 0.2% as a preservative agent in cosmetics.

* Recapitulation of the studies of toxicity

ORAL TOXICITY

Acute : LD50 (rat) = 1.43 g/kg bw

In a 3-wk oral test in rats with 20, 50, 100 and 200 mg/kg/day, the top dose caused mortality ; with 100 mg/kg there was reduced erythropoiesis. No changes were seen at lower levels.

In a one year study, rats received 0.02, 0.05 or 0.1% in the diet. Growth retardation and decreased food concumption were seen in the top dose group. No treatment-related changes were found upon microscopic examination. (Only summary report is available).

In a sub-chronic (90-day) oral study, rats received 0, 5,15 or 50 mg/kg bw in the diet. There were no deaths or clinical signs of toxicity. Growth depression, decreased intake of food and water,changes in biochemical blood parameters and increased organ-to-body weight ratios for several organs,occurred in the top-dose group.Microscopically, the top dose rats showed signs of systemic lipidosis. No detailed report of the pathological findings was available. It is stated that 300 ppm was a no-effect level (or 27 mg/kg bw).

An oral dose of ^{14}C-hexetidine in rats (20 mg/kg) and dogs (10 mg/kg)was rapidly excreted. After 72 hours 60-70% was recovered in the faeces and 20-25% in the urine, while 1-2% was found in the liver and kidneys.

CUTANEOUS AND MUCOUS TOLERANCE

DERMAL ROUTE TOXICITY

No skin irritation or sensitization was observed in 200 volunteers examined for primary irritation with 1% in an ointment. A repeated insult test in 50 humans with 1% in oil caused mild irritation ;10 and 5% in oil caused strong primary irritation but no sensitization.

Very slight eye irritation was seen in rabbits with 0.1% ointment, moderate irritation with 4% aqueous dispersion, and with the undiluted substance, and severe irritation with 25% ointment and 50 % dispersion.

A 0.1% aqueous solution used twice daily as a mouth wash by 200 volunteers for 3 months did not induce clinical signs, primary irritation or sensitization.

In a study in 327 patients with cervicovaginal infections 0.1% as a gel and 0.5% solution did not induce irritation, sensitization or systemic toxicity.

In a skin sensitization test in guinea pigs, induction with 10 intradermal injections of 0.001, 0.005 and 0.01% in acidic saline,followed by one intradermal injection of 0.001% and topical application of a

25% in absolute ethanol after 14 days, no signs of sensitization occurred. Repeating the challenge treatment after one week, and also after 2 weeks, was likewise negative.

Daily application of 50, 100,200 or 500 mg/kg to rats for 3 weeks (5 days/week) caused mortality in the top-dose group. No changes occurred at lower levels.
Dermal treatment of rabbits with 0.0625 up to 4.0 ml/kg/day for 90 days caused mortality and growth retardation with 0.25 ml/kg and above. No effects on haemopoietic and urogenital systems were seen at any treatment level (Summary reports only).

Dermal LD50 in rats : 1.86 ml/kg bw
in rabbits : 4 ml/kg.

MUTAGENICITY

An Ames test with 5 strains of S.typhimurium and dose levels up to 5 mg/plate was negative. A chromosomal aberration test with CHO-cells *in vitro* was negative (The exposure concentrations used in the latter test are not clearly indicated in the report ; it is stated in the German text that the highest test concentration was a 6×10^{-5} dilution (Colipa subm. I,ref. 1).

TERATOGENICITY

In a teratogenicity study in rats ,groups of 20 females were treated orally with 12.5, 25 or 50 mg hexetidine/kg bw/day by gavage on days 6-15 of pregnancy. Forty control females were treated with the vehicle (corn oil) only. The top-dose induced growth retardation and decreased food intake of the dams. There was no indication of foetotoxicity, embryolethality or of a teratogenic response, in this well conducted and reported study. (Colipa subm I, ref.2).

PARENTERIC TOXICITY

LD50 (intraperitoneal in mice) : 0.03-0.085 g/kg; intravenous administration of 5 mg/kg was lethal for cats.

CONCLUSION

The substance showed considerable systemic toxicity in animals and comparison of oral and dermal toxicity in rats indicates high dermal absorption.
No teratogenic or mutagenic properties were detected.
The oral NEL in rats was 27 mg/kg bw/day.
Dermal exposure in humans from use in cosmetics will be less than 1 mg/kg.

Classification : A

Information

- Data Sheet Council of Europe
- Environmental Safety Laboratory, Unilever Research, Colworth House, England.
  Dossier Safety Evaluation of Hexetidine. Vol. I and II
  Document : D 83/037, August 1983.
- TOXLINE 1974 - 1984, MEDLARS 1981 - 1984 and RTECS have been searched.
- Colipa submission I, May 1987.

SUB I - May 1987

1. Test report of the laboratory of Dr. Leimbeck, dated April 2, 1987 Bad Kissingen

2. W.E. Parish
   "Hexetidine rat teratology study : effects on pregnancy and post-natal development"
   Vol I and II
   Environmental Safety Laboratory Unilever Research - England
   24 July 1986

(2) BENZYLFORMAL

* Formula

EEC n+ 16

Colipa P 21

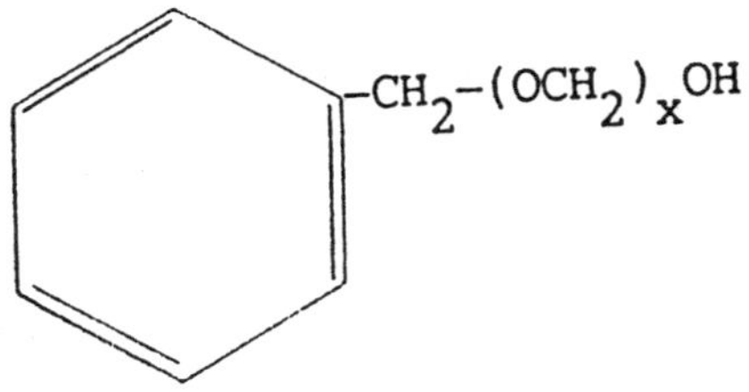

Benzylformal

Mixture of benzyloxymethanol and benzyloxymethoxymethanol

Preventol

* Characteristics

Soluble in organic solvents

Solubility in water 25 gr/l

* Use

Used up to 0,2% as a preservative agent in all types of cosmetics.

* Recapitulation of the studies of toxicity

LD50	oral (rats)	:	1700 mg/kg bw
	I.V. (rats)	:	153 mg/kg
	dermal (rats)	:	> 1000 mg/kg
	(rabbit)	:	1429-2000 mg/kg

ORAL TOXICITY

Acute ; LD50 oral (rat) : 1700 mg/kg

There is no information on short- or long-term oral toxicity.

CUTANEOUS AND MUCOUS TOLERANCE

DERMAL ROUTE TOXICITY

A skin irritation test in rabbits with 500 mg undiluted substance applied to the intact skin of the ear for 8 hours induced redness and oedema ; when applied for only 2 hours, slight redness was observed. A 0.2% aqueous solution applied for 24 hours did not induce any changes.

In an eye irritation test in rabbits 50 mg undiluted substance caused erythema and oedema and an opaque cornea.A 0.2% aqueous dilution only produced erythema.

A sensitization test by the Landsteiner-Draize method with 0.1% of the test substance in saline both for the induction and for the challenge treatment did not reveal signs of sensitization.

A sub-chronic (90-day) dermal study was conducted in groups of 10 rabbits/sex which received 0, 1, 4 or 16 mg /kg b.w. daily,5 days/ week. With 4 and 16 mg,the skin showed dose-related changes at the site of application. In the top dose group, growth depression occurred in females and decreased cholesterol values in both sexes. The weight of the pituitary was decreased in males of the intermediate- and high-dose group, however, no treatment related pathological changes were found in the internal organs . One mg/kg was a clear no-effect level in this well conducted dermal study.

MUTAGENICITY

An Ames test with up to 500 g/plate was positive for S.typhimurium TA 100. This result is attributed to the presence of 29.7% formaldehyde in the product. In a micronucleus test, male and female mice received 2 x 500 mg/kg with an interval of 24 hours. No increase in the incidence of micronuclei was observed.

TERATOGENICITY

There is no information on reproduction or teratogenicity.

CONCLUSION

The substance liberates formaldehyde (at a maximum of 0.004% under test conditions, according to D r. Crisp of the U.K. Laboratory of Governmental Chemists).
There is no information on short- or long-term oral toxicity, on reproduction or teratogenicity. A chromosomal aberration test is also lacking. Although studies on dermal absorption are not available high uptake through the skin is suggested by comparing oral and dermal LD50-values. In a sub-chronic dermal study in rabbits, 1 mg/kg bw/day was a no-effect level. One mg/kg bw/day might be also the dermal exposure in humans if 0.2% is used in all types of cosmetics. It is doubtful, therefore, whether continued use of this substance is justified.
Therefore, informations are required on :

- dermal absorption in humans
- a chromosomal aberration test in mammalian cells in vitro
- a short-term oral toxicity study to establish systemic effects

Classification : B

Information

Data Sheet Council of Europe

Sub I - Avril '84

1. Bayer AG, test of Toxicology, report n° 1002
   18.10.1968

2. Study n° TO 004.574
   "90-Tage-Hauttoxizität aus Kaninchen unit Preventol D2"
   von Heyden GmbH, Toxicologie, Werk Regenisburg 13.1.1983

3. Bayer AG, Institut f. Toxikologie, report n° 10.619 5.2.1982

4. " " " " report n° 10.497 12.1.1982

Sub II - Mai '87

5. Mobay Chem. Corp. Stilwell Kansas (report 67442)

Sub III - Sept '90

6. Bayer AG. Preventol D2 - Subakute toxikologische
   Untersuchunger an Wisar Ratten (Verabreichung mit der
   Mogensonde über 28 bzw. 29 Tage)
   Bencht n° 19293 20.7.90

(3) 2 - CHLOROACETAMIDE

* Formula

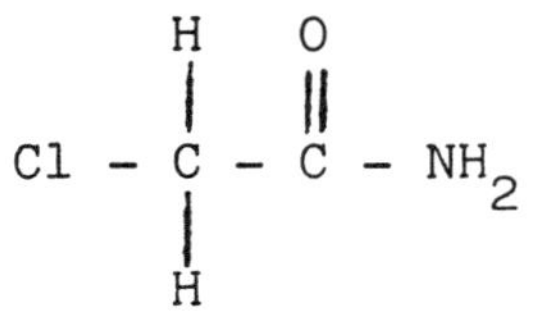

C_2H_4ONCl

EEC n° 22

Colipa P 27

* Characteristics

The product manufactured in the EEC is 99.5-99.8% pure (impurities are ammonium-chloride 0.1-0.2% and monochloracetic acid 0.08 %).
Moderately soluble in water (5%)
Easily soluble in ethanol

* Use

Used mainly as a mixture of chloroacetamide and sodiumbenzoate (70/30), in rinsed off products at the level of 0.15%, in non-rinsed off products at the level of 0.3%.

* Recapitulation of the studies of toxicity

LD50 (of a mixture of 70/30 mixture of chloroacetamide and sodium benzoate)

Oral (rats) = 0.37 gr/kg bw
(mice) = 0.15 gr/kg

LD50 (intraperitoneal) = 50 mg/kg.

ORAL TOXICITY

Acute : LD50 (rat) = 0.37 gr/kg (mixture 70/30)
(mice) = 0.15 gr/kg

In a 13-wk feeding study in rats with levels providing 12.5 and 50.0 mg/kg bw/day, marked testicular atrophy occurred at each dose level.

Increased weight of the thyroid was also observed (no further details provided). A recent 90-day oral study in rats was conducted with a 99.7% pure substance at levels of 0,20,100 and 500 mg/kg diet.
The top-dose induced growth depression, increased leucocyte counts, reduced weight of testicle and liver, suppression of spermiogenesis and proliferation of Leydig cells. No treatment-related changes occurred with 20 and 100 mg/kg diet (Colipa subm. IV).

CUTANEOUS AND MUCOUS TOLERANCE

DERMAL ROUTE TOXICITY

A 1% solution was not irritating to the skin of guinea pigs.

Single patch testing of 209 dermal patients with 0.5% of the 70/30 mixture (containing 99% pure chloroacetamide, solvent not mentioned) did not result in any skin reaction, even after exposure of several persons to sunlight (ref. 1, subm. II; only summary report).

In another single patch test with 0.1% aqueous solution of a similar mixture in 200 dermal patients, no skin reactions were seen.Repeated application of the same daily for 14 days to 10 healthy persons and to 15 persons with allergy for benzoic acid, p-hydroxybenzoic acid ester, or peruvian balsam,did not induce irritation (ref. 2,subm. II; only summary report).

Single patch testing of 296 eczema patients with 0.2% aqueous chloroacetamide (99% pure) induced skin reactions in 7 patients (2.3%) (ref. 3, subm. II, summary report). Single patch tests with the 70:30 mixture of chloroacetamide (99% pure) and sodium benzoate did not induce irritation in 14 humans treated with 1%, in 10 humans treated with 0.5% or in 102 humans treated with 1% (ref. 10, subm. III).
Daily patch testing of 25 subjects with 0.1 or 0.2% in formulations was likewise negative (ref. 10, subm. III).

A 5% aqueous dilution of the 70:30 mixture applied to the rabbit eye was found to be non-irritating.
A single application of a 10% solution/suspension to the rabbit eye induced redness of the conjunctivae in 1/4 animals. Slight redness was observed upon daily application of 1% in an ointment for 12 days.

A similar response ,however, occurred in controls treated with the ointment only.

A 0.2% preparation was well tolerated in the eye of 5 humans.

In a sensitization test by a modified Bhehler method with the mixture of chloroacetamide (99% pure) and Na-benzoate (70:30) guinea pigs received topical induction treatments with 0.5 ml 0.3% in water, once a week, for 3 weeks. After a 14-day rest period, a challenge treatment with 0.3% in water,did not induce any sign of sensitization.

In a second sensitization test, a 1% aqueous solution of the mixture (containing the 99% pure compound) was painted on the skin of guinea pigs nine times every other day. After a 14-day rest period the same solution applied as a challenge treatment did not induce any positive response.

A third test was conducted with 0.1% of the mixture in a skin cleaning formulation, by rubbing it into the intact and scarified skin of guinea pigs in an amount of 0.1 ml, three times weekly for three weeks. This induction treatment caused erythema and oedema. Since the challenge treatment with 0.1 ml, 0.1% solution after 14 days did not produce more severe changes, the test substance was not considered a sensitizer.

A fourth test conducted in a similar way,however,with a 1.0% (instead of a 0.1%) solution, also induced erythema and oedema. These findings were not considered the result of sensitizing properties.

In a fifth test, guinea pigs were treated daily for 4 weeks with 1% and 3% of the pure compound in an aqueous mixture. The challenge treatment with 0.2% after 10 days rest did not produce sensitization (ref.3; subm. II; summary only).

In a maximization test in guinea pigs with the 99% pure compound (using for induction a 9% aqueous concentration intradermally and a concentration of 9% in vaseline topically) a challenge treatment with 3, 1 and 0.3% in distilled water given after 14 days rest did not produce sensitization (ref. 11 ; subm. III).

A sensitization study was conducted in 147 humans by topical application of a 0.5% aqueous solution of chloroacetamide of unknown purity, on every other day for 3 weeks. The challenge, also with 0.5% after

14 days, consisting of two consecutive patch applications for 48 hours each (on the same site, but away from the induction site) resulted in a high incidence of sensitized subjects, viz. 19/33 females and 28/114 males (ref. 4, subm. II).
In another study in humans with 1.25% in a cream base used both for induction and challenge treatment, 35/205 subjects showed a positive response (ref. 5, subm. II).
Positive reactions were obtained with 2/18 humans treated with 0.18% in a cream. A positive result was obtained also with 0.18% in water. In another patch test on 200 humans with 0.18% in a cream no positive reactions occurred (ref. 6, subm. II).
Tests with 1832 patients of a dermatological clinic showed 30 clearly positive reactions (1.6%) (Colipa sumb. V : May 1987).

Dermal treatment of rats with 12.5 and 50.0 mg/kg bw for 13 weeks did not induce fross or microscopical changes (no further details provided).

MUTAGENICITY

Mutagenicity tests were conducted with the 70:30 mixture of chloro-acetamide and Na-benzoate. An Ames test with up to 1000 µg/plate was negative.
Chinese hamsters treated with up to 50 mg/kg intraperitoneally did not show chromosomal aberrations or an increased incidence of micro-nucleated erythrocytes.
A dominant lethal test in male mice treated intraperitoneally with 114 and 123 mg/kg was likewise negative.

TERATOGENICITY

Rats treated with an intraperitoneal dose of 20 mg/kg on day 7 of gestation,or on days 11 and 12 showed no signs of toxicity to the mothers or the foetuses (no further details were provided).

Single administration (route ?) of 50 mg/kg to pregnant rats did not affect foetal development (no further details were provided).

CONCLUSION

Chloroacetamide has shown considerable sensitizing potency in humans at concentrations in the range of those present in cosmetics. This property might be due to a contaminant in certain batches of the substance.

The Committee agrees with provisional acceptance if information is provided on the results of ongoing surveillance.

Classification : A + mention "contains chloroacetamide"
(annexe VI n° 41)
Maximum level use : 0.3%

REFERENCES

1. Prof. Dr. Schmidt-La Baume in Zusammenarbeit
Prof. Dr. R. Pfister
Städt, Hautklinik, Karlsruhe
Baden-Baden, 25.5.1966

Copy of Expert report

2. Prof. Dr. med. H. Röckl
Universitäts-Klinik und Poliklinik für Hautkrankheiten
Würzburg 16.11.1970

Copy of Expert report

3. Prof. Dr. med. K.H. Schulz
Universitäts-Hautklinik
Abteilung für Allergologie
Hamburg 18.12.1984

Copy of Expert report

4. W.P. Jordan, Ir. und S.W. King
"Delayed hypersensitivity in females"
Contact Dermatitis 3, 19-26 1977

5. Marzulli, Maibach 1973 - Antimicrobials :
experimental contact sensitization in man
Journal of the Society of Cosmetic Chemists 24.399

6. Nater 1971
Allergie reactions due to chloracetamide
Dermatologica 142. 191-192

7. Meynadier 1982
Allergie aux conservateurs
Ann. Dermatol. Venrol. 109 1017-23

8. IBR-Forschung-GmbH, Hannover 1981

Copy of Expert report

9. Prof. Dr. Gloxhuber, HENKEL & Cie
GmbH, Düsseldorf 1966

Copy of Expert report
page 2

10. Prof. Dr. med. Franz Klaschka
Berlin 27.7.1985 Ref. 10

11. B.4. IBR - Forschungs GmbH, Walsrode - 15 August 1985
Delayed contact hypersensitivity in guinea pigs,
mod. Method of B. Magnusson and A.M. Kligman Ref. 11

12. Drei Monate Fütterung an der Wistar-Ratte
September 9 1985
Unpublished Company data

4. PHENOXYPROPANOL

. Formula and Synonymes

EEC n° 56

COLIPA P54

1 - phenoxy propane - 2 - ol : chemical name

phenoxy propanol = propylene phemoxetol

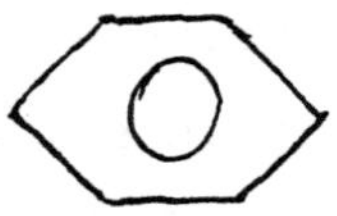 $- O - CH_2 - CH_2OH$

$C_8H_{10}O_2$

MW 138,17

. CHARACTERISTICS

Soluble in water (1%)

Mixable with alcohol, ether and chloroform

. USE

Used only in rinsed-off cosmetics.

- up to 1% as a perservative
- up to 2% for other purposes

. RECAPITULATION OF THE STUDIES OF TOXICITY

LD50 oral (rat) : 2554 - 2720 mg/kg/bw

ORAL TOXICITY

In a 2-wk oral test in rats with 0, 200, 500 or 1250 mg/kg/day by gavage, the top-dose was lethal. The two lower dose levels induced clinical signs and gross pathological changes.

A 2-wk oral test in rabbits with 0, 100 and 500 mg/kg/day showed paralysis and death in the top-dose group. No effect were noted with 100 mg/kg.

In a 4-wk oral study in rats with 0, 40, 120 and 400 mg/kg/day, the top-dose group showed clinical signs including sedation and abnormal gait, but no pathological changes. No abnormalities were seen with 120 mg/kg.

In a 4-wk oral test in rabbits with 0, 50, 100 and 200 mg/kg/day, the top-dose induced a general decline of condition, and increased liver weights. With 100 mg/kg growth rate and food intake were reduced. No change occurred with 50 mg/kg.

CUTANEOUS AND MUCOUS TOLERANCE
DERMAL TOXICITY

A skin irritation test in rabbits with the undiluted substance produced only minimal irritation.

In an eye irritation test in rabbits the undiluted substance elicited moderate to marked reactions of cornea, conjunctivae and iris. With 5% and 2% aqueous dilutions there were no eye reactions attributable to the substance.

A maximization test in Guinea pigs with 0.5% in water for induction and 75% aqueous preparation for the challenge did not produce evidence of sensitization.

Dermal exposure of rabbits to 0, 20, 50 or 1000 mg/kg/day for 90 consecutive days induced transient signs of toxicity only in the top-dose group (peripheral hypothermia, paralysis). Gross and microscopic examination were negative. (In this study the stratum corneum was removed at the site of treatment to enhance penetration).

In a subsequent 90-day study in rabbits with several exposure levels of 50 up to 1000 mg/kg/day, similar clinical signs of toxicity (including decreased motor activity, immobility, paralysis and hypothermia) were seen with 400, 500 and 1000 mg/kg. The signs which occurred shortly after the daily dosing, became noticeable at the end of the first week of treatment, decreased in frequency and intensity as the treatment progressed, and were no longer apparent after week 6.
The "no-toxic effect level" was 300 mg/kg. Local skin changes, largely confined to exfoliation, occurred in all treatment groups (Colipa submission IV, ref. 24).

A dermal dose of 2 ml of a 2% solution of the radioactively labelled compound, caused a maximum blood level in rabbits of 4-5 ug/ml. An average of 54% of the applied radioactivity was excreted in the urine and an additional 23% was excreted in the faeces. Two radioactive compounds (a glucuronic acid conjugate and the aglycone) were identified in the urine, and an additional metabolite was extracted from the faeces. Absorption of the labelled compound through the skin of hairless mice amounted up to 73% if applied in 0.5% aqueous solution (Subm. IV, ref. 21). Rabbit skin was found to be twice as permeable for the compound as abdominal human skin (ref. 22).
The penetration through grafted abdominal human skin in vivo was 70% when applied unrinsed in a liquid soap, but only 9% if rinsed off shortly after application (ref. 23).

MUTAGENICITY

An Ames test with 4 strains of S. typhimurium, using up to 5 mg/plate, did not reveal mutagenic activity (ref. 26). A chromosomal abberration test in human lymphocytes in vitro using up to 0.8 mg/ml culture medium showed no clastogenic activity (ref. 27).

TERATOGENICITY

In teratogenicity studies, rats and rabbits were treated orally during pregnancy with dose levels of 0, 80, 160 or 320 mg/kg/day for rats and 0, 25, 50 or 100 mg/kg/day for rabbits. In neither of these two studies were there adverse effects on foetal development.

CONCLUSION

The substance has shown considerable toxicity upon oral administration to animals, and dermal absorption is high if not rinsed off. No mutagenic or teratogenic properties were detected. Oral and dermal exposure levels above 200 mg/kg body weight/day in rats and rabbits were found to effect the central nervous system. No-effect levels varied between 50 and 300 mg/kg in different studies in rats and rabbits. If used in rinsed-off cosmetics only, the margin of safety is considered sufficient.

CLASSIFICATION A.

SECTION II - REFERENCES

1. Official US protocol - US Fedral Register 191.12
17 September 1971

2. Official French method - Journal Officiel de la République Française
Arrêté du 5 Avril 1971 (J.O. du 9.2.73)
complété par l'Arrêté du 17 Janvier 1973 (J.O. du 5.6.73)
complété par l'Arrêté du 16 Avril 1973 (J.O. du 5.6.73)

3. Irritants de Référence - Interprétation de l'irritation oculaire provoqueée par des substances étalon.

4. VICK International Ltd Report N° 73/UK 18/072 -
Life Science Research - Stock Essex CH4 - 9 PE - 2nd October '73
Propylene Phenoxetol. Irritance towards Rabbit Skin and Eye Mucosa, and Acute Oral Toxicity to Rats
Life Science Research (LSR) October 1973

5. HUNTINGDON Research Center - Report 3398/10 D/75 -
22 November 1974 - R. Davies - Irritant effects of Propylene phenoxetol on Rabbit Skin

6. HUNTINGDON Research Center - Report 3399/11 D/74 -
22 November 1974 - R. Davies - Irritant effects of Propylene phenoxetol on Rabbit eye and mucosa

7. HUNTINGDON Research Center - Report 3559/D 9/74 -
6 January 1975 - R. E. Davies - Acute toxicity to rats of Propylene phenoxetol

8. HUNTINGDON Research Center - Propylene phenoxetol delayed dermal sensitization in the albino guinea pig - Report of December 1974
January 1975

9. VICK International Report 76/VNL 16/018 - Propylene phenoxetol in Topexane Base : Toxicity in repeated Oral administration to Rats for 14 days
(Range-finding test) LSR January 1976

10. VICK International Report 76/VNL 17/023 - Propylene phenoxetol in Toxpexane Base : Toxicity in repeated Oral administration to Rats for 4 weeks
Life Science Research - Stock Essex LSR February 1976

11. VICK International Report 76/VNL 18/016 - Propylene phenoxetol in Topexane Base : Toxicity in repeated Oral Administration to Rabbits for 2 weeks
(Range-finding test)
Life Science Research - Stock Essex LSR January 1976

12. VICK International Report 76/VNL 19/017 - Propylene phenoxetol in Topexane Base : Toxicity in repeated Oral administration to Rabbits for 4 weeks
Life Science Research - Stock Essex - LSR January 1976

13. VICK International Report 76/VNL 21/020 - Propylene phenoxetol : Toxicity to Rabbits by Dermal application for 13 weeks
Life Science Research - Stock Essex - LSR February 1976

14. VICK International Report 76/VNL 15/067 - The absorption of Propylene phenoxetol through intact and abraded rabbit skin
Life Science Research - Stock Essex - LSR April 1976

15. VICK International Report 76/VNL 22/052 - Propylene phenoxetol : Effects upon Pregnancy in the Rats
Life Science Research - Stock Essex - LSR March 1976

This second submission on P 54 consists of :

- Ref. 17 : A safety assessment of P 54 as used in rinse-off cosmetic products (doc. 85/270)
- Ref. 18 : CUNLIFFE W.J., Experiment to determine the amount of Topexane left on the face after washing
  Internal. Report May 27 1977
- Ref. 19 : Protocol AMES-Test - Life Science Research, September 10 1985
- Ref. 20 : Protocol Lymphocyte Cytogenetic Study -
  Life Science Research, September 11 1985

REFERENCES

21. Short J. Phenoxypropanol - in-vitro permeation through hairless mouse skin. November 1986
Internal report 86.5052.01

22. Short J. Penetration of propylene phenoxetol in-vitro - comparison of diffusion rate through human and rabbit skin. September 1985
Internal report

23. Gowland G. The penetration of phenoxypropanol through grafted human skin in vivo - February 1987 - Internal report UK 195

13. * Life Science Research. Propylene phenoxetol : Toxicity to rabbits by dermal application for 13 weeks
February 1976 - report 76/VNL 21/020

24. Life Science Research. Phenoxypropanol : 13 week percutaneous toxicity study in rabbits. February 1987 - report VNL 065/PP

25. Meynadier J. Study of the effects of Biactol on acne lesions
September 1985 - Report F-117

26. Life Science Research. Propylene phenoxetol : assessment of its mutagenic potential in histidine auxotrophs of Salmonella typhimurium (Ames test) December 1985 - Report 85/VNL 063/746

27. Life Science Research. In-vitro assessment of the clastogenic activity of propylene phenoxetol in cultured human peripheral lymphocytes - March 1986 - Report 86/VNL 064/056

28. Ashby J. The prospects for a simplified and internationally harmonised approach to the detection of possible human carcinogens and mutagens. Mutagenisis 1, 3-16 (1986)

14. * Life Science Research. The absorption of propylene phenoxetol through intact and abraded rabbit skin. April 1976
Report 76/VNL 15/067

* See submission I, February 1980

5. SUTTOCIDE A (P84)

(opinion expressed on February 20th, 1987, 33th reunion)
Conclusion : classification C.

New informations : Colipa Subm. II (8/87)

Micronucleus test in mice

The results of MNT indicate that Suttocide A did not induce a statistically significant increase in micronucleated PCEs or change in the PCE/NCE ratio at doses of 750, 1250 and 1750 mg/kg of body weight administered in single oral doses with sacrifice times of 30, 48 and 72 hours. (Inconsideration of Suttocide A as a 50% aqueous solution, the actual calculated doses were 375, 625 and 875 mg/kg). These findings are based upon the absence of a significant increase in the incidence of micronucleated PCE per 1000 PCEs in the Suttocide A treated animals as compared with those treated only with the vehicle control. Therefore, Suttocide A is considered negative under the experimental conditions of this protocol.

This new data doesn't justify a modification of opinion.

CLASSIFICATION C.

REFERENCE :

1. Heddle, J.A., M. Hite, B. Kirkhart, K. Mavournin, J.T. MacGregor, G.W. Newell and M.F. Salamone. Induction of Micronuclei as a Measure of Genotoxicity. A report of the U.S. Environmental Protection Agency Gene-Tox Program, 1983, Volume 123, pages 61-118.

2. Schmid, W., The Micronucleus Test, Mutation Research, 31 (1975) 9-15.

3. EPA Guidelines, Federal Register Vol. 50, N° 188, Friday, September 27, 1985.

4. Organisation for Economic Co-operation and Development Guidelines for Testing Chemicals (OECD), ISBN 92-64-12221-4, adopted by the council at its 535th meeting on May 12th, 1981.

5. Salamone, M., J. Heddle, E. Stuart and M. Katz. Towards An Improved Micronucleus Test, Mutation Research 74 (1980) 347-356.

REPORTS OF THE SCIENTIFIC COMMITTEE ON COSMETOLOGY ON THE USE OF CERTAIN COLOURING AGENTS IN COSMETIC PRODUCTS

. THE COMMITTEE'S MANDATE

To give its opinion on the question whether the use of certain colouring agents in cosmetics is admissable from the health point of view.

Consult Reports of See seventh series EUR 11303 for preceding informations and opinions.

CI	13065	P. 23 - 25	
CI	42535	P. 90 - 91	
CI	42555	P. 92 - 93	
CI	44045	P. 97 - 98	
CI	61554	P. 117 - 118	
CI	73900	P. 103 - 104	(CI 46500)
ACID RED	195	P. 138	

1. CI 13065

. FORMULA AND SYNONYMES

CAS Reg. N° 587-98-4

Acid Yellow 36
Ext. D. and C. Yellow N°1
C. Ext Gelb 10
Metanil Yellow

m(p-anilinophenylazo) benzene sodium sulfonate

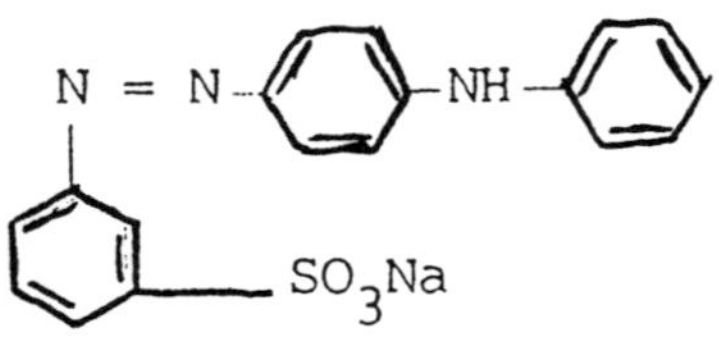

$C_{18}H_{15}N_3O_3S,Na$

Mw 376.41

. USE

Used at levels up to 0.5% in non rinsed off products and up to 0.1% in rinsed off products.

. PRECEDING CONCLUSIONS (1st of July 1986)

The Committee saw no objection to maintaining the use of this colouring agent for the time being in rinsed off products. No opinion could be expressed for non rinsed off products.

Required informations :

- a clear NEL by oral administration
- a sensitization test
- a chromosomal aberration test

. NEW INFORMATIONS

In vivo studies in mice given orally 2mg/kg/day for 30 days, or i.p. a single dose of 50 mg/kg showed strong clastogenic activity. Positive in vitro tests for chromosomal aberrations had already been reported in 1978. An in vivo test in mice given a single i.p. dose varying from 2.5-200 mg/kg bw showed dose-related increases of sister chromatid exchanges in bone marrow.

CONCLUSIONS

The substance possessed considerable systemic toxicity and may adversely affect spermatogenesis. A no-effect level has not been established, but it is less than 25 mg/kg bw/day. The substance was found to induce chromosomal aberrations and sister chromatid exchanges both in vivo and in vitro. The secondary amine group of the molecule may lead to formation of nitrosamines which pass the skin. In view of the above properties and the gaps in the available information, e.g. on sensitization, the continued use of this colourant in cosmetics is not considered justified.

CLASSIFICATION D

REFERENCES

Colipa Subm. III , June 1987.

- Giri et al. J. Fd Sci. Technol. (India) 21 (1984)
- Giri et al. Cancer Letters 30 (1986a) 315-320
- Giri et al. Cancer Letters 31 (1986a) 299-303

2. CI 42535

. FORMULA AND SYNONYMES

CAS Reg. N° 8004-87-3

- Basic Violet 1
- Methyl Violet

bis-4,4' - dimethylamino - 4'' - methylamino - triphenyl - carbenium chloride

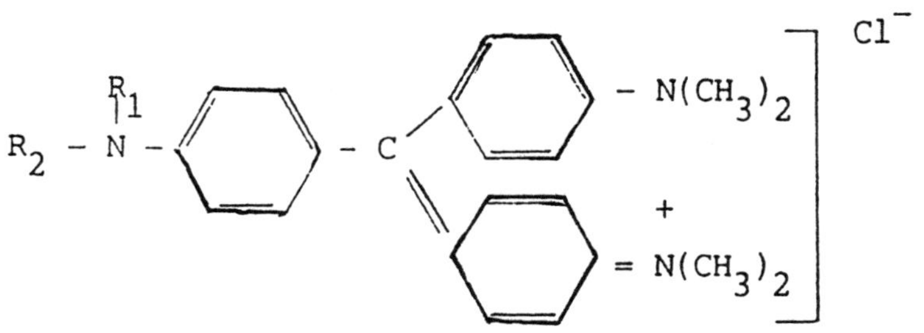

	$R_1, R_2 = H$	/ $R_1 = H, R_2 = CH_3$	/ $R_1, R_2 = CH_3$
	$C_{23}H_{26}N_3$	/ $C_{24}H_{28}N_3$	/ $C_{25}H_{30}N_3$ $^{+}$ Cl^{-}
MW	344	358	372

This colour is called Methyl Violet or Gentian Violet but it is actually a poorly defined combination of violet rosanilins comprising a series of dyes including Crystal Violet (Mut. Res. 98 (1982) p. 157).

. USE

Use level up to <u>0.1%</u> in rinsed off and non rinsed off cosmetics, and up to <u>2.5%</u> in hair setting lotions.

. PRECEDING CONCLUSION (1st of July 1986)

Because this dye is not clearly defined, and the results of the genotoxicity test depend on the source of the sample an evaluation is not possible. An opinion can not be given until information is available on the level of CI 42555 in CI 42535.

CONCLUSION

This colourant is chemically not well defined, which might explain why the results of genotoxicity tests depend on the source of the sample. One sample was found to contain 5% CI 42555 a related colourant with genotoxic and carcinogenic properties.

CLASSIFICATION D

Bibliography 1

Huntingdon Research Centre Report n° 6704/D7/76
Acute Oral Toxicity to Rats of Substanz n° 11
4'-(N-Dimethyl)-Amino-4"-(N-Dimethyl)-Amino-N-Methylfuchsoniumchloride (NBL)
29.10.1976

Bibliography 2

Huntingdon Research Centre Report n° 6682/9D/76
Irritant Effects of Substanz n° 11
4'-(N-Dimethyl)-Amino-4"-(N-Dimethyl)-Amino-N-Methylfuchsoniumchloride (NBL)
on Rabbit Eye Mucosa
21.10.1976

Bibliography 3

Huntingdon Research Centre Report n° 6311/8-D/76
Irritant Effects of Substanz n° 11 On Rabbit Skin
26.7.1976

Bibliography 4

Reprotox Report n° 188
Screening-Test zur Hautsensibilisierung am Meerschweinchen mit
4'-(N-Dimethyl)-Amino-4"-(N-Dimethyl)-Amino-N-Methylfuchsoniumchlorid (NBL)
17.7.1979

Bibliography 5

N. Simon, A. Dobozy, 8.4.1982
Clinica Dermato-Venerologica / Szeged, Hungaria
Bericht über die Prüfung von Bis-4,4-dimethylamino-4-methylaminotriphenyl-carbeniumchlorid auf reizende und sensibilisierende Eigenschaften am Menschen

Bibliography 6

Chr. Gloxhuher et al ; J. Soc. Cosmetic Chemists 23 (1972) 259-269

Bibliography 7

Huntingdon Research Centre Report n° WLA5/77433
Ames Metabolic Activation Test to Assess the Potential Mutagenic Effect of 4'-Dimethylamino-4"-Methylamino-N-Dimethylfuschsine Ammonium Chloride Basic Violet 1 (compound n° 12)
4.7.1977

Bibliography 8

Hill Top Research Project n° 79-0226-11
The Salmonella/Microsomal Assay for Bacterial Mutagenic Activity of Sample n° 6
14.3.1979

Bibliography 9

K.-T. Chung, G.E. Fulk, and A.W. Andrews
Mutagenicity Testing of Some Commonly used Dyes
Applied and Environmental Microbiology 42 (1981) 641-648

Bibliography 10

A.M. Bonin, J.B. Farquharson and R.S.U. Baker
Mutagenicity of Arylmethane Dyes in Salmonella
Mutation Research 89 (1981) 21-34

Bibliography 11

Tong-Man Ong
Use of the spot, plate and suspension test systems for the detection of the mutagenicity of environmental agents and chemical carcinogens in Neurospora Crassa
Mutation Research 53 (1978) 297-308

Bibliography 12

Huntingdon Research Centre Report n° SWF 19/78966
Micronucleus Test on 4'-(N-Dimethyl)-Amino-4"-(N-Dimethyl)-Amino-N-Methylfuchsoniumchloride (NBL)
25.10.1978

Bibliography 13

Ch. Gloxhuber, F. Wingen ; 3.4.1981
Basic Violet 1, Methylviolett CI 42535, Resorptionsscreening an Humanhaut in vivo

Reference 14 : M.M. SHAHIN (1988)
"Mutagenic evaluation of the compound CI 42535 in AMES Salmonella typhimurium plate-test"
Division "Recherche Fondamentale" - Department of Mutagenesis - L'OREAL Aulnay-sous-bois - France
Report 24.10.1988

Reference 15 : A.T. NATARAJAN & F.H. SOBELS (1988)
"Evaluation of Compound CI 42535 in the Chromosome Aberration Test with Chinese Hamster Ovary Cells (In-Vitro)"
Department of Radiation Genetics and Chemical Mutagenesis, University of Leiden, THE NETHERLANDS
Report 9 November 1988

3. CI 42555 see /284/87

. FORMULA AND SYNONYMES

CAS Reg. N° 548-62-9

- Basic Violet 3
- C - Ext. Violet 6
- Crystal Violet
- Gentian Violet

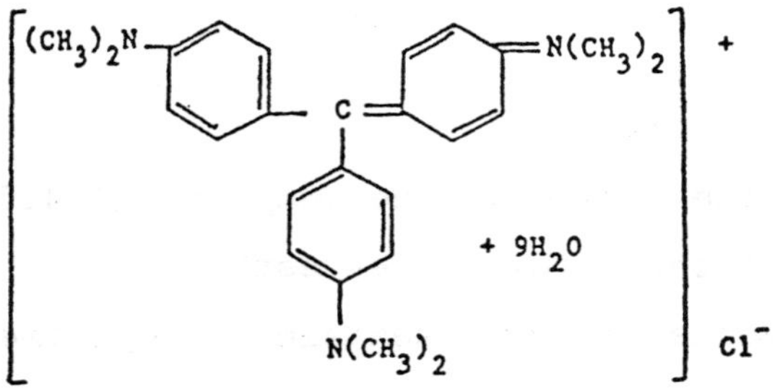

$C_{25}H_{30}N_3.Cl$

MW 408

This dye is the major component of CI 42535 which is a mixture of dyes.

. USE

Used up to 0.1% in rinsed off and non rinsed off cosmetics and up to 0.5% in hair colourants. Industry proposed to restrict its use to hair colourants up to 0.2% (Colipa Subm II, June 1987).

. PRECEDING CONCLUSIONS (1st of July 1986)

Although Gentian Violet is used extensively in medical practice, the Committee came to the conclusion that this substance should not be used as a colourant in cosmetics. However, the Committee noted that Gentian Violet has therapeutic uses. Industry was to be asked to define the extent to which the toxic compound was present in the mixtures.

. NEW INFORMATIONS

A chronic toxicity and carcinogenicity study in mice with dietary levels of 0, 100, 300 and 600 ppm (equivalent with approximately 0, 13, 36 and 68 mg/kg/day) resulted in dose-related increase in mortality, in hepatocellular carcinoma in both sexes, and in reticulum cell sarcomas in uterus, vagina, ovaries and bladder of females after 2 years. These effects were noticeable even in the low-dose group. The authors calculated the virtual safe dose (VSD) to be 1 ppb for males and 2 ppb for females (equivalent with 0.0001 and 0.0002 mg/kg bw/ day respectively) (Littlefield et al, 1985). This chronic mouse study was initiated by NTP. Oral carcinogenicity and reproduction studies in rats which also formed part of this project, have not yet been noticed in the literature.

The metabolism of the ^{14}C-labelled dye was examined after oral administration by gavage of a single dose to rats and of multiple doses to rats and mice on 7 days. After a single dose, tissue residues were maximal at 4 hr in liver, kidney, muscle and gonads ; in fat, a maximum was reached after 24 hrs. The highest residue occurred in fatty tissue. Urinary excretion of the label was 1.6-2.2% in rats and 6-8% in mice (McDonald et al. 1984a).

Gentian Violet is metabolized to leucogentian violet by the intestinal microflora of man, rat and chicken, and by several genera of anaerobic bacteria (McDonald and Cerniglia, 1984b).

In vitro metabolism studies showed demethylation of the compound by liver microsome preparations from mice, rats, hamster, guinea pig and chicken. The pattern of demethylated metabolites was comparable among the species with little difference between the sexes (McDonald et al. 1984c).

Dermal absorption, examined in vitro after 30 minutes contact time, was 0.03% through hairless rat skin, and not detectable through human epidermis (Colipa Subm. II, ref. 28).

CONCLUSIONS

This substance is highly toxic. Oral dose levels as low as 2.5 mg/kg bw/day in rats, and 0.5 mg/kg bw/day in rabbits produced signs of maternal toxicity in teratogenicity studies. Mutagenic properties have been well established. In an oral mouse study the substance was found to be carcinogenic. There are indications of teratogenicity in animals and humans. Although gentian violet is used in medical practice the Committee had insufficient arguments to maintain the substance in cosmetics, even if used only at 0.2% in hair colourants.

CLASSIFICATION D

REFERENCES

- JECFA (1984) WHO Technical Report Series 710
- Littlefield et al. (1985) Fundam. Appl. Toxicol. 5, 902-912
- McDonald et al (1984a) Fd. Chem. Toxicol. 22, 331-336
- McDonald and Cerniglia (1984b) Drug. Metab. Dispos. 12, 330-336
- McDonald et al (1984c) J. Agric. Food Chem. 32, 596-600

REFERENCES

Reference 1 : Dr. LEUSCHNER (1971)
Laboratorium für Pharmakologie und Toxikologie, Hamburg
Report from 15.2.1971

Reference 2 : HODGE et al (1972)
"Acute Oral Toxicity of Methylrosaniline Chloride"
Toxicol and Applied Pharmacol, 22, p. 1-5

Reference 3 : BROCK - ERHARDT (1951)
"Zur Pharmakotherapie der Oxyuriasis I"
Arzneimittel-Forschung (Editio-Cantor KG, 796 Aulendorf, Germany), Vol. 1, p. 5-21

Reference 4 : Dr. LEUSCHNER (1971)
Laboratorium für Phamakologie und Toxikologie, Hamburg
Report from 10.2.1971

Reference 5 : Dr. LEUSCHNER (1970)
Laboratorium für Pharmakologie und Toxikologie,Hamburg
Report from 20.12.1970

Reference 6 : BJORNBERG et al (1972)
"Necrotic Skin Reactions Caused by 1% Gentian Violet and Brillant Green"
Acta Dermatovener (Stockholm), 52, p. 55-60

Reference 7 : GUILLOT et al
"Evaluation of the sensitizing potential of a test substance by topical applications in the Guinea-pig" :
IFG 68-79 - Colorant (Prelt : 8-25112)"
Laboratories of "Institut Français de Recherches et Essais Biologiques (IFREB) - Saint Germain s/L'Arbresle"
Report IFREB-R 911304 of 5th November 1979

Reference 8 : IVANCOVIC S. PREUSSMANN R. (1973)
Deutsches Krebsforschungszentrum, Heidelberg
Report from 4.11.1973

Reference 9 : Riojun KINOSITA
"Studies on the cancerogenic azo and related compounds" : a lecture delivered under the auspices of the Jane Coffin Childs Fund for Medical Research,
at Yale University School of Medecine, New Haven
27.9.

Reference 10 : N.H. CROMWELL et al (1982)
"Hair Coloring Carcinogenesis Study in Mice"
Eppley Institute for Research in Cancer
University of Nebraska Medical Center
Omaha, Nebraska 68105
Final Report May 4, 1982, reviewed and approved August 9 and 10, 1982

Reference 11 : COMBES et al 1982
"A review of the genotoxicity of food, drug and cosmetic colours"
Mut. Res. 98, p. 101-102-156-243

Reference 12 : ROSENKRANZ et al 1971
"Possible Hazard in Use of Gentian Violet"
Br. Med. J. 3, p. 702-703

Reference 13 : CHURCHILL et al 1977
Unpublished observations - not available

Reference 14 : KADA et al 1972
"In-Vitro Host-Mediated "Rec-Assay" Procedures for screening chemical mutagens ; and phloxine, a mutagenic red dye detected"
Mut. Res. 16, p. 165-174

Reference 15 : CLEAVER 1969
"Repair of Mammalian Cell DNA : Effects of Compound that Inhibit DNA Synthesis or Dark Repair"
Radiation Res. 37, p. 334-348

Reference 16 : NAGAI 1959
"Induction of the Respiration-Deficient Mutation in Yeast by Various Synthetic Dyes"
Science, 130, p. 1188-1189

Reference 17 : SUGIMURA et al (1976)
"Overlapping of Carcinogens and Mutagens"
in P.N. MAGEE et al (Eds)
"Fundamentals in Cancer Prevention"
Univ, Tokyo Press, Tokyo p. 191

Reference 18 : LEVIN et al (1982)
"Light-Enhanced Genetic Toxicity of Crystal Violet"
Mut. Research 103, p. 283-288

Reference 19 : BONIN et al (1981)
"Mutagenicity of Arylmethane Dyes in Salmonella"
Mut. Research, 89, p. 21-34

Reference 20 : AU et al (1978)
"Cytogenetic Toxicity of Gentian Violet and Crystal Violet on Mammalian Cells In-Vitro"
Mut. Research 58, p. 269-276

Reference 21 : AU et (1979)
"Further Study of the Genetic Toxiicty of Gentian Violet"
Mut. Research, 66, p. 103-112

Reference 22 : AU et al (1979)
"Studies on Clastogenic Effects of Biological Stains and Dyes "
Environ. Mutagen. 1, p. 27-35

Reference 23 : SHANIN et al (1978)
"Comparisons of Mutation Induction in Reversion Systems of Saccharomyces Cerevisiae and Salmonella Typhimurium"
Mut. Research 53, p. 1-10
Cited in LEVIN et al (1982)

Reference 24 : R. WOLKOWSKY-TYL et al (1982)
"Teratologic evaluation of Gentian Violet (CAS N° 548-62-9) in CD rats"
Research Triangle Institute - North Carolina 27709 - USA
Report PB83-155754 - 31 May 1982

Reference 25 : R. WOLKOWSKY-TYL (1983)
"Teratologic Evaluation of Gentian Violet (CAS N° 548-62-9) in New Zealand White Rabbits"
Research Triangle Institute - North Carolina 27709 - USA
Report PB83-182519 - 15 March 1983

Sub II

Reference 26 : Division of Chemical Toxicology
National Center for Toxicological Research, Jefferson, Arkansas 72079 - USA
"Chronic Toxicity and Carcinogenicity Studies of Gentian Violet in Mice"
NCTR Technical Report, Experiment 304 - September 1984

Published in :
N.A. LITTLEFIELD, B.N. BLACKWELL, C.C. HEWITT, D. GAYLOR (1985)
Fund Appl. Toxicol. 5, (1985) p. 902-912

Reference 27 : S.M. THOMAS & D.G. MAC PHEE (1984)
"Crystal Violet : A direct Acting Frameshift Mutagens Whose Mutagenicity is enhanced by Mammalian Metabolism"
Mutation Research 140, (1984) p. 165-167

Reference 28 : CHALAYE & BAUER (1985)
"In vitro absorption of Crystal Violet through the Total skin of the Hairless Rat, the Human Epidermis"
Laboratories de Recherche Fondamentale - L'OREAL
Aulnay-sous-bois - FRANCE
Report February 1985

Reference 29 : J.J. Mc DONALD et al (1984)
"Synthesis and Disposition of ^{14}C-Labelled Gentian Violet in F344 Rats and B6C3F1 Mice"
Fd Chem. Toxic Vol. 22, n° 5, (1984) p. 331-336

Reference 30 : J.J. Mc DONALD et al (1984)
"Species and Strain comparison of the metabolism of Gentian Violet by Liver Microsomes"
Journal of Agricultural and Food Chemistry 32, (1984) p. 596-600

Reference 31 : J.J. Mc DONALD & C.E. CERNIGLIA (1984)
"Biotransformation of Gentian Violet to Leucogentian Violet by Human, Rat and Chicken intestinal Microflora"
Drug Metabolism and Disposition Vol. 12, n° 3, (1984) p. 330-336

4. CI 44045

. FORMULA AND SYNONYMES

CAS Reg. N° 2580-56-5

Basic Blue 26

L - Ext. Blau 2

C - WR Blau 8

Bis - 4,4' - dimethylaminophenyl - 4'' - phenylaminonaphthyl - carbenium chloride

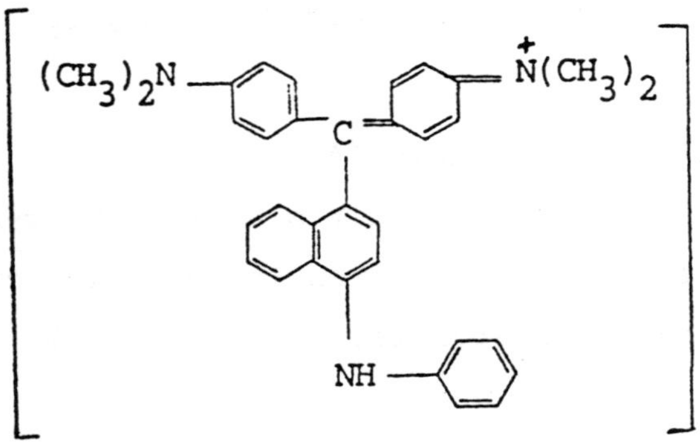

Cl$^-$

MW 506

$C_{33}H_{32}H_3$,Cl

. USE

Use level 0,1% in rinsed off and non rinced off cosmetics, and up to 0,5% in hair dye formulations.

. PRECEDING CONCLUSIONS

No opinion could be expressed because a lack of data.

Were required : - an adequate short-term oral toxicity study
- an Ames test
- a chromosomal aberration test

. NEW INFORMATIONS

In a short-term (13-wk) oral toxicity study 10 rats/sex were treated by gavage with 10 mg/kg b.w./day on 5 days/week. Ten rats/sex served as control. No treatment-related changes were observed at this low exposure level, which was only c. 1% of the LD_{50}. (Colipa subm. III,ref. 7).

An Ames test with up to 1 mg/plate was negative.
A chromosomal aberration study in CHO-cells in vitro up to 1.5 ug/ml culture medium was also negative. (Colipa subm. III, ref. 8 and 9).

CONCLUSION

In the only short-term oral study available, no treatment-related changes were observed at an exposure level of 10 mg/kg b.w. on 5 days/wk. Because higher levels were not examined this study does not meet the requirement that several levels should be administered of which at least one should exert an effect.

Classification : C.

REFERENCES

1. U.HACKENBERG, D.KUHN,Inbifo,Institut für biologische Forschung,Köln Report from 18.01.1978.

2. G.RONDOT. Laboratories of Institut Français de Recherches et Essais biologiques (IFREB) Report IFREB.R 801227.10.1.78

3. U.HACKENBERG, D.KUHN,Inbifo,Köln,Report from 17.01.1978

4. " " 13.01.1978.

5. J.P. GUILLOT, Report IFREB.R 802264. 02.03.1978.

6. J.P. GUILLOT et al.,Report IFREB.R 911365. 27.11.1979.

7. P.E. FOURNIER, "Etude de toxicité chronique chez le rat d'un produit dénommé Bleu Victoria" Société de Recherches Biologiques Bonnez-aux-Loire. FRANCE. Report 13.01.1987.

8. M.M. SHAHIN. L'Oréal. Division Recherche Fondamentale. Departement de Mutagenèse. Report 27.11.1986.

10.C. BOUTEMY et al.(1988) Laboratoires d'Etudes et de Recherches SYNTHELABO (LERS).GARGENVILLE. FRANCE. Report N°1310 TMR/030.88. 06.06.1988.

11.H. SCHEFER (1988) "Penetration In Vitro du Colorant CI 44045" CIRD. VALBONNE. FRANCE. Report 28.08.1988.

5. CI 61554

. FORMULA AND SYNONYMES

CAS Reg. N° 12769-17-4

Solvent Blue 35

C - Ext. Blau 12

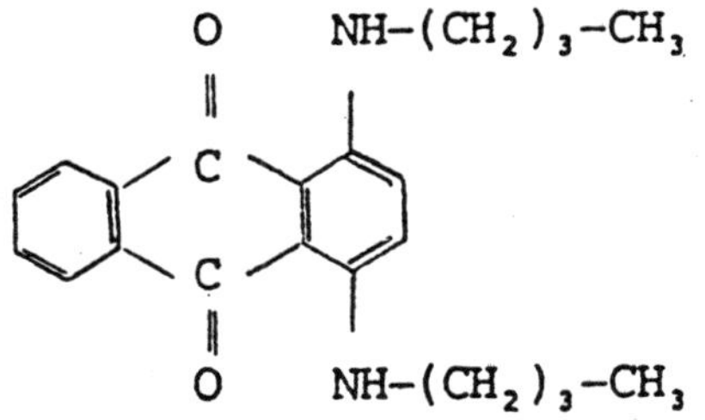

. USE

Before used at level up to 0,1% in rinsed off and now rinsed off products. Actual use level up to 50 ppm only for hair care products.

. PRECEDING CONCLUSIONS (1st of July 1986)

In view of the high systemic toxicity and the absence of a NEL the Committee recommends not to use this colouring for cosmetics.

CONCLUSION

After 5 years no further information had been provided, the Committee confirmed its opinion in October '87.

Classification : D.

REFERENCES

1. Dr. Leuschner, Hamburg - Laboratorium für Pharmakologie U. Toxikologie. Hamburg. Report dated 15.02.1971.
2. " " " " 10.02.1971.
3. " " " " 21.12.1970.
4. S. Ivanovic, R.Prenssmann. Deutches Krebsforschungszentrum. Report from 04.11.1973.
5. Ch. GLOXHUBER. F. WINGEN/ HENKEL INSTITUT fur TOXIKOLOGIE, 12.05.1980

6. CI 73900 (Formerly 46500)

. FORMULA AND SYSONYMES

CAS Reg. N° 1047-16-1

Pigment Violet 19

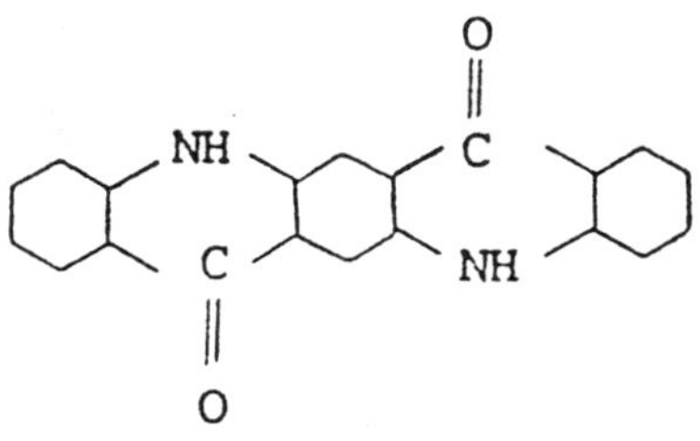

$C_{20}H_{12}N_2O_2$

MW 238

. USE

Use level 0,4% in rinsed off products and 0,003% in non rinsed off products.

. PRECEDING CONCLUSIONS

The Committee saw no objection to maintaining the use of this colouring agent is cosmetic products for the time being. Information on possible sensitizing properties, and a detailed report on the short-term oral study and a chromosome aberration test were required.

CONCLUSION

In the recent submission, it is stated that studies with this colourant are in progres, including 28-days oral toxicity, in vivo mutagenicity and bio-availability.

Classification : B.

7. CI ACID RED 195

. FORMULA AND SYNONYMES

CAS Reg. N° 12220-24-5

4 - (2' - oxo - 4' - sulphonaphthylazo) - 1 - phenyl - 3 - methyl - 5 - oxo - pyrazole, chromiumcomplex (1/1) monosodiumsalt

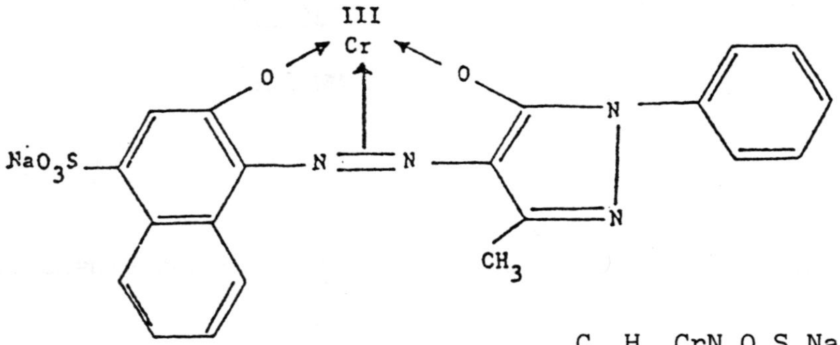

MW 496

$C_{20}H_{13}CrN_4O_5S$ Na

. USE

Used up to 0.03% in rinsed off cosmetics and up to 0.02% in non rinsed off products.

. PRECEDING CONCLUSIONS

Information was required on sensitization, on genotoxicity and on short-term oral toxicity. The Committee had also required information on the incidence of deletorious effect (Stevens - Johnson syndrome) in patients using pyrazole containing drugs which are also exposed to cosmetics containing this colourant.

. NEW INFORMATIONS

ORAL TOXICITY

A sub-acute (28-day) oral study was conducted in groups of rats/sex, treated with 0, 250, 500 or 1000 mg/kg bw/day by gavage. All test groups of males showed lower body weights than did the controls, but there was no relationship with the dose. In females, albumin levels were higher-, and globulin levels were lower in all test groups than in controls, but no dose-related response was noticeable. Liver weights of females were decreased in the intermediate-, and high-dose group. These changes were not accompanied by microscopical abnormalities (Colipa subm. III, ref. 11).

CUTANEOUS AND MUCOUS TOLERANCE

Sensitization potency was examined by a modified maximization test in guinea pigs which first received one intradermal injection of 0.1 ml 50% Freund's complete adjuvans in saline on day one. Then they were treated topically 7 times in 15 days with 0.5 g aqueous mixture with 18.5% a.i. under occlusion for 48-72 hrs. After a 12 day rest period the topical challenge treatment with 0.5 g of test product containing 37% a.i. induced a slight response in 2/20 animals. Therefore the colourant was classified as a week sensitizer (Colipa subm. III, ref. 9).

A dermal test with the colourant in 20 users of pyrazolon drugs (4 of which were allergic to Metamizol and 5 to Propyphenazon) did not induce skin reactions. These results do not support the possibility of cross reactions caused by the colourant in users of pyrazolon drugs (Summary report only, in Colipa subm. III, ref. 10).

MUTAGENICITY

An Ames test with up to 5 mg/plate was negative.
A chromosomal aberration test was conducted in CHO - cells with concentrations up to 0.8 mg/ml. In the absence of the metabolic activation system there was an increase only in gaps at the two high concentrations. It was concluded that the test substance did not induce any dose dependent increase in chromosomal aberrations (Colipa sumb. III, ref. 12 and 13).

CONCLUSION

Although the target organ and a no effect level have not been clearly established, there was little if any toxicity upon very high oral exposure of rats. Because, in addition, the use levels in cosmetics are relatively low, the use of this colourant seems to be justified, if sensitization is not a problem.

Classification : A.

References

1. Médicine, Hygiène, Sécurié du travail - Bureau International du Travail (in collaboration with the World Health Organisation) Geneva 1, p. 407 (1973)

2. F.A. PATTY
   Industrial hygiene and Toxicology - 2nd edition 2, p. 1017-1022 (1963)

3. K. AKATSUKA, L.T. FAIRHALL
   J. Ind. Hyg., 16 - 1 (1934)

4. Pigment Handbook - Vol 1 (1973) Edited by T.D. PATTON

5. Dr. Med. Vet. R. BATHE
   Dr. Med. Vet. K. SACHSSE
   "Acute Oral LD 50 in the Rat of FAT 20042/B"
   Laboratories of CIBA-GEIGY Ld, Basle, Switzerland
   Project n° 785415, Report of January 2nd, 1979

6. L. Ullmann, Dr. Med. Vet. K. Sachsse
   "Eye irritation in the Rabbit after single application of FAT 20042/B"
   Laboratories of CIBA-GEIGY Ld, Basle, Switzerland
   Project n° 785414, Report of January 18th, 1979

7. L. Ullmann, Dr. Med. Vet. K. Sachsse
   "Skin irritation in the rabbit after single application of FAT 20042/B"
   Laboratories of CIBA-GEIGY Ld, Basle, Switzerland
   Project n° 785416, Report of January 18th 1979

8. Company Data
   Test conducted by Dr. C.D. Calnan
   Consultant Dermatologist 20.12.83

9. C. CLEMENT (1987)
"Test to Evaluate the Sensitizing Potential by Topical Applications in the Guinea-Pigs"
Hazleton-IFT - Saint-Germain-sur-l'Arbresle - FRANCE
Report n° 703355 - 13 March 1987

10. Letter 23/07/1985
Pr. Dr. K.H. SCHULZ - Pr. Dr. H. IPPEN
Georg - August - Universität GOTTINGEN

11. P.H. ELLIOTT et al (1987)
"Twenty-eight Day Oral Toxicity Study in Rats with CI Acid Red 195"
Huntingdon Research Centre - Hungtingdon - ENGLAND

12. M.M. SHAHIN (1986)
"Mutagenic Evaluation of the Dye CI Acid Red 195 in Ames Salmonella typhimurium/Microsome Plate Test"
Division Recherche Fondamentale - Département de Mutagénèse - L'OREAL - Aulnay - sous - bois - FRANCE
Report 31.12.1986

13. A.T. NATARAJAN & F.H. SOBELS (1987)
"Evaluation of Compound CI Acid Red 195 in the Chromosome Aberration with Chinese Hamster Ovary Cells (in vitro)"
Department of Radiation Genetics and Chemical Mutagenesis - University of Leiden - THE NETHERLANDS
Report 27 April 1987

REPORT BY THE SCIENTIFIC COMMITTEE ON COSMETOLOGY

ON THE USE OF 1,2-DIAMINO-4-NITROBENZENE

. THE COMMITTEE'S MANDATE

To give its opinion on the question whether the use of 4. NOPD at the maximal level of 3,5% is admissible from the health point of view.

. FORMULA AND SYNONYMES

CI 76020
CAS Reg. 99-56-9
Colipa B24

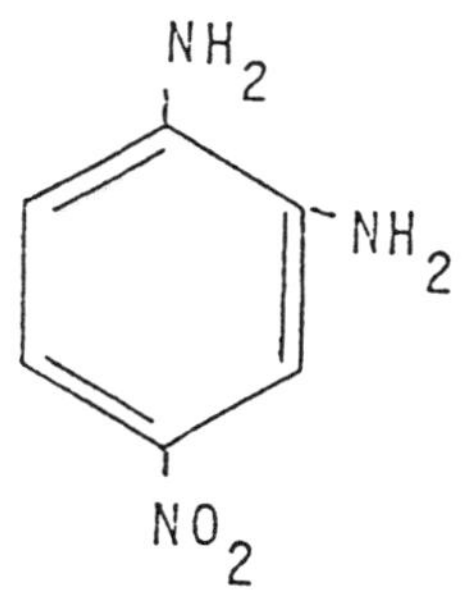

1,2-DIAMINO-4-NITROBENZENE

4-NITRO-o-PHENYLENEDIAMINE (4-NOPD)
2-AMINO-4-NITROANILINE
4-NITRO-1,2-DIAMINOBENZENE
4-NITRO-1,2-PHENYLENEDIAMINE
p-NITRO-o-PHENYLENEDIAMINE

Mol.w. 153.1

$C_6H_7N_3O_2$

. USE

It is used in direct or semi-permanent hair colouring products in combination with oxidant dyes at the maximal use concentration of 3,5% and at normal use concentration of 0,6%. It produces brown, red and blonde shades on the hairs without any chemical reaction.

. PRECEDING CONCLUSIONS

SEE, SECOND SERIES, EUR 8634 (1983)

"In view of the absence of conclusive carcinogenic effects in animals. The SCC saw no reasons for prohibiting 4 NOPD at present but wished to obtain additional information concerning percutaneous resorption and the repetition of more realistic carcinogenicity tests and in the meantime it could accept its continuing use on a provisional basis. The implementation of this recommendation will be reviewed each year". (Hair dye which was temporarily acceptable for use in cosmetic products until 31 december 1985 : EUR 8634, p.1. 1980).

17-10-86 : Banned in Italy and Denmark ; recommended for banning in F.R.G.

. RECAPITULATION OF THE STUDIES ON TOXICITY

LD50 oral (rat) : 2100-3720 mg/kg
(NIOSH : 681 mg/kg)
i.p. (rat) 1600 mg/kg

ORAL TOXICITY

Subacute toxicity : RAT Fisher 344 & Mouse B6C3F1. NCI bioassay.
5 males and 5 females/group received in the diet for 7 weeks period these doses of 4-NOPD : 0 (2 groups) - 681 - 1000 - 1430 - 2160 - 3150 - 4600 - 6800 and 10000 ppm for rats and 0 (2 groups) - 1470 - 2160 - 3150 - 4600 - 6800 - 10000 - 14700 and 21500 for mice.
The maximum tolerated dose was 750 ppm for rats and 7500 ppm for mice.

CUTANEOUS AND MUCOUS TOLERANCE

DERMAL TOXICITY

Dermal irritation : the compound applied to intact and abraded skin of rabbit as a 2.5% (w/v) preparation resulted non irritating.

Eye irritation : the compound as a 2.5% (w/v) suspension on rabbit's eyes caused a mild conjunctival inflammation and did not persist for more than 24 hours.

Sensitization was tested in Guinea pigs treated with 3% of 4-NOPD solution containing 2% Natrosol, 2% Tween 80, 0.05% Sodium sulphite and 10% isopropanol (pH = 7) applied 6 days/week for 3 weeks, sensitization was evaluated 2 weeks later. The results showed a relatively strong reaction (18/20 animals have an allergic effect).

Dermal absorption : 4-NOPD (120 µg/cm2 as a 0.6% (w/v), 200 ul as 3-H labelled hair dye solution) was applied on 10 cm2 of the skin in a 50% solution of a semi-permanent hair colourant shampoo base for 20 min. before rinsing off. Absorption was evaluated from the levels of 3-H present in the excreta and carcasses of the animals 48 hours after application. The results showed that 2.2 µg/cm2 (1.83%) of 4-NOPD apparently penetrated in the skin.

Short term dermal toxicity : 4-NOPD containing formulation (0.25% in water) tested on shaven intact skin of New Zealand by topical applications produced no toxic effects at 3-7-13 weeks after treatment at the histopathological analyses.

Human : In a repeated insult patch test in human (previous sensitized to p-phenylenediamine) with the compound no positive reactions were observed.

MUTAGENICITY

Studies have shown that 4-NOPD is mutagenic in vitro :

(1) on B.subtilis ;

(2) on Salmonella, in several experiments and in different experimental conditions ;

(3) on E.coli ;

(4) on Mouse lympoma test ; and in vivo :

(5) on Salmonella (urinary assay on rats) ;

(6) on Drosophila (X-recessive lethals and visible mutation tests systems by microinjection to adult and in SLRL test system by oral administration for 3 days) for gene mutation induction ;

(7) on 3 different cells line of chinese hamster (fibroblast line CHL, prostat gland and A(T1) CI-3 cell line) for chromosome aberrations in vitro ;

(8) on rat (until to 100 mg/kg/day i.p. × 10 wks.) by dominat lethal test (weak response) in vivo ;

(9) on B.subtilis (rec-assay) ;

(10) on E.coli by umu-test and differential killing tests system ;

(11) on S.cerevisiae for the induction of mitotic recombination ;

(12) on CHO cells and in mammalian cells culture (SCE in vitro).

Other mutagenicity studies have shown that the compound is negative for the induction of gene mutation in vitro in A.nidulans for forward mutation in two genetic markers ; in S.cerevisiae for the induction of reversions at three genetic loci ; and in N.crassa in ad-3 reversion mutations system ; and negative in vivo on Drosophila melanogaster in a minute loci and SLRL tests by fed for 21 hours.
The compound did not induce chromosome aberrations in vivo on rats (5000 mg/kg, g.i.) and mice (300 mg/kg, i.p.) by micronucleus test on bone marrow and dominant lethal test on rat (20 mg/kg, i.p. for 8 weeks). The compound has resulted negative for genotoxicity by DNA repair test on HeLa human cell line, and in E.coli for the induction of SOS function on primary rat hepatocytes culture in vitro, and in vivo on primary rat hepatocytes culture in vitro, and in vivo for the induction of SCE in bone marrow and in intestinal epithelium of Chinese hamster cells.

CARCINOGENICITY

Studies were carried out on mice and rat by a NCI bioassay, the compound fed in the diet at 750 and 375 ppm for rats for 103 weeks and 7500 and 3750 ppm for mice for 102 weeks showed no significant evidence of carcinogenicity. A formulation containing 4-NOPD at 0.6% level tested in A and DBAf mice by repeated topical application in aqueous acetone solution showed lymphoid tumours in both strains, but only in DBAf the incidence was statistically significant ; this report is incomplete and the experimental protocol is not adequate. Other studies carried out on mousse and rat (formulation 7403 containing 0.25% of 4-NOPD) and dog (Dye/Base composite containing 0.16% of 4-NOPD) have produced negative results.

CONCLUSION

- Negative evidence of carcinogenicity on two species (mice & rats).
- Sufficient evidence for mutagenicity in vitro, but negative in vivo mutagenicity and DNA damage.
- Some evidence of systemic toxicity on rat.
- Negative evidence of teratogenicity with doses lower than 128 mg/kg/day on rabbits.

The SCC in his plenary meeting of October 13th, 1987 requested a short-term oral toxicity study to determining the NEL.

Classification : B.

BIBLIOGRAPHY

N. LOPRIENO & G. BONCRISTIANI : A review on toxicology of 1,2 - Diamino - 4 - nitrobenzene.
Pisa, Italy, sept. 1986.

CONTENTS

36th meeting of the SCC - January 19, 1988

37th meeting of the SCC - April 12, 1988

2nd ADDITIONAL OPINION OF THE SCIENTIFIC COMMITTEE ON COSMETOLOGY RELATING TO THE USE OF BENZOYL PEROXIDE

(Opinion delivered on 19.1.1988)

The Committee has examinded the new data sent in by the Member States which are opposed to the Commission's proposal to regulate the use of benzoyl peroxide in cosmetics in accordance with the opinion delivered on 22.11.1983 and 17.12.1984.

For preceding informations, consult :
- Reports of SCC, fourth series, EUR 10305, P9, P23.

. FORMULA AND SYNONYMES

Benzoyl peroxide
Dibenzoyl peroxide
Benzoyl superoxide
Benoxyl (T.N.)
Oxy - 5
Cas reg n° 94-36-0

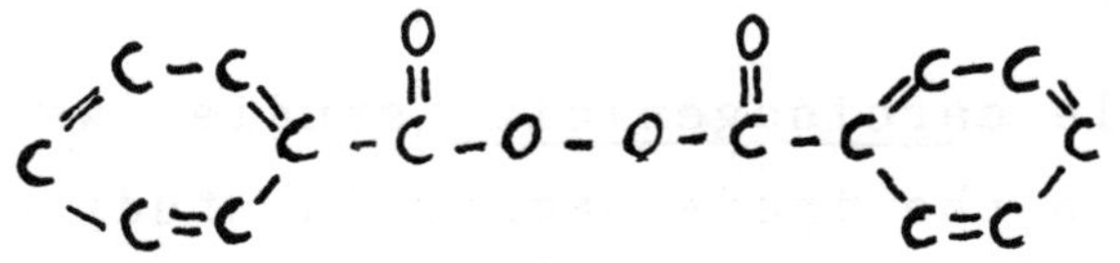

. USE

Used in anti-acne formulations.

In the medical practice BP use-concentrations in anti-acne preparations are 5 or 10%. (37) (38)
The Scientific Committee on Cosmetology (SCC) of the EC in 1983 has approved of a use concentration of 3% in cosmetic formulations. (20)
In a recent meeting of the SCC a proposed use concentration of 2% is mentioned. (33)

. <u>RECAPITULATION OF ADDITIONAL INFORMATIONS</u>

- Skin <u>irritation</u> after application of BP formulations has been demonstrated to occur in humans at concentrations of 2.5% and higher. At lower concentrations the data are scarce. In a limited study in humans absence of skin irritation at 1.2% BP was found but in rabbits concentrations as low as 0.1% were skin riitating.
- The available data on <u>phototoxicity</u> are too limited to allow drawing conclusions on this point.
- In controlled human studies high <u>sensitisation</u> rates have been found for 5 or 10% BP formulations. These results are not refuted by the mostly anecdotal reports on low sensitisation rates after clinical use. Almost all sensitisation data refer to BP use concentrations $\geq$ 5%. Controlled sensitisation studies in humans with the currently proposed use conentrations of 2 or 3% are lacking.
- The available <u>carcinogenicity</u> studies were of limited design. Given the inadequacies in study duration, number of dose levels tested and application frequency, the results of the studies, while not showing a carcinogenic effect by BP, cannot be considered conclusive evidence for the absence of such an effect.

- In tests in hamsters for tumour promotion by BP after chemical initiation the result was positive. In the sensitive mouse strain Sencar BP clearly exerts a tumour promotive effect after chemical initiation, an effect for the induction of which relative low and infrequent exposure suffice. Other mouse strains appear less sensitive.

  The Limited design of all available tumour promotion/cocarcinogenicity studies (in all cases duration $\leq$ 60 weeks) reduces the value of the negative results obtained in some of the studies. Thus, the absence of an effect in the studies with UV-treatment - (such studies are particularly relevant with regard to anti-acne use of BP) - cannot be considered as conclusive evidence for the absence of an effect by BP when UV-radiation is used as initiator or cocarcinogen. A study of adequate design could resolve this issue.

- On the basis of the available evidence BP cannot be qualified as a mutagenic agent. The relevance of the (positive result in) in vitro studies for DNA-damage/strand breaks or SCE's for possible mutagenicity or carcinogenicity is unclear.

CONCLUSION

The overall toxicological profile of benzoyl peroxide is incompatible with the level of human exposure resulting from cosmetic products.
In particular, the Committee cannot accept the risk of the widespread utilisation of a substance which promotes skin tumours in a low dose, whether the reservations on concentration, formulation, scope and labelling. The Committee recommends the banning of the use of benzoyl peroxide in cosmetic products.

Classification : D.

REFERENCES

1. Benzoyl peroxide. Monograph from National Institute of Public Health - The Netherlands. June 1983. Doc/Tox 300/137. (Summarized by P.J.C.M. Janssen).

2. IARC-Monographs on the evaluation of the carcinogenic risk of chemicals to humans, volume 36. WHO International Agency for Research on Cancer, Lyon 1985, pp. 267-284.

3. Anonymous, Effectiveness of 2.5% Benzoyl Peroxide in the treatment of acne, Herbert Laboratories Technical Report #57, 1980.

4. Aanvullend Advies van het Wetenschappelijk Comité voor Kosmetologie over het gebruik van benzoyl peroxide - Advies uitgebracht op 17 december 1984. Document n° XI/744/84-NL.

5. W.J. Cunliffe, Leeds Foundation for Dermatological Research, Letter to Dr. Tackray, dated 9th June 1983.

6. A.M. Kligman, Department of Dermatology, University of Pennsylvania. Letter to W.J. Cunliffe, dated May 20, 1983.

7. J.S. Strauss, Department of Dermatology, The University of Iowa, Letter to W.J. Cunliffe, dated May 6, 1983.

8. P.E. Pochi, Department of Dermatology, Boston University Medical Center, Letter to W.J. Cunliffe, dated May 10, 1983.

9. A.R. Shalita, Department of Dermatology, State University of New York, Downstate Medical Center, Letter to W.J. Cunliffe, dated May 6, 1983.

10. Statement of the Canadian and American Nonprescription Drug Industry on the Safety of Benzoyl Peroxide in non-prescription drug products for the treatment of acne vulgaris. Report dated July 9, 1987.

11. Y. Kurokawa et al. Studies on promoting and complete carcinogenic activities of some oxidizing chemicals in skin carcinogenesis. Cancer Letter, 24, (1984), 299-304.

12. W.C. Easting, Public Health Service of the U.S. Department of Health and Human Services. Research Triangle Park. Personal Communication to D. Hughes, dated July 2, 1986. Protocol outline and tables with results of the NTP initiation/promotion dermal study to compare susceptibility of B6C3F1 mice with Swiss CD-1 and Sencar mice.

13. J.A. McDermott, Note on cosultation with T. Slaga in Cincinnatti on May 2, 1986. Note dated 7/1/1987 (Abstract of study results).

14. P.C. Merker, Vicks Research Center. Note on information given by M. Hergenhahn, Cancer Research Institute, Heidelberg. Note dated January 26, 1987 (Abstract of study results).

15. O.H. Iversen, Skin tumorigenesis and carcinogenesis studies with UV-light, benzoyl peroxide (Panoxyl gel 5%) and ointment gel II. Institue of Pathology, University of Oslo, Norway. Manuscript submitted for publication in the Journal of Investigative Dermatology. Sent as personal communication to P.C. Merker, Vicks Research Center, 12/5/1987.

16. J.Schweizer et al. Benzoyl peroxide promotes the formation of melanotic tumors in the skin of 7,12-dimethylbenz (a) anthracene-initiated Syrian golden hamsters, Carcinogenesis, 8 (n° 3), (1987), 479-482.

17. R.A. Cartwright et al., Malignant melanoma benzoyl peroxide epidemiologica case-control investigation, Draft manuscript as attachment to letter from G. Haase to P. Merker and R. Wilson, dated April 30, 1987.

18. J.F. O'Connell et al., Enhanced malignant progression of mouse skin tumours by Free-radical generator benzoyl peroxide, Cancer Res. 46, (1986), 2863-.

19. T.J. Slaga and S. Nesnow, Sencar mouse skin tumorigenesis. In : H.A. Milman and E.K. Weisburger (eds). Handbook of carcinogen testing, Park Ridge, New Jersey, USA. Noyes Publications, 1987, 230-250.

20. EEC Environment and quality of life. Reports of the Scientific Committee on Cometology (fourth series). Report n° EUR 10305 EN. 1986. pp. 9-11.

21. S. De Flora et al. Genotoxicity and potency of 135 compounds in the Ames reversion test in a bacterial DNA-repair test, Mutat. Res. 133, (1986), 161-198.

22. S. De Flora et al. Mutagenicity testing with TA97 and TA102 and 30 DNA-damaging compounds, negative with other Salmonella strains, Mutat. Res. 134, (1984), 159-165.

23. B. Trainer, Personal Communication to E.B. Siegel, July 23, 1985 (Summary of Salmonella data on benzoyl peroxide ; NTP unpublished results). (Table of results only, no further information).

24. T.I. Matula et al., Mutagenicity studies of benzoyl peroxide in bacteria, Environ. Mutat. 9 (supplement 8) (1987), p. 69 (Abstract n° 177). (Summary only).

25. S.H.H. Swierenga et al., An evaluation of cytotoxicity genotoxicity of benzoyl peroxide, Environ. Mutat. 9 (supplement 8) (1987), p. 105 (Abstract n° 273) (Summary only).

26. H.L. Gensler and G.T. Bowden, Evidence suggesting a dissociation of DNA strand scissions and late stage promotion of tumour cell phenotype, Carcinogenesis, 4 (n° 2), (1983), 1507-1511.

27. A.J. Saladino et al. Effects of formaldehyde, acetaldehyde, benzoyl peroxide and hydrogen peroxide on cultured normal human bronchial epithelial cells, Cancer Res. 45, (1985), 2522-2526.

28. H.C. Birnboim, DNA strand breaks in human leukocytes induced by superoxide anion, hydrogen peroxide and tuour promotors are repaired slowly compared to breaks induced by ionizing radiation, Carcinogenesis, 7, (n° 9) (1986), 1511-1517.

29. J.A. Hartley et al. Association of DNA strand breaks with accelerated terminal differentiation in mouse epidermal cells exposed to tumour promotors, Cancer Res. 45, (1985), 4864-4870.

30. W.J. Cunliffe and B. Burke, Benzoyl peroxide : Lack of sensitization, Acta. Derm - venereol. 62 (n° 5), (1982), 458-459.

31. B. Bandenir and M. Pambar, Zur hautirrifativen Wirkung von Antiseptika, Z. Klin. Ned. 40 (1985), 1893-1895.

32. G. Angelini et al. Allergic contact dermatitis to some medicaments, Contact. Dermatitis 12, (1985), 263-269.

33. Anonymous, Compte-rendu de la 35ème réunion du Comité Scientifique de Cosmetologie - Bruxelles, le 13 octobre 1987. Document n° XI/932/87-FR.

34. T.W. Kensler, Division of Experimental Pathology and Toxicology, The John Hopkins University, School of Hygiene and Public Health, Letter P.C. Merker dated April 13, 1987. (Summary of test result).

35. U.F. Haustein et al. Allergic and irritant potential of benzoyl peroxide, Contact Dermatitis, 13 (1985), 252-257.

36. H.J. Bandmann and M. Agathos, Die posttherapeutische Benzoylperoxid-kontaktallergie, Der Hautarzt, 36 (1985), 670-674.

37. Anonymous, Richtlijn 1984-11 Benzoylperoxidebevattende produkten. College ter Bevordering van Geneesmiddelen 278e College vergadering. Bijlage 2.

38. Martindale Pharmacopee ed. 1977 p. 1201.

REPORTS BY THE SCIENTIFIC

COMMITTEE ON COSMETOLOGY

37th REUNION

APRIL 12, 1988

LIST OF PARTICIPANTS

Present

Mr. AGACHE
Mr. DE GROOT
Mrs. DONY
Mrs. ENJOLRAS
Mr. FIELDER
Mr. GOULDING
Mr. KAPOULAS
Mr. LOPRIENO
Mr. O'MAHONY
Mr. RAMOS MORGADO
Mr. SHOU
Mr. STUTTGEN

Absent

Mr. GRANADOS JARQUE
Mrs. KNAAP
Mr. MUSCARDIN

Commission

Mrs. MASSE
Mr. GONTIER (DG XI/C/2)
Mr. COLLIN (expert)

SUMMARY

Opinions expressed on April 12th, 1988 concerning the use of

- HAIR DYES (XI/827/87 - XI/151/88)
  1. 2 AMINO 4 NITROPHENOL (B26)
  2. 2 AMINO 5 NITROPHENOL (B27)
  3. ORTHO - AMINOPHENOL (A14)
  4. META - AMINOPHENOL (A15)

- PRESERVATIVE AGENTS (XI/204/88 - XI/189/88)
  1. CHLORPHENESIN (EEC n° 3 P4)
  2. BENZETHONIUM CHLORIDE (EEC n° 53 P70)
  3. BENZALKONIUM CHLORIDE (EEC n° 54 P71)
  4. CHLORQUINADOL (P76)

- U.V. FILTERS
  1. GLYCERYL PARAAMINOBENZOATE (EEC n° 2.4)
  2. ETHYL HEXYL - 4 - METHOXYCINNAMATE (EEC n° 2.13)
  3. TERT - BUTYL - 4 - METHOXYDIBENZOYLNETHANE (EEC n° 2.31)

REPORTS BY THE SCIENTIFIC COMMITTEE ON COSMETOLOGY ON THE USE OF CERTAIN HAIR DYES

. THE COMMITTEE'S MANDATE

To give its opinion on the question whether the use of certain hair dyes :

- 2 - amino - 4 - nitrophenol
- 2 - amino - 5 - nitrophenol
- ortho - aminophenol
- meta - aminophenol

is admissible from the health point of view.

1. 2-AMINÓ-4-NITROPHENOL

1. FORMULA AND SYNONYMES

CI 76530
CAS Reg. 99-57-0
Colipa B26

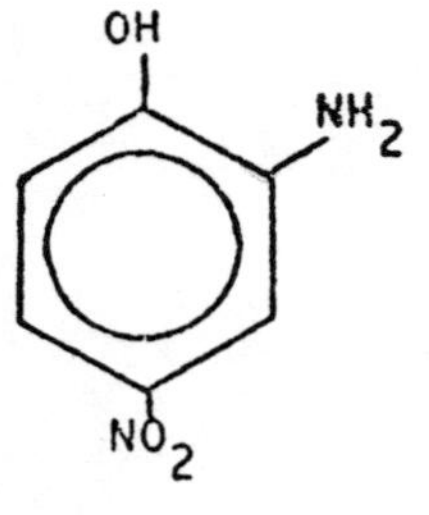

$C_6H_6N_2O_3$ MW:154,1

2 - amino - nitrophenol
2 - amino - 4 - nitrophenol
1 - hydroxy - 2 - amino - 4 - nitrobenzene

. PRECEDING CONCLUSIONS

From See, Second Series, EUR 8634 p.12

"Insufficient data for carcinogenic assessment ; additional information concerning carcinogenicity in animals after dermal application of adequate doses".

. NEW INFORMATIONS

- Cancerogenicity ; The compound (98%) was tested by NCI for the study of carcinogenicity on Fischer 344 rat and B6C3F1 mice by oral gavage administrations of 0 - 125 - 250 mg/kg for 2 years (5-d/wk). The results showed evidence of carcinogenicity in male rats (renal cortical (tubular cells) adenoma (vehicle : 0/50 ; 125 mg/kg ; 1/48 (2.1%) ; 250 mg/kg 3/50 (6%). Nonneoplastics effects were observed in male and female rats (chronic nephropaty and pigmentation of the small and large intestines). No evidence of carcinogenicity in female rats and male and female mice.
- Mutagenicity and genotoxicity ; A pure preparation of the compound resulted positive for gene mutation in vitro by reverse mutations on Salmonella typhimurium (4 studies), Sordaria brevicollis and on mouse lymphoma L5178Y/TK+/- assay ; positive for chromosome aberrations in vitro on CHO and CHL cells system ; negative for chromosome aberrations in vivo by micronucleus test on rat at a dose of 5000 mg/kg and dominant lethal on rat at dose of 20 mg/kg (inadequate studies) ; and positive for genotoxicity by DNA-repair liquid test on E.COLI and for the ability to induce sister chromatid exchange in CHO cells in vitro.

BIBLIOGRAPHY

(1) NTP. Toxicology and Carcinogenesis studies of 2-Amino-4-Nitrophenol (CAS No. 99-57-0) in F344/N rats and B6C3F1 mice (gavage studies). Tech. Rep. Ser. No. 139, July 1987.

(2) HOSSACK D.J.N., J.C. RICHARDSON. Examination of the potential mutagenicity of hair dye constituents using the micronucleus test. Experientia 33, 377-379, 1977.

(3) BURNETT C. et al., Dominant lethal mutagenicity study on hair dyes. J.Toxicol.Environ.Health 2, 657-662, 1977.

(4a) M.M.SHAIN et al., Mutagenicity of aminonitrophenol compounds in Salmonella typhimurium: a study of structural-activity relationships. Int.J.Cosm.Scien. 4, 25-35, 1982.

(4b) M.M.SHAIN et al., Relationships between the chemical structure and mutagenic activity of monocyclic aromatic amines. In "Chemical Mutagens - Principle and Methods for their detection" Vol. 8, ed. F.J. de Serres, Plen.Pres., NY & L., cap.5, pp.151-181, 1983.

(5) AMES B.N. et al., Hair dyes are mutagenic: identification of a variety of mutagenic ingredients. P.N.A.S. 72, 2423-2427, 1975.

(6) GARNER R.C., C.A. NUTMAN, Testing of some azo dyes and their reduction products for mutagenicity using Salmonella typhimurium TA1538.

(7) JENSSEN D., C.RAMEL, The micronucleus test as part of a short-term mutagenicity test program for the prediction of carcinogenicity evaluated by 143 agents tested. Mutat. Res. 75, 191-202, 1980.

(8) YU-SUN C.C. et al., Mutagenicity of 4-nitroquinoline 1-oxide and three heir dye components in *Sordaria brevicollis*. Neurospora Newsl 28(2), 22, 1981.

(9a) NISHIOKA H., Mutagenic studies on coloured make-up cosmetics as index of carcinogenicity. Meeting of Environmental Mutagenic Society of Japan, Japan - Sept. 27, 1975.

(9b) NISHIOKA H. et al., Detection of carcinogenicity of color cosmetics in bacterial test systems. Mutat. Res. 38(5), 1976. Abstract.

(10) ISHIDATE M., Chromosomal Aberrations Test in Vitro. Realiz. Inc., Japan, 1983.

(11) INOUE T. et al., Purification and properties of a plant desmutagenic factor for the mutagenic principle of triptophan pyrolysate. Agric.Biol.Chem. 45(2), 345-353, 1981.

2. 2-AMINO-5-NITROPHENOL

. FORMULA AND SYNONYMES

CI 76535
CAS Reg. 121-88-0
Colipa B27

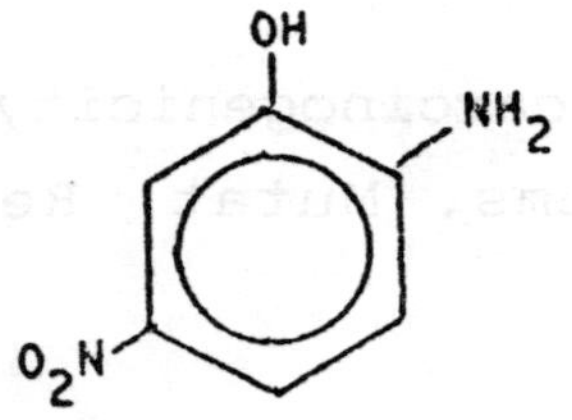

$C_6H_6N_2O$

MW : 154,1

2 - amino - 5 - nitrophenol

1 - hydroxy - 2 - amino - 5 - nitrobenzene

. PRECEDING CONCLUSIONS

From see, second series, EUR 8634 p13

"Insufficient data available for a conclusive toxicological assessment ; additional data are required for the assessment of its potential for carcinogenicity in animals after dermal applications".

. NEW INFORMATIONS

- Carcinogenicity : The compound was tested by NCI for the study of carcinogenicity on Fisher 344 rat and on B6C3F1 mice. 2-Amino-5-nitrophenol (98% purity) has been administered in corn oil by oral gavage, 5-day per week, at dose levels of 0 - 100 - 200 mg/kg (5ml/kg) to groups of 50 male and 50 female mice, and 0 - 400 - 800 mg/kg (10 ml/kg) to 50 males and 50 female mice for 2-years. The results showed evidence of carcinogenicity in male rats : pancreatic acinar cell adenomas (vehicle : 1/50 ; 100 mg/kg : 10/50 ($P < 0.002$) ; 200 mg/kg : 3/49). Nonneoplastic effects were observed either in male and female rats. In male and female mice : inflammation and pigmentation of the large intestine. No evidence of carcinogenicity were observed in female rats and female and male mice.

- Mutagenicity and genotoxicity. A pure preparation of the compound resulted positive for gene mutation in vitro by reverse mutations on Salmonella typhimurium (6 studies/1 study negative, but this latter was inadequate because performed only in the absence of metabolic activation), and by forward mutations on mouse lymphoma L5178Y/TK+/-assay ; positive for chromosome aberrations in vitro on two different cells line of chinese hamster (CHO and CHL) ; negative for chromosome aberrations in vivo by micronucleus test on rat at a dose of 3000 mg/kg and dominant lethal on rat at dose of 20 mg/kg (inadequate studies) ; positive for genotoxicity by DNA repair liquid test on E.coli and for the ability to induce sister chromatid exchange in CHO cells in vitro.

CONCLUSION

Classification : D.

BIBLIOGRAPHY

NTP Technical Report on the Toxicology and Carcinogenesis studies of 2-Amino-5-Nitrophenol (CAS N° 121-88-0) in F344/N rats and B6C3F1 mice (gavage studies).
Tech. Rep. Ser. N° 334, March 4th, 1987/

3. o-AMINOPHENOL

. FORMULA AND SYNONYMES

CI 76520
CAS Reg. 95-55-6
Colipa A14
o - aminophenol
1 - hydroxy -2- aminobenzene
2 - aminophenol
2 - hydroxy - aniline
o - hydroxy - aniline

MW : 109,129

. USE

Oxidative hair dye
Max. use : up to 1%
0,5% in combination with H_2O_2

. RECAPITULATION OF THE STUDIES OF TOXICITY

LD50 oral (rat) : 1052 - 1058 mg/kg

ORAL TOXICITY

Rat : 150 mg/kg : no nephrotoxicity and no hepatotoxicity.

- Short term oral : o-AP administered oral to rat (50 mg/kg in 1% propylene glycole x 3 months) showed slightly toxic effects : bronco-pulmonary injuries at anatomopatological analyses in several rat and slight and isolated increase in alcaline phosphatase. No mortality occured.

CUTANEOUS AND MUCOUS TOLERANCE

DERMAL TOXICITY

- Dermal irritation : it resulted slightly irritating (0.1 ml, 1% in propylene glycole) on rabbit.
- Eye irritation : it resulted not irritating (0.5 ml, 1% in propylene glycole) on rabbit.
- Sensitization : it showed negative results on guinea pigs (0.5 g 3 times/wk X 3 wks and at the start of 4 wk ; challenge reaction on day 36 with 0.5 g).
- Human-sensitization : A sensitized man with dermatitis related to tires and other rubber products, IPPD and PPD positive, showed at patch-test focal flare after 1 wk, in the site of application of o-AP(27 aromatic amines tested).
- Dermal absorption : In a formulation (0.8%) mixed 1:1 with H_2O_2 (0.32% of o-AP, i.e. 587 nmol or 64 ug) applied on the back of female Hairless rat (1 cm2) for 30 min. showed these absorption values : 4.55% with human hair ; 4.66% with m-AP (0.8%) and human hair ; 6.62% with m-AP (0.8%).
- Short term dermal : o-AP containing formulation (0.3%, mixed 1:1 H_2O_2) showed negative results after dermal topical application (1 ml/kg, 2 time/wk. X 13 wks) on shaven or not shaven skin of rabbit.

MUTAGENICITY

The compound tested for gene mutation resulted positive in vitro on Salmonella (2 studies) and in vivo on mice with Sperm-Head abnormality ; positive for chromosome aberrations using micronucleus test on mouse (1 study) ; and positive for genotoxicity in vitro on human fibroblasts and lymphocytes using SCE analyses and on E.coli for DNA damage.

The compound resulted also negative for gene mutation in vitro on Salmonella (3 studies) and S.pombe ; negative for chromosome aberrations in vitro on CHO cells and in vivo by chromosome aberrations and micronucleus test on mice ; and negative for genotoxicity on CHO cells in vitro and hamster chinese (bone marrow) in vivo for induction of SCE.

CARCINOGENICITY

Long term studies was carried out with a hair dye composite containing test compound (0.3%, 1:1 with H_2O_2) by dermal topical applications on mice (0.05 ml/wk. X 2y) and rats (0.2 ml increased by 0.1 ml to 0.5 ml 2 appl/wk, X 2y - to Fo generations from the time of weaning to the weaning of young (Fl generation) : no biologically significant difference was observed between treated and control groups.

TERATOGENICITY AND REPRODUCTION

The compound resulted teratogenic without maternal toxicity on Syrian golden hamster (100-150-200 mg/kg i.p. in acified isotone saline on the morning of day 8 of gestation (8 days after the evening breeding) showing at day 13 of gestation several malformations in the fetuses : neural tube defects (exencephaly, enocephale and spina bifida), eye defects, limb defect, rib defects, tail defects and umbilical hernia. The compound containing formulation (0.3% in water, 1:1 with H_2O_2) tested on rats (2 mg/kg bw day to the shaven skin on day 1-4-7-10-13-16-19 of gestation) showed negative results, but the report resulted incomplete, because no detail on histopatology and body weight were presented.

Embryotoxicity : the compound showed moderate acute embryoletality (LD50) by HET test : 0.91 mg/egg at day 1 (ca. 20 ppm) and 0.75 at day 5 (ca. 15 ppm), with a no-effects of 0.2 mg/egg (1 d.) and 0.1 mg/egg (5 d.).

Reproduction : the compound containing formulation (0.3%, with 6% H_2O_2) showed negative results in a reproduction study on raty by dermal topical applications (0.5 ml/appl. start to 0.2 ml/appl. increased of 0.1 ml/appl/week) twice weekly through growth, mating, gestation and lactation to the weaning of the F1b, F2b and F3c litters of the respective generation.

MISCELLANEOUS INFORMATIONS

Protein binding : o-Aminophenol after oxidation with ferricyanide formed adducts with proteins.

Blochemical studies in vitro : o-Aminophenol had no effects on mitochondrial respiration and phosphorilation with different oxidative substrate, only slightly reduction of the P/O ratio at the highest doses (>0.5 mM). It had no effects on mitochondrial integrity structure, only a physiological swelling increase of mitochondria observed.

Effects on hemoglobin : in Vitro. The compound rapidly produced ferrihemoglobin formation on both human hemoglobin and washed red cells in vitro, more rapidly in washed red cells. The oxidized o-Aminophenol, by the activated oxygen, preferred phenoxazone condesation at high dose (<100 uM) and binding with hemoglobin at low dose (0-80 uM) ; it is oxidized by the activated oxygen in oxyhemoglobin. Glutathlone diminished the covalent binding of o-Aminophenol to hemoglobin and pre-treatment of hemoglobin with p-chloromercuribenzolc acid (block the reactive groups SH) reduced to 1/5 the amount observed with native hemoglobin. The fetal hemoglobin is more susceptible than adult hemoglobin for methemoglobin formation, it has been reported that this result was also obtained at different values of pH in Vivo. The dose of 50 mg/kg bw in aqueous by i.p. injections on 2 adult Japanese quail showed a peak of 25% of methemoglobin after 10 min. (0.4 umol/ml blood) no undetectable after 30 min.

CONCLUSIONS

The SCC requires additional data for defining the no effect level in an oral subacute toxicity study (28 days), the potential for the induction of UDS or DNA binding in vivo and conclusive data for teratogenicity.

Classification : C.

BIBLIOGRAPHY - ACUTE TOXICITY
MUCOUS MEMBRANES TOLERANCE
SKIN IRRITATION
SENSITIZATION

1. B. COQUET, (1977)
"Hydroxy-1 Amino-2 Benzene (Ref : AH 8242/46)
- Oral Acute Toxicity in the Rat
- Tolerance Tests in the Rabbit
. Cutaneous Irritation
. Mucous Irritation"
Laboratories of "Institut Français de Recherches et d'Essais Biologiques" (IFREB)
Report IFREB-R 705271, May 27th 1977

2. B. COQUET, (1978)
"IFG.60/78-Colorant (Ref : AH 8242/46)
Evaluation of the Sensitizing Potential of a Test Substance by Topical Application in the Guinea-Pig"
Laboratories of "Institut Français de Recherches et d'Essais Biologiques" (IFREB)
Report IFREB-R 807311, July 5th 1978

3. RUDZKI et al, (1980)
"Active Sensitization to Ortho-and Para-aminophenol with negative patch test to Meta-aminophenol"
Contact Dermatitis 6, n° 7, p. 501 (Dec. 1980)

4. PROF. P.E. FOURNIER, (1981)
"Etude de toxicité chronique chez le rat d'un poroduit dénommé o-Amino-phenol"
Société de Recherches Biologiques - Asnières/Seine
Report of 25th June 1981

5. BURNETT et al, (1976)
"Teratology and Percutaneous Toxicity Studies on Hair Dyes"
J. of Toxicol. and Environ. Health 1, p. 1027-1040

6. MILLER et al, (1948)
"The carcinogenicity of certain derivatives of p-Dimethylamino-azobenzene in the rat"
J. Exp. Med. 87, p. 139-156

7. BURNETT et al, (1980)
"Evaluation of the Toxicity and Carcinogenicity of Hair Dyes"
J. of Toxicol. and Environ. Health 6, p. 247-257

MUTAGENICITY STUDIES - IN-VITRO STUDIES

8. E. LAVOIE et al, (1979)
"Mutagenicity of AMinophenyl and Nitrophenyl Ethers, Sulfides and Disulfides"
Mutation Research 67, p. 123-131

9. YOSHIKAWA et al, (1976)
"Studies on the Mutagenicity of Hair-Dyes"
Bull. Nation. Inst. Hygien. Scien. 94, p. 28

10. M. DEGAWA et al, (1979)
"Mutagenicity of Metabolites of Carcinogenic Aminoazodyes"
Cancer Letters 8, p. 71-76

MUTAGENICITY STUDIES - IN-VIVO STUDIES

11. PROF. SOBELS and DARROUDI, (1982)
"Sex-linked Recessive Lethal Test"
O-Amino-Phenol
Sylvius Laboratories
University of Leiden (NL)
Report March 5th, 1982

12. PROF. SOBELS and DARROUDI, (1981)
"Sister Chromatid Exchanges Test"
O-Amino-phenol
Sylvius Laboratories
University of Leiden (NL)
Report October 23rd, 1981

LONG TERM TOXICITY/CARCINOGENICITY STUDY

Method : Bladder implantation testing

Species : Albino mice

Number of animals : 37 (treatment group)
55 (control group)

Method of administration : Implantation of cholesterol pellet containing the test colorant in the bladder of the mice

Dose level : 12.5 %

Total duration : 40 weeks

Negative control : Pellets of Cholesterol alone

Results : The implantation of Cholesterol pellet cuased 5 carcinomas in 55 mice (9.0 %) while the pellet containing o-Aminophenol caused 2 carcinomas in 37 mice (5.4 %) and no benign tumors

Conclusion : o-Aminophenol is considered inactive

13. D.B. CLAYSON et al, (1958)
"The Testing of Ortho Hydroxy-Amines and Related Compounds by Bladder Implantation and a Discussion of their Structural Requirements for Carcinogenic Activity"
Brit. J. Cancer, 12, (1958), p. 222-230

24. E.I. GODENTHAL et al, (1979)
"Lifetime Chronic TOxicity/Carcinogenesis Study in Rats"
International Research and Development Corporation, Michigan, USA
Report February 1979

MUTAGENICITY STUDY

14. S. HAWORTH et al, (1983)
"Salmonella Mutagenicity Test Results for 250 Chemicals"
Environ. Mutagenesis Suppl., 1, (1983), p. 3-142

15. N. LOPRIENO, (1983)
"Forward Mutation Induction on the Yeast Schizosaccharomyces pombe (Strain P1) : In Vitro"
Instituto di Biochimica, Biofisica e Genetica, Università di Pisa, ITALY
Report 4.3.1983

16. F. DARROUDI & F.H. SOBELS, (1982)
"Evaluation of Compound Ortho Amino Phenol in the Chromosome Aberrations Test : Chinese Hamster Ovary (CHO) Cells (In Vitro)"
Department of Radiation Genetics and Chemicals Mutagenesis - University of Leiden - THE NETHERLANDS
Report 5 March 1982

17. J.L. WILMER et al, (1981)
"Sister-Chromatid Exchanges Induction and Cell Cycle Inhibition by Aniline and its Metabolites in Human Fibroblasts"
Environemental Mutagenesis, 3, (1981), p. 627-638

18. G. KIRCHNER & U. BAYER, (1982)
"Genotoxic Activity of the Aminophenols as Evidenced by the Induction of Sister-Chromatid Echanges"
Human Toxicol., 1, (1982), p. 387-392

19. D. WILD et al, (1981)
"Mutagenic Activity of Aminophenols and Diphenols, and Relations with Chemical Structure"
Abstract presented at a GUM Meeting (Darmstadt 1981)
Mutation Research, 85, (1981), p. 456

20. K.G. DOSSOU, (1983)
"Micronucleus test : o-Aminophenol"
Laboratoires de Recherche Fondamentale, L'OREAL,
Aulnay-sous-bois, FRANCE
Report 4 March 1983

21 N. LOPRIENO, (1983)
"In Vivo Cytogenetic Test for the ANAlysis of Chromosome Aberrations Induced in Bone Marrow Cells of CD-1 Swiss Mice Treatment with o-Aminophenol"
Instituto di Biochimica, Biofisica e Gentica della Università di Pisa, ITALY
Report 1st September 1983

EMBRYOTOXICITY STUDY

22. F.H. KEMPER & N.P. LUEPKE, (1984)
"Bericht zur Toxikologischen Prüfung am Bebrüteten Hühnerei (HET - Embryotoxicity - Test)"
Institut für Pharmakologie und Toxikologie - Münster - GERMANY
Report 1.9.1984

TERATOGENICITY STUDY

23. J.V. RUTKOWSKI & V.H. FERM, (1982)
"Comparison of the Teratogenic Effects of the Isomeric Forms of Aminophenol in the Syrian Golden Hamster"
Toxicol. and applied Pharmacol., 63, (1982), p. 264-269

REPRODUCTION STUDY

25. F.X. WAZETER et al, (1977)
"Multigeneration Reproduction Study in Rats"
International Research and Development Corporation, Michigan, USA
Report November 1977

EFFECT ON HEMOGLOBIN (IN VITRO STUDY)

27. M. WIND & A. STERN, (1977)
"Comparison of Human Adult and Fetal Hemoglobin : Aminophenol-Induced Methemoglobin FOrmation"
Experientia, 33, (5 February 1977), p. 1500-1501

28. K.G. ECKERT & P. EYER, (1983)
"Differences in the Reactions of Isomeric Ortho- and Para-Aminophenols with Hemoglobin"
Biochemical Pharmacology, Vol. 22, n° 6, (1983), p. 1019-1027

IN VIVO TOXIC EFFECTS AND FERRIHAEMOGLOBIN FORMATION

29. B.J. BLAAUBOER et al, (1980)
"Effects of Phenylhydroxylamine and Aminophenols in Japanese Quail in Vivo"
Xenobiotica, Vol. 10, n° 7/8, (1980), p. 495-498

TOXICITY ON CYANOBACTERIA

30. C.E. CERNIGLIA et al, (1981)
"Biotransformation and Toxicity of Aniline and ANiline Derivatives in Cyanobacteria"
Arch. Microbiol., 130, (1981), p. 272-275

NEPHROTOXICITY STUDY

31. J.F. NEWTON et al, (1982)
"Nephrotoxicity of p-Aminophenol, a Metabolite of Acetaminophen, in the Fischer 344 Rat"
Toxicol. and applied Pharmacol., 65, (1982), p. 336-344

IN VITRO PROTEIN BINDING

32. "The Differential Reactivity of the Oxidation Products of o-Aminophenols Towards Protein and Nucleic Acid"
Biochimica et Biophysica Acta, III, (1965), p. 147-153

MUTAGENICITY STUDY
IN VITRO ASSAY

33. DE FLORA et al, (1984)
"Genotoxic Activity and Potency of 135 Compounds in the AMES Reversion Test and in a Bacterial DNA-Repair Test"
Mutation Research, 133, (1984), p. 161-198

34. J.S. YOON et al, (1985)
"Chemical Mutagenesis Testing in Drosophila. IV. Results of 45 Coded Compounds Tested for the National Toxicology Program"
Environmental Mutagenesis, 7, (1985), p. 349-367

EFFECTS ON RAT LIVER MITOCHONDRIA

35. A. ALIFANO et al, (1965)
"Effeto di Alcune Amine Aromatiche e dei loro Derivati Idrossilati su Mitocondri di Fegato di Ratto"
G. Biochim, 14 (6), (1965), p. 722-742

4. m-AMINOPHENOL

. FORMULA AND SYNONYMES

CI 76545

CAS Reg. 591-27-5

Colipa A15

m-Aminophenol

1-hydroxy-3-aminobenzene

3-amino-phenol

1-3-amino-phenol

m-hydroxy-aniline

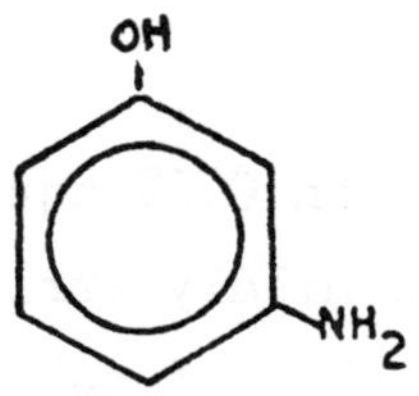

MW : 109,29

. USE

oxidative hair dye

max. use : 2%

1% in combination with H_2O_2

. RECAPITULATION OF THE STUDIES OF TOXICITY

LD50 oral (rat) : 812 - 1660 mg/kg

(mouse) : 330 mg/kg

ORAL TOXICITY

The compound did not show nephrotoxicity on Fisher 344 rat orally treated with 1.37 mmol/kg in 0.5 mM HCl in isotone saline.

Short term oral : m-Aminophenol administered by oral intubation to rats (20 males and 20 females) for 12 weeks (5 times/wk) showed that the dose of 50 mg/kg does not represent a toxic cumulative dose.

CUTANEOUS AND NUCOUS TOLERANCE

DERMAL TOXICITY

Skin irritation : The compound (2.5% (w/v) in 0.5% aqueous gum tragacanth with 0.05% sodium sulfite resulted mildly irritating (reading at 24 and 72 h) on abraded and intact skin of rabbit (1/3 animals at 72 h with very slight oedema).

Eye irritation : 2.5% (w/v) in 0.5% aqueous gum tragacanth with 0.05% sodium sulfite, instilled into one eyes of 3 rabbit resulted non irritating ; only minimal conjunctival irritation was observed 1 hour after instillation in all 3 rabbits.

Sensitization : m-Aminophenol (3% in water with 2% Natrosol, 2% Tween 80 and 0.05% sodium sulfite) showed no allergic reaction in guinea pig by open epicutaneous method. In another study on guinea pigs it has been reported that m-Aminophenol (0.1 ml, dose not reported) showed no sensitized animals after challenge reaction.

Human : m-Aminophenol resulted negative in a patch test on a sensitized man (positive to IPPD and PPD) and it was unable to produce cross-reactions.

Dermal absorption : m-Aminophenol HCl (ring ^{14}C, radiochemical purity 98%) administered dermally (2mg/kg = 0.3 ml of an 8% aqueous solution on the shaved skin (9 cm^2) for 30 min, before shampooing and rinsing), subcutaneously or orally (37.5 mg/kg bw, 1 ml of 1% aqueous solution) to OFA Sprague Dawley rats showed that after dermal study 0.41% (B.1 µg/cm2) penetrated in the skin ; 89-95% of the dose absorbed was revealed in the urine after 24 hours and reached the maximum level in the organs after 35 min. in subcutaneous and oral administration treatment it was mainly excreted in the urine. The labelled compound (ring [14C]) containing formulations (1%, 2% and 3%, 1:1 with 6% H_2O_2) applied on the shaven skin of OFA Sprague-Dawley rats for 30 min. before shampooing and rinsing showed these values of dermal absorption : 0.14% (1% of m-AP) ; 0.16% (2% m-AP) ; 0.15% (3% m-AP) ; and 0.053% (2% m-AP) when skin was not shaven. After rinsing 92-96% of the radioactivity was found in the rinsing water.

Short term dermal : The compound containing formulations (0.04%-0.7%), mixed 1:1 with H_2O_2 by topical application for 13 wks (twice weekly) on abraded and intact skin of rabbit showed negative results.

MUTAGENICITY

Studies have been demonstrates that m-Aminophenol does not induce gene mutation in Salmonella, E.coli, S.pombe and S.cerevisiae in vitro and in vivo by Sperm-head abnormality on Mouse and SLRL on Drosophila ; chromosome aberrations in vitro on CHO cells and in vivo on bone marrow cells on mouse by micronuleus test on mouse and rat ; genotoxicity effects in vitro on E.coli (DNA Repair by plate-test), S.cerevisiae (mitotic gene-conversion) and V79 (SCE) genotoxicity effects in vivo on mouse (SCE, s.c. treatment) lymphocytes (SSB), human fibroblasts (SCE) and S.cerevisiae.
Positive results were obtained in one study on Salmonella TA1538 with co-factors for gene mutation in vitro and for genotoxicity in vitro on mouse (SCE, i.p. treatment) ; the compound is resulted positive for ability to produce aneuploid products of meiosis in Neurospora. Formulations containing m-Aminophenol resulted negative ; (a) for heritable translocation test on rat (1%),

(b) chromosome aberrations and SCE in human volunteers (dyed the hair : 3-6% H_2O_2 solutions, lymphocytes analysis).

CARCINOGENICITY

Dermal topical application : Four oxidative formulations (7403, 7406, P-25, P-26, mixed 1:1 with 6% H_2O_2) containing 0.7%, 0.09% and 0.04% m-Aminophenol tested on Swiss Webster mouse by dermal-topical application (0.05 ml/cm2 X 21-23 months) showed negative results. m-AP containing in two formulations (0.7%, 1:1 with H_2O_2) tested on Charles River rat (Fo generation) from the time of weaning to the weaning of their young (Fla generation) by dermal topical applications (0.2 ml increased by 0.1 ml to 0.5 ml, 2 times/week X 2 years) showed negative results.
m-AP in other two formulations (0.09% nad 0.02%) tested with the same previous treatment schedule on rats showed hyperkeratosis and/or acanthosis of stomach mucosa in several males and females and hepatocellular hypertrophy/hyperplasia or hyperplastic/hypertrofic nodules in livers of several rats, especially males treated with formula at 0.09% m-AP level (possible compound related).

TERATOGENICITY AND REPRODUCTION

Reproduction : In a multireproduction study on rat with formulations (0.02%-0.9%) applied dermally (0.5 ml/rat) twice week for through, growth, mating, gestation and lactation to weaning at the Flb, F2b and F3c litters of respective generations given negative results.

Teratogenicity study performed on Syrian golden hamster by i.p. injection (100-150-200 mg/kg in 10% DMSO aqueous solution) on the morning of day 8 of gestation (8 days after the evening breeding) showed (females killed at 13 days of gestation) that the compound induced some malformation (type not reported) at 150 mg/kg (6/84=7.1% malformed fetuses) not statistically significant,

without maternal toxicity, m-AP (0-0,10-0,25-1.00% in the diet) administered to female rats for at least 90 consecutive days showed at the dose of 1% maternal toxicity (some effects also at 0.25%) without teratogenicity or embryotoxicity effects.
m-AP containing formulations (0.04%-0.7%, 1:1 with H_2O_2) were applied (2 mg/kg/day) to the shaven skin on rat on day 1-4-7-10-13-16-19 of gestation, or mouse (0.7% ; 0.05 ml/mouse 2 times week for 4 weeks prior to mating through day 18 of gestation), or rabbit (0.7% ; 2 ml/kg 2 times week from 4 weeks prior to mating through 30 day of gestation).
In rat only significant reduction of the mean live fetal weight was noted. In mouse a retarding effect of ossification process (bones of the feet and the cervical and caudal vertebral center) and slightly lower fetal weights. In rabbit no signs of maternal toxicity ; only focal alopecia until to the last 3rd of gestation) and a reduction of fetal survival as possible embryotoxic effects were observed. These results are not significant for teratogenicity potency.

MISCELLANEOUS INFORMATIONS

Hemoglobin effects : m-Aminohenol did not cause metahemoglobin formation in either fetal hemoglobin or adult hemoglobin at differents walues of pH (6.35-7.20).

Immunosuppressive : THe compound (2.5 mg, 4 times with 1/4 maximal tolerated dose, sc. inject.) resulted negative for immunosuppressive action on mice.

CONCLUSION

Classification : A.

BIBLIOGRAPHY - ACUTE TOXICITY
MUCOUS MEMBRANES IRRITATION
SKIN IRRITATION
SENSITIZATION

1. KYNOCH et al, (1975)
"Acute Oral Toxicity to rats of m-Amino-Phenol"
Huntigdon Research Centre - Cambridgeshire - ENGLAND
Report 4858/D1/75, 14th October, 1975

2. LLOYD et al, (1977)
"Assessment of the Acute Toxicity and Potential Irritancy of Hair Dye Constituents"
Fd. Cosmet. Toxicol. 15, p. 607-610, 1977

3. PROF. SEGR2, (1977)
"Rapport concernant les essais de toxicité aigüe chez la souris"
Università di Siena ITALY - Instituto di Farmacologia
Report of 1st February, 1977

4. KYNOCH et al, (1975)
"Irritant effects of m-Amino-phenol on Rabbit Eye Mucosa"
Huntingdon Research Centre - Cambridgeshire - ENGLAND
Report 4994/2D/75, 12th November 1975

5. KYNOCH et al, (1975)
"Irritant effects of m-Amino-phenol on Rabbit Skin"
Huntingdon Research Centre - Cambridgeshire - ENGLAND
Report 4983/3D/75, 12th November, 1975

6. PROF. Dr. SCHULZ, (1976)
"Vergleichende Untersuchungen über das Sensibilisierungs-vermögen verschiedener Haarfarbstoffe"
Universitäts -Hautklinik - Hamburg DEUTSCHLAND
Report of 24th May, 1976

6a. RUDZKI et al, (1980)
"Active sensitization to ortho and para-aminophenol with negative patch-test to meta-aminophenol"
Contact dermatitis, 6, n° 7, 501, (Dec 1980)

IMMUNOSUPPRESSIVE ACTION

6b. GERICKE, (1977)
"Bericht über die Prüfung auf Immunosuppressive Wirkung von elf Substanzen der Hans Schwarzkopf GmbH"
Krebsforschungslabor Hoechst AG Frankfurt/M
Report 4 January 1977

SHORT TERM TOXICITY

7. PROF. Dr. GLOXHUBER et al, (1978)
3-Amino-phenol
"Subakute orale Toxizität an Ratten bei einer Versuchsdauer Von 12 Wochen"
HENKEL Kommanditgesellschaft auf AKtien
Hauptabteilung TOxikologie - Düsseldorf - DEUTSCHLAND
Report of 17th March, 1978

8. BURNETT et al, (1976)
"Teratology and Percutaneous Toxicity Studies on Hair Dyes"
J. of Toxicol. and environ. Health, 1, p. 1027-1040, 1976

LONG TERM TOXICITY

9. BURNETT et al, (1980)
"Evaluation of the toxicity and carcinogenicity of Hair Dyes"
J. of Toxicol. and environ. Health, 6, p. 247-257, 1980

MUTAGENICITY STUDIES - IN-VITRO STUDIES

10. AMES et al, (1975)
"Hair Dyes are mutagenic : Identification of a variety of mutagenic ingredients"
Proc. Nat. Acad. Scien. 72, p. 2423, 1975

11. NISHIOKA, (1975)
Data presented to the Japanese Study Group of Environmental Mutagens

12. NISHIOKA, (1976)
"Detection of Carcinogenicity of Color Cosmetics in bacterial test Systems"
Mut. Research 38, n° 5, p. 345, 1976

13. YOSHIKAWA et al, (1976)
"Studies on the mutagenicity of hair dyes"
Bull. Nation. Inst. Hygen. Scien. 94, p. 28, 1976

14. LAVOIE et al, (1979)
"Mutagenicity of Aminophenyl and Nitrophenyl Ethers, Sulfides and Disulfides"
Mut. Research 67, p. 123-131, 1979

15. HOPE, (1978)
"Bacterial Plate Incorporation Mutagenicity Assay (Ames test) : m-Aminophenol"
Unilever Research, Colworth House
Project n° 30960, 25 May 1978

16. HASTWELL & Mac GREGOR, (1977)
"Testing for Mutagenic Activity of 22 Hair Colourant Components"
Inveresk Research International (IRI)
IRI Project n° 407485, August 1977

17. MULLER, (1976)
"Prüfung des Farbstoffes 1-Hydroxy-3-amino-benzol auf Mutagenität im Bakterienstest"
Battelle Institut Frankfurt/Main
Report of January 1976

MUTAGENICITY STUDIES - IN-VIVO STUDIES

18. HOSSACK et al, (1977)
"Examination of the potential mutagenicity of hair dyes constituents using the micronucleus test"
Experientia 33, p. 377-378, 1977

. Report Huntingdon Research Center - Cambridgeshire - ENGLAND SWF6/76 174, 26th March, 1976

TERATOLOGY STUDIES

8. BURNETT et al, (1976)
"Teratology and Percutaneous Toxicity Studies on Hair Dyes"
J. of Toxicol. and Environ. Health, 1, p. 1027-1040, 1976

19. BIO - DYNAMICS INC., (1977)
"A modified segment II - Teratology Study of Hair Dyes in Mice"
Submitted to : Clairol Inc. Stamford Connecticut
Report Project n° 76 - 1667
7th September 1977

SENSITIZATION

20. RAO et al, (1981)
"A Collection of Guinea Pig Sensitization Test Results grouped by Chemical Class"
Drug and Chemical Toxicology, 4(4), 1981, p. 331-351

LONG TERM TOXICITY/CARCINOGENICTY STUDY

21a. E.I. GOLDENTHAL et al, (1979)
"Lifetime CHronic Toxicity/Carcinogenesis Study in Rats"
International Research and Development Corporation,
Michigan, USA

21b. E.I. GOLDENTHAL et al, (1979)
Report February/March 1979

MUTAGENICITY STUDY
IN VITRO ASSAY

22. N. LOPRIENO, (1984)
"Forward Mutation Induction on the Yeast Schizosaccharomyces pombe (Strain pl) : In Vitro"
Istituto di Biochimica, Biofisica e Genetica - Univeristà di Pisa - ITALY
Report June 1984

23. N. LOPRIENO, (1984)
"Mitotic Gene Conversion Induction on the Yeast Saccharomyces cerevisiae (Strain D4) : In Vitro"
Istituto di Biochimica, Biofisica e Genetica - Università di Pisa - ITALY
Report June 1984

24. W.G.H. BLIJLEVEN, (1983)
"Evaluation of Compound m-Aminophenol in the Chromosome Aberrations Test : Chinese Hamster Cells (In Vitro)"
Department of Radiation Genetics and Chemical Mutagenesis - University of Leiden - THE NETHERLANDS
Report 31 December 1983

25. G. KIRCHNER & U. BAYER, (1982)
"Genotoxic Activity of the AMinophenols as Evidenced by the Induction of SIster CHromatid Exchanges"
Human Toxicol. 1, (1982, p. 387-392

26. D. WILD et al, (1981)
"Mutagenic Activity of Aminophenols and Diphenols, and Relations with CHemical Structure"
Abstract presented at a GUM Meeting (Darmstadt 1981)
Mutation Research, 85, (1981), p. 456

MUTAGENICITY STUDY

IN VIVO ASSAY

27. K.G. DOSSOU, (1983)
"Micronucleus Test : m-Aminophenol"
Laboratories de Recherche Fondamentale - L'OREAL - Aulnay-sous-bois - FRANCE
Report 11.7.1983

28. H. HOFER et al, (1983)
"Analysis of Human Chromosomes After Repeated Hair Dyes"
Fd Chem. Toxic. Vol. 21, n° 6, 1983, p. 785-789

29. K. TURANITZ et al, (1983)
"Investigations on the Effect of Repeated Hair Dyeing on Sister Chromatid Exchanges"
Fd Chem. Toxic., Vol. 21, n° 6, 1983, p. 791-793

30. W.G.H. BLIJLEVEN, (1983)
"Evaluation of Compound m-Aminophenol in the Sex-Linked Recessive Lethal Test with Drosophila melanogaster"
Department of Radiation Genetics and Chemical Mutagenesis - University of Leiden - THE NETHERLANDS
Report 31 August 1983

31. C. BURNETT et al, (1981)
"Heritable Translocation Study on Two Hair Dye Formulations"
Fundam. and applied Toxicol., 1, (1981), p. 325-328

TERATOGENICITY STUDY

32. BIO-DYNAMICS INC, (1982)
"A modified Segment II Teratology Study on Hair Dyes in Rabbits"
Project n° 76-1666

33. J.V. RUTOWSKI & V.H. FERM, (1982)
"Comparison of the Teratogenic Effects of the Isomeric Forms of Aminophenols in the Syrian Golden Hamster"
Toxicol. and applied Pharmacol., 63, 1982, p. 264-269

34. T.A. RE et al, (1984)
"Results of Teratogenicity Testing of m-Aminophenol in Sprague-Dawley Rats"
Fundamental and applied Toxicol., 4, 1984, p. 98-104

35a. F.X. WAZETER et al, (1977)
"Multigeneration Reproduction Study in Rats"

35b. International Research and Development Corporation, Michigan, USA
Report November 1977

REPRODUCTION STUDY

36. H. HOFER et al, (1985)
"Studie zur Resorption und Toxikokinetik von m-Aminophenol"
Institut für Biologie - Seibersdorf - AUSTRIA
Report OFZS Ber. n° A0651 - BL-503/85 - March 1985

NEPHROTOXICITY STUDY

37. J.F. NEWTON et al, (1982)
"Nephrotoxicity of p-Aminophenol, a Metabolite of Acetaminophen, in the Fischer 344 Rat"
Toxicol. and applied Pharmacol., 65, 1982, p. 336-344

EFFECT ON HEMOGLOBIN

38. M. WIND & A. STERN, (1977)
"Comparison of Human Adult and Fetal Hemoglobin : Aminophenol - Induced Methemoglobin Formation"
Experientia, 33, (15.11.1977), p. 1500-1501

TOXICITY ON CYANOBACTERIA

39. C.E. CERNIGLIA et al, (1981)
"Biotransformation & Toxicity of Aniline and Aniline Derivatives in Cyanobacteria"
Arch. Microbiol., 130, 1981, p. 272-275

MUTAGENICITY STUDY
IN VIVO ASSAY

40. F. NOSER? M. BRACHER & J. SWISTAK, (1982)
"Ermittlung der Potentiellen In Vivo Induktion von "Sister-Chromatid Exchanges" in Knochenmarkzellen nach Intraperitonealer und Peroraler Applikation von 1-Hydroxy-3-aminobenzol"
Cosmital Sa - MARLY - SWITZERLAND
Report n° 9396/46 - 14 July 1982

REPORT OF THE SCIENTIFIC COMMITTEE ON COSMETOLOGY ON THE USE OF CERTAIN PRESERVATIVE AGENTS IN COSMETIC PRODUCTS

. THE COMMITTEE'S MANDATE

To give its opinion on the question whether the use of certain preservative agents in cosmetic products is admissable from the health point of view.

1. CHLORPHENESIN CSC/204/88

. FORMULA AND SYNONYMES

EEC N°3
Colipa P4
CAS N° 104-29-0
3-(p-chlorophenoxy) propane -1,2 - diol (chem.name)
chlorphenesin
p-chlorophenyl-glycerol ether

$C_9H_{11}CLO_3$

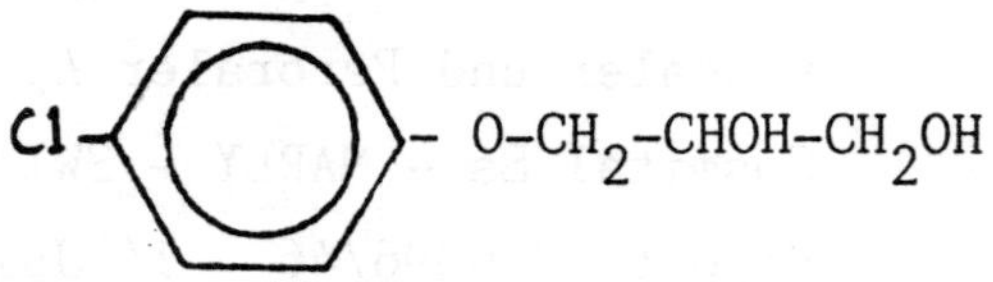

MW : 202,64

. CHARACTERISTICS

Slightly soluble in water (0.6%)
Moderately soluble in glycerol (9.5%) and alcohol (15%)

. USE

As a preservative in cosmetics at a concentration up to 0.2%.

. RECAPITULATION OF THE STUDIES OF TOXICITY

LD50	oral	(rat)	: 1400 mg/kg
		(mouse)	: 1060 mg/kg
		(guinea pig)	: 820 mg/kg
	i.p.	(rat)	: 520 mg/kg
		(mouse)	: 675 - 911 mg/kg
		(guinea pig)	: 425 mg/kg
	S.C.	(mouse)	: 930 mg/kg

ORAL TOXICITY

An oral dose of the labelled compound given to rats was rapidly absorbed and reached a peak concentration in the blood in 30 min. The half life in serum was 140 min. More than half of an oral dose was excreted in the urine in 4 hr, partly as the unchanged compound. Four metabolites have been identified : 3-p-chlorophenoxylactic acid, p-chlorophenoxyacetic acid, a conjugate of chlorophenol, and a conjugate of chlorphenesin.

In an oral 13-week study in rats given doses of 50, 100 or 200 mg/kg bw/day by gavage, no effect on growth rate or food intake was observed. Examination of vaginal smears provided no evidence of interference with oestrus. No gross changes were observed at autopsy (a detailed report is not available).

CUTANEOUS AND MUCOUS TOLERANCE

A skin irritation test in rabbits was negative (no details).
In repeated insult patch tests with 18 humans, application of 0.05 ml of 0.2% in hand cream, skin lotion and skin smoothing milk on 5 successive days was negative, or produced slight erythema in some cases.

An eye irritation test in rabbits with 1% in glycerine, did not provoke corneal irritation.

MUTAGENICITY

No evidence of mutagenic potential was obtained in a well-conducted Ames test with up to 0.5 mg/plate. Mutagenicity was examined also by the CHO/HGPRT locus bioassay.
Treatment of the cells *in vitro* with up to 1.5 mg/ml did not demonstrate mutagenic potential (Colipa subm. III).

MISCELLANEOUS DATA

Dogs given 75 or 150 mg/kg/day, (route not specified) 5 days a week for 18 weeks, did not show any significant changes in behavior or growth, in haematology or clinical chemistry, and in urine composition (summary report only).

Chlorphenesin may affect the immune system : both stimulating and inhibiting properties have been reported (Colipa subm. II). Lymphocyte function *in vitro* was found to be suppressed by 20-50 µg/ml culture medium (Colipa subm. II, ref. 19). Tumour inhibiting properties have also been reported. It is one of the glycerol ethers which are known to cause paralysis and to act as anticonvulsants. In a repeated intramuscular injection test in mice, with 0.5 ml of a 0.6% aqueous solution daily for 40 days there were no observable effects on growth or on the state of the organs.

CONCLUSION

ALthough short-term studies have been conducted, insufficient information on these studies is available.

Are required :
- a repeated dose oral study (with special attention for the immune system.
- a chromosomal aberration test in vitro
- information on dermal absorption
- a teratogenicity study (depending on the degree of dermal absorption).

Classification : C.

. FORMULATIONS AND SYNONYMES

EEC N° 15

Colipa P70

4'-(1,1,3,3-tetramethylbutyl) phenoxy-ethoxyethylene-dimethyl-benzyl-ammonium chloride.

Hyamine 1622

phenerol chloride

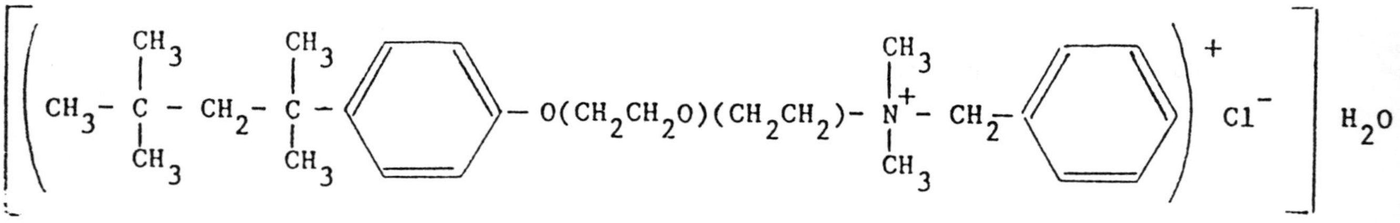

$C_{27}H_{42}NO_2CL$

. CHARACTERISTICS

Soluble in water, alcohols and other organic solvents.

. USE

Used in cosmetics as a preservative agent at level of 0.1%.

. RECAPITULATION OF THE STUDIES OF TOXICITY

LD50	oral	(rat)	:	420 mg/kg
	i.p.	"	:	33 mg/kg
	i.v.	"	:	19 mg/kg

Intranasal administration of 0.06 ml of a solution of 0.25% or more was lethal to rats.

ORAL TOXICITY

In a 28-day feeding study, rats received diets with 0, 20, 100, 500 or 2500 ppm, providing intake levels of 0, 1.7, 8, 40 or 200 mg/kg bw/day. The changes in the top-dose group included growth retardation, caecum enlargement, signs of liver damage and decreased serum levels of inorganic phosphorus in males. The latter finding was the only effect considered treatment-related in males fed 500 ppm. The diet with 100 ppm (8 mg/kg bw/dày) was a clear no-effect level (Colipa subm. III, January 1988).

In a one year study, groups of 3 dogs were fed 0, 5, 100 and 500 ppm in the diet. No changes were observed in growth rate, haematology or in gross - or microscopic pathology.

A two year study has been conducted with groups of 5 rats/sex, fed diets containing 0, 50, 200, 1000, 2500 and 5000 ppm. The top dose induced mortality. With 2500 and 5000 ppm testicular atrophy and caecal enlargement occurred. With 1000 ppm there was only caecal enlargement.

CUTANEOUS AND MUCOUS TOLERANCE

DERMAL TOXICITY

Various dilutions applied to the eye of rabbits produced barely perceptible irritation at concentrations of 0.01 and 0.03%.

Skin irritation in rabbits did not occur when 2 ml of a 0.1% dilution were applied daily 5 days a week for 4 weeks. In humans, 0.1 ml of a 5% aqueous solution applied under patches for 48 hours, was irritating (C.I.R. 1985).

A sensitization test in humans with 0.12% in formulations applied to the skin under closed patches was negative (C.I.R. 1985).

Upon sub-acute, dermal application of 2 ml 0.1% solution to the skin of rabbits daily, 5 days/week for 4 weeks no systemic effects were observed (summary report).

Sub-chronic (13-wk) dermal studies in rats and mice are being conducted by the NTP.

Dermal absorption was examined by applying 1.0 ml of a 10% aqueous solution of the ^{14}C-labelled compound under occluded patches to the skin of two rabbits on 4 consecutive days. One rabbit had the skin abraded. Blood samples taken on each day, showed an average concentration of 0.2 ppm, which corresponds to 0.003% of the amount applied. No mention is made of analyses in urine, faeces or carcasses (ref. 3, sumbm. II).

MUTAGENICITY

An Ames test with up to 100nmoles/plate was negative. Another Ames test with up to 7500 µg Hyamine 1622/plate was likewise negative.
It was stated that in an in vitro assay with CHO-cells no evidence was found of sister chromatid exchange or chromosomal aberrations but a report is not available (Colipa subm. III, January 1988).

CARCINOGENICITY

Several subcutaneous injection studies have been conducted in rats and mice. In one study in rats, a dose-related increase in the incidence of granulomatous reactions (mainly fibrosarcomas) occurred at the injection site (Cosmetic Ingredient Review 1985). It is not clear wheter this result is an indication of carcinogenic properties of the test substance.

TERATOGENICITY

An oral teratogenicity study in rabbits with 1, 3 and 10 mg/kg/day revealed signs of maternal toxicity with 3 and 10 mg, increased mortality of mothers and pups with 10 mg, and an increased incidence of supernumerary ribs with 3 and 10 mg. The latter finding was attributed to stress (ref. 4, subm. I).

In a second teratogenicity study in rabbits with oral dosing of 1.125, 3.558 and 35.576 mg/kg/day, the high dose induced maternal and foetal mortality. A dose-related increase in foetal resorptions occurred in all treatment groups although the change was statistically significant only in the high-dose group. The mid-dose was not clearly without effect (ref. 8, subm. I).

In a teratogenicity study in rats with oral dosing of 1.125, 3.558 and 35.576 mg/kg/day the high-dose group showed decreased maternal body weight and an increased number of smaller pups. An increased variation in ossification occurred in all treated groups. Skeletal malformation was increased in the high-dose group. Slight hydrocephalus was seen in one pup of the mid-dose group and in 5 pups (in 2 litters) of the high-dose group (ref. 6, subm. I).

A second oral rat teratogenicity study with 0.059, 1.125, 3.558 and 35.576 mg/kg showed lower maternal body weights, increased variation of skeletal ossification and increased incidence of skeletal malformations (wavy ribs) in the top-dose group only. The latter finding was considered to be within the limits for historical controls (ref. 7, subm. I).

Fertility and reproductive performance were examined in rats treated orally with 1.125, 3.558 and 35.576 mg/kg/day prior to and during mating and during the gestation and lactation period. The high-dose produced growth depression, increased irritability, respiratory signs in the parents and decreased viability and decreased body weight of pups at birth. Fertility and general reproductive performance were not affected (ref. 5, subm. I).

Peri - and postnatal effects were examined in rats dosed orally with 1.125, 3.558 and 35.576 mg/kg/day from day 15 of gestation through day 20 of lactation. A slight decrease in foetal viability occurred in all dose groups and in postnatal survival in the mid- and top-dose group (ref. 9, subm. I).

Maternal and foetal absorption of the ^{14}C- labelled compound was examined in pregnant rats treated orally with 1.125 and 3.558 mg/kg/day on days 6 through 15 of gestation. Average blood levels in the two groups were 1.5 and 0.97 ng/g respectively. In urine, the maximum levels were 52 and 149 ng/ml after a single oral dose. Virtually all radioactivity was recovered in the maternal faeces and carcass. Results of foetal analyses varied between not-detectable and 6.8 ng/g foetus (ref. 4, subm. II).

MISCELLANEOUS DATA

Concentrations as low as 0.002% inhibited the motility of the isolated ileum of rats and rabbits. Blood pressure measurements in the dog indicated nearly complete blockage of sympathetic ganglions at an i.v. dose of 2 mg/kg.

CONCLUSIONS

The available data from teratogenicity studies in rats indicate that benzethonium chloride produces both maternal toxicity and adverse effects on the developing embryo at oral doses of 3 mg/kg and above. The NEL was around 1 mg/kg. The safety factor resulting from cosmetic use and assuming a NEL of 1 mg/kg and 10% dermal absorption is very low (around 20). Although the actual extent of absorption through intact skin is likely to be low, it is essential to have adequate information on dermal absorption.

Classification : C.

3. BENZALKONIUM CHLORIDE

SCC/204/87-88

. FORMULA AND SYNONYMES

EEC n° 54

Colipa P71

Alkyl C_8-C_{18}dimethylbenzylammoniumchloride, -bromide and -
-saccharinate

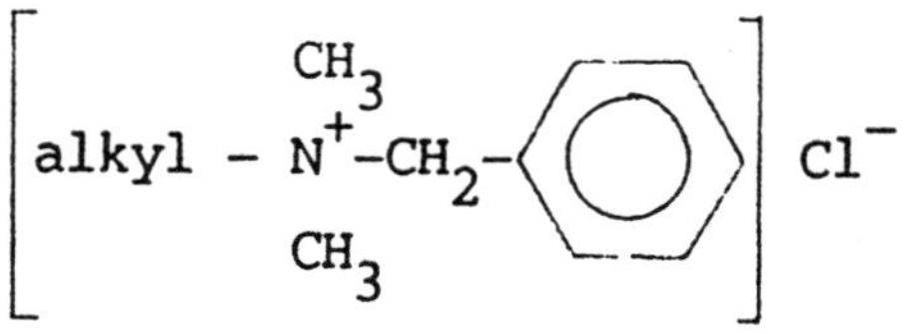

dodécyl : $C_{21}H_{38}N.Cl$

. CHARACTERISTICS

Soluble in water and alcohols.

Poorly soluble in hydrocarbons, oils and fats.

. USE

Used as a preservative at levels of 0.25%

. RECAPITULATION OF THE STUDIES OF TOXICITY

LD 50 oral (rats and mice)	:	0,5 - 1,0 gr/kg bw (obtained for commercial products with different alkyl groups)
i.v. (mice)	:	12,8 - 26 mg/kg intranasal administration of 0,06 ml of a 0,125% solution was lethal for rats.

ORAL TOXICITY

Sub-chronic (13-wk) oral studies in rats revealed toxicity and mortality at dose levels of 25 mg/kg bw and above. With 25 mg/kg bw, administered to dogs daily for 52 weeks, mortality and gastrointestinal damage was observed.

Further, sub-chronic (90-day) oral studies in rats and dogs, conducted in 1968, became available recently (Colipa subm. IV, ref. 20 and 21). In both studies, dose levels of 5, 12.5 and 25 mg/kg bw/day were administered ; in the rat study by stomach tube in the dog study by capsule. No changes attributed to treatment were observed. Because the studies showed several deficiencies, the results do not justify to establish a no-effect level.

In a 2 year rat study, 0.5% in the diet (250 mg/kg bw) caused high mortality and pathological changes in the gastrointestinal tract. Microscopical changes of the intestinal tract were seen also in a 2-year study with a second commercial product at dose levels of 25 and 12.5 mg/kg, and in a 2-year study with a third product at a dose level of 30 mg/kg bw. Dogs given 50 mg/kg bw/day by gavage (at a concentration of 5%) showed changes in the intestinal tract after one year.

The distribution of the compound was studied after oral, rectal and intramuscular administration of the 10-fold lethal dose to rabbits, dogs and cats. Most of the dose remained at the application site. After oral and rectal administration, small amounts were detected in blood and liver. Upon rectal administration a small amount was found also in the kidneys (ref. 16, subm. II).

CUTANEOUS AND NUCOUS TOLERANCE

DERMAL TOXICITY

Skin irritation tests in rabbits with 0.1% solutions, and in humans with 1.0% solutions were negative. With extended contact period in the rabbit, or repeated application in humans these concentrations produced distinct irritation. In rabbits, repeated application 0.3% induced only mild erythema.

Eye irritation in rabbits may occur upon a single application of 0.01% solution and above and upon repeated application of 0.004%. Concentrations of 0.01% and above caused eye irritation in guinea pigs when applied repeatedly on the same day. Single treatment of human eyes with 0.1%, or daily treatment with 0.03 - 0.04% caused irritation.

Soft contact lenses disinfected daily with 0.0025% benzalkonium-chloride + 0.01% EDTA induced severe irritation when brought into contact with the rabbit eye for 6 hr/day.

A sensitization test in 100 male and 100 female volunteers with 0.1%, applied daily for 5 days, followed by a challenge treatment with 1% after 3 weeks, was negative. In the literature only a few cases of sensitization in humans have been reported. Short-term oral administration to several animal species in the diet or the drinking water containing concentrations of 0.02% or more induced toxic effects.

A dermal 90-day study was conducted in rats with a formulation containing 1% stearyldimethylbenzylammoniumchloride and 0.2% benzalkoniumchloride 50%. Once daily, 5 days/week for 13 weeks the rats received topically 2.4 ml/kg (2.4 mg benzalkoniumchloride/kg). It is stated that no significant local or systemic effects occurred. However, the report is confusing and incomplete (ref. 7, subm. II).

Dermal life-time studies in mice and rabbits, treated topically with 0.02 ml of 8.5 or 17.0% solutions twice weekly showed local skin damage in both species, but no skin tumours.

Skin penetration tests in vitro with pieces of human skin were conducted in aqueous solutions of 0.005M to 0.1M benzalkoniumchloride (i.e. 0.17 to 3.4%). No penetration into the dermis was detected when the solution was unbuffered or acid. Measurable penetration occurred when the epidermal barrier was damaged or with intact skin in solutions of pH 11 (ref. 13, subm. II).
No penetration was found in vitro with skin from hairless rats exposed to 2.5% ^{14}C-dimethylbenzylammoniumchloride for 4.5 hours (ref. 14, subm. II).
In a similar in vitro test with human epidermis the mean penetration was 1.47% of the dose applied (ref. 15, subm. II).

MUTAGENICITY

A Ames test with S.typhimurium His G 46-uvr B exposed to 10-100 µg/plate was negative.

A micronucleus test in mice treated i.p. with 20 mg/kg bw, twice, with an interval of 24 hours did not reveal increased numbers of micronuclei (ref. 11, subm. II).

The substance was found to induce repairable DNA damage in the E.coli DNA polymerase A assay, but no mutagenic properties were observed (ref. 10, subm. II). No forward mutations were induced in Schizosaccharomyces pombe P_1 with or without metabolic activation (ref. 18, subm. III).

A chromosome aberration test with CHO-cells in vitro was negative (ref. 19, subm. III).

TERATOGENICITY

In an oral teratogenicity study, groups of 15 pregnant rabbits were treated by gavage with 0, 10, 30 or 100 mg/kg/day (in aqueous solu- of 0.5, 1.5 and 5.0% respectively) from day 7 through day 19 of gestation. All rabbits of the high dose group died. The intermediate dose caused maternal and embryotoxicity. Signs of maternal toxicity occurred also in the low-dose group. There were no indications of teratogenic properties. (Colipa subm. I, ref. 1).

A dermal teratogenicity study was conducted in rats treated topically with 0.5 ml aqueous solutions of 1.6, 3.3 and 6.6%, (estimated to be about 30, 60 and 120 mg/kg) once daily from day 6 to day 15 of pregnancy. No embryopathic effects were observed (ref. 12, subm. II).

CONCLUSION

Benzalkoniumchloride possesses considerable irritative properties for the eye and the gastrointestinal tract, and was highly toxic under certain conditions of acute exposure. In short-, and long-term studies changes have been observed in rats and rabbits with oral doses of 12.5 mg/kg/day, while mortality occurred in dogs and rats with 25 mg/kg/day. The substances is used in cosmetics for preservation and for other purposes. From the estimated maximal daily use of cosmetics and the levels used in the different types of product, a potential daily exposure to 5 mg/kg bw/day is calculated. A no-effect level for systemic toxicity has not been firmly established, and dermal absorption under practical use conditions is not known.

In view of the low safety margin it is essential therefore that detailed information be provided on skin absorption, and an in vivo study in rabbits (using radiolabelled material, single dose, back skin) is requested. It was also noted that 0.1% produced marked eye irritation, and the use of significantly higher concentrations in cosmetic areas where some eye contact was likely to occur, was of concern.

Classification : C.

REFERENCES

Colipa subm. I sept '83.
Colipa subm. II sept '85.

---

1. SHELANSKI ans SHELANSKI, (1953)
   Proceedings in Toilet Goods Association N° 19 NOT AVAILABLE
   May 1953

2. FINNEGAN and DIENNA, (1953)
   "Toxicological Observations in Certain Surface-Active Agents"
   Proc. Sci. Sec. Toilet Goods Assoc., 20, (1953), p. 16

3. WARF Institute, (1976)
   Submission of data by CTFA : "Rabbit eye irritation study - Stearalkonium Chloride" - Cited in the C.I.R. Final Report (11.1.1980) (see ref. 17)

4. GRIFFITH et al, (1980)
   "Dose-Response Studies with Chemical Irritants in the Albino Rabbit Eye as a Basic for Selecting Optimum Testing Conditions for Predicting Hazard to the Human Eye"
   Toxicol. and Applied Pharmacol., 55, (1980), P. 501-513

5. Th. KALLESEN (1981)
   "Investigation of the Eye Irritation Properties of a Preparation Performed on Albino Rabbits (Benzalkonium Chloride C14 - Cryst, Batch 8730)"
   SCANTOX Biological Lab. Ltd - Ejby - DENMARK - Report 6 January '81

6. COULSTON et al, (1961)
   "Toxicology of Benzalkonium Chloride Given Orally in Milk or Water to Rats and Dogs"
   Toxicol. and Applied Pharmacol., 3, (1961), P. 584-594

7. A.A. STEIN, (1978)
   AVON Toxicology Department, Suffern, New-York, U.S.A.
   Report February 1978

8. ROCKEN and SCHULTE, (1978)
"Prüfung von Brauerei-Desinfektionsmitteln auf Mögliche Mutagene Nebenwirkungen - I. Quaternäre Ammoniumverbindungen, Chlorpräparate, Jodophore, Kationogene Amine und Amphotere Verbindungen"
Monatsschrift für Brauerei, 31, (28 February 1978), P. 55-59

9. SUESSMUTH et al, (1979)
"Mutagenic Effect of 1,1' -Hexamethylene-bis (5-p-chlorophenyl) -biguanide"
Chem. Biol. Interactions, 28, (1979), P. 249-258

10. LOVELY and al, (1982)
"Light-Induced Genetic Toxicity of Thimerosal and Benzalkonium Chloride in Commercial Contact Lens Solutions"
Mutation Research, 101, (1982), P. 11-18

11. DOSSOU, (1985)
"Test de Micronucleus In-Vivo : Chlorure de Benzalkonium"
Laboratoires de Recherche Fondamentale - L'OREAL - Aulnay-sous-bois - FRANCE
Report 25 March 1985

12. PALMER et al, (1983)
"Absence of Embryotoxic Effects in Rats with Three Quaternary Ammonium Compounds (Cationic Surfactants)"
Toxicology, 26, (1983), P. 313-315

13. BLANK et al, (1964)
"Penetration of Cationic Surfactants into Skin"
Journal Invest. Dermatol., 42, (1964), P. 363-366

14. D. BAUER, (1985)
"Pénétration in vitro du Chlorure de benzalkonium à travers la peau totale de rat Hairless (cellules de diffusion)"
Laboratoires de Recherche Fondamentale - Département Métrologie et Calculs - L'OREAL - Aulnay-sous-bois - FRANCE
Report 6 September 1985

15. D. BAUER, (1985)
"Pénétration in vitro du Chlorure de benzalkonium à travers l'épiderme humain (cellules de Franz)"
Laboratoires de Recherche Fondamentale - Département Métrologie et Calculs - L'OREAL - Aulnay-sous-bois - FRANCE
Report 6 September 1985

16. U. BOGS and E. LOHSE, (1971)
"Zur Verteilung von Invertseifen im Körper von Säugetieren"
Arch. Toxikol., 28, (1971), P. 68-71

17. C.I.R., (1980)
"Final Report of the Safety Assessment for Stearalkonium Chloride"
Washington - USA
Report 1 November 1980

Colipa subm. III Oct '85.

---

18. MUTAGENICITY
N. LOPRIENO et al (1985)
Life Science Research Romo Toxicol. Centre Italy
LSR - RTC Report N° 088099-M-07185

19. Idem.
Report N° 088100-M-06685

4. CHLORQUINALDOL SCC/204/88

. FORMULA AND SYNONYMES

Colipa P86

Sterosan

5,7 - dichloro - 2 - methylchinolin - 8 - ol

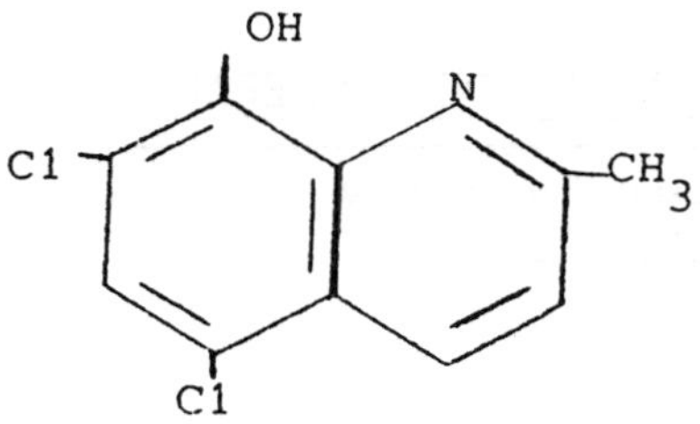

$C_{10}H_7Cl_2NO$

MW 228.08

. CHARACTERISTICS

Moderately soluble in organic solvents (chloroform acetone, ether)
Practically insoluble in water.

. USE

Used as a preservative in cosmetics up to 0.05% (also used at 3% in ointments and creams in medical and veterinary drugs as antiseptic and fungistatie).

. RECAPITULATION OF THE STUDIES OF TOXICITY

LD50 oral	(rats)	:	2,9 gr/kg
	(mice)	:	0,583 gr/kg
	(dogs)	:	> 1 gr/kg
LD50 subent	(mice)	:	2,9 - 3,1 gr/kg
	(rats)	:	> 4,0 gr/kg
	(dogs)	:	> 0,5 gr/kg
i.p.	(mice)	:	0,185 gr/kg
	(rats)	:	1,39 gr/kg

ORAL TOXICITY

A short-term (28-day) oral study with 0, 7 and 21 mg/kg bw/day administered in maize oil by gavage to groups of 7 rats/sex did not reveal any changes which were attributed to treatment. Both the design of the study and the report showed deficiencies (Colipa subm. II, ref. 43 ; summary report only).

CUTANEOUS AND MUCOUS TOLERANCE

DERMAL TOXICITY

An eye irritation test in rabbits with 0.1 ml cleansing milk containing 0.02, 0.1 or 0.5% did not reveal clearly more eye changes than did the cosmetic without the test compound (ref. 22).

The substance is known to possess sensitizing potency for humans. In a study with 5558 patients, 2.4% were positive to a mixture of Vioform + Sterosan (ref. 36). Cross sensitization of Sterosan with other halogenated hydroxyquinolines has been observed (ref. 39).

Repeated dermal application of 0.05 ml of a 3% cream to the intact and scarified skin of hairless mice, daily for 28 days, only induced some reddening in 2/10 mice on a few days, while thickening of the skin was seen in 1/10 mice (ref. 21). More sever dermal reactions occurred in 90-day topical application studies in rabbits and dogs (see below, ref. 12-15).

The sub-chronic (90-day) dermal toxicity was examined in groups of 5 rabbits/sex by applying 0.7 g/kg bw of an ointment containing 0, 1, 3 or 10% of the test substance each day for 13 weeks (or 0, 7, 21 or 70 mg/kg bw/day). In the top-dose group all animals died or were killed when moribund in the first 2 weeks. In the top-, and mid-dose group there were haematological changes, and blood was found in urine.

Increases in blood urea-N, absence of spermiogenesis, and microscopical renal changes occurred in all treated groups (ref. 12, 14). In this study, oral uptake was not fully excluded, but in a repeat study with the 10% ointment under occlusion similar changes were found though less severe (ref. 13).

A 90-day dermal study with the 0, 1, 3 and 10% ointments (or 0, 7, 21 and 70 mg/kg bw/day) was also conducted in groups of 2 beagle dogs/sex treated with 7.0 g/dog on 7 days/week.
In all treated groups the dogs showed dermal changes at the application sites, enhanced blood sedimentation rate and changes in white blood cell counts. In addition the 3 and 10% groups showed weight loss, increased numbers of leucocytes, increased activity of some blood enzymes and microscopical liver changes (ref. 15).

Dermal absorption studies with the $[^{14}C]$-labelled compound in guinea pigs in vivo showed that 96-98% of the dose, applied in various formulations, could still be removed from the skin after a 7-hr contact period (ref. 23). In humans treated with a cream, 96% was found not to have penetrated after 8 hrs (ref. 23).

In vitro tests with intact skin from guinea pigs and from nude mice showed more penetration (20-30%) especially when the compound was applied in ethanol (40 and 50% respectively into skin of guinea pigs and nude mice) (ref. 23).

In a study in humans (exposed to Sterosan in ointments and creams), up to 19.5% of a dermal dose was recovered in the urine, while after oral administration of an equal dose (30 mg) the mean urinary excretion (mainly in the form of the glucuronide) was 67.6% (ref. 40, Dermatologica 159 (1979) 239-244.

MUTAGENICITY

An Ames test with up to 100 µg/plate was negative, but 5 ug and above already caused growth depression or complete inhibition of the bacteria (ref. 18).

In the mammalian cell culture HGPRT system with Chinese hamster ovary cell line V79 exposed to up to 15 µg/ml, no mutagenic activity was detected. The highest concentration tested was distinctly toxic (ref.19).

A micronucleus test in mice given once intraperitoneally 37.5, 75 or 150 mg/kg bw, with bone marrow counts after 24, 48 and 72 hours did not reveal an increase in micronucleated cells (ref. 20).

Embryotoxicity studies were conducted in rats and rabbits by dermal application of 0.5, 5.0, 50.0 or 250.0 mg/kg bw, in a 10% containing ointment under occlusion for 5 hours daily on days 7-16 of pregnancy in rats and on days 7-19 of pregnancy in rabbits.

Apart from local irritation of the skin in the highest dose group of rats and in the two higher dose groups of rabbits, there were no other signs of intolerance. The reproduction data were not influenced by treatment. In the rat study several abnormalities of the skeleton were more frequent in the treatment groups than in controls, but it is stated that the incidence were in the normal range (ref. 16).

The two higher dose groups in the rabbit study showed growth retardation which was attributed to the severe local skin damage. The higher incidence of a few bone abnormalities in the treated groups were not considered to represent embryotoxic or teratogenic effects (ref. 17).

CONCLUSION

The substance shows sensitizing potency in humans. Results of dermal absorption studies vary widely. In humans, 19.5% of a dermal dose was recovered in the urine. In short-term studies in rabbits and dogs, dermal application of 21 mg/kg bw/day induced haematological effects, and microscopical changes in kidneys and liver have been reported.

Slight systemic changes were seen also upon dermal treatment with 7 mg/kg.

However, no changes were observed upon short-term oral treatment of rats with 21 mg/kg bw/day.

Interpretation of these studies is difficult in view of the apparent inconsistencies between species, but a no-effect level of below 7 mg/kg must be assumed on the present, together with good absorption through the skin.

The safety factors arising from cosmetic use, making such assumptions, were relatively low, and it was felt that more data, from an adequate 90-day study, using the oral route and a sensitive species (preferably the dog), were needed to obtain more definite information on a NEL.

Classification : C.

REFERENCES

Colipa Subm. II Oct. 1985

43. ADAMIKER & HÖGER
Versuchstierzucht und - haltung ;
Gemeinsame Einrichtung der Medizinischen
Fakultät der Universität Wien : Chlorquinaldol
Trial according to OECD
Guidelines for testing chemicals 407 ;
Himberg 3-7-1987

REPORTS OF THE SCIENTIFIC COMMITTEE ON COSMETOLOGY

CONCERNING THE USE OF CERTAIN U.V. FILTERS

. THE COMMITTEE'S MANDATE

To give its opinion on the question whether the use of certain U.V filters in cosmetics is admissible from the health point of view.

- Glyceryl paraaminobenzoate up to 5%
- 2 - Ethylhexyl - 4 - methoxycinnamate up to 10%
- 4 - tert - butyl - 4' - methoxydibenzoylmethane up to 5%

1. GLYCERYL PARAAMINOBENZOATE SCC/414/88

. FORMULA AND SYNONYMES

EEC N° 2.4
Colipa S6
Glyceryl - 1 - (4-aminobenzoate)
Glyceryl PABA
"ESCALOL 106"

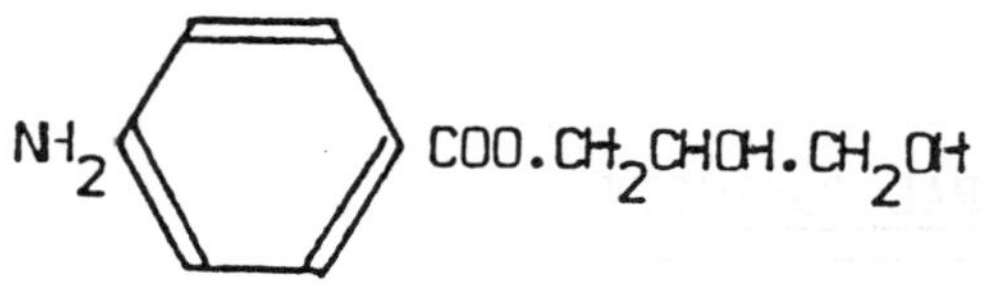

$C_{10}H_{13}O_4N$
MW = 211,2

. CHARACTERISTICS AND COSMETIC QUALITY

Creamy-white powder.
Insoluble in water and mineral oils.
Soluble in ethanol and propylene glycol.
Powder 80-90% pure, containing as impurities glyceryl di-para-amino-benzoate and traces of p-aminobenzoic acid and glyceryl-tri-paraamino-benzoate.
Stated to be "free of benzocaine".

In a test for purity of samples of glyceryl-PABA, 2 from one manufacturer in 1983 showed substantial amounts of benzocaine present.
On the other hand, there was a smaller concentration of benzocaine in another sample (from a different manufacturer) in 1978, and in 1983, a sample from the latter source showed no benzocaine (limit of detection, 0.05%). It is understood, however, that analyses of this sunscreen agent in the EC has frequently shown evidence of the presence of benzocaine.

. USE

Up to 5%

. RECAPITULATION OF THE STUDIES OF TOXICITY

LD50 oral (rat) :> 8 gr/kg

ORAL TOXICITY

A marketed lotion containing 3% a.i. and 3% amyl p-dimethylaminobenzoate was non-toxic to rats and mice in a single oral dose of 50 ml/kg (=1.5 g/kg of a.i.).

CUTANEOUS AND MUCOUS TOLERANCE

DERMAL TOXICITY

Irritation of the mucous membrane.

Rabbit : Twelve animals were treated with 2 commercial solutions, applied to one or other eye. Each lotion contained 3.15% a.i. and 3.15% amyl p-dimethyl-aminobenzoate. Half the animals had rinsing after 2 mins. Slight conjunctivitis subsided in 1 hr in the rinsed eyes and in 24 hrs in the unrinsed eyes.

In 8 rabbits 0.5 ml of a 5% solution of a.i. in propylene glycol was instilled into the eye. Washing was carried out in 5 animals. No detectable response was found in any animal.

Irritation of the skin

Rabbit : A 5% solution of a.i. in propylene glycol was applied to intact and abraded skin of 6 rabbits under occlusion for 24 hrs.
No irritant response was found.

Two commercially available lotions were tested. Each contained 3.15% a.i. and 3.15% amyl p-dimethylaminobenzoate. Six animals received 0.5 ml of one preparation, the remaining 6 received 0.5 ml of the other. Applications were made to both intact and abraded skin, and the sites in 3/6 animals in each group were occluded. Contact time was 24 hrs, and readings were made up to 72 hrs. There was slight initial erythema in all animals which quickly cleared.

A similar study which used 0.2 ml instead of 0.5 ml had the same results.

Man : Fifty human volunteers received 15 applications, each of 24 hrs duration, with occlusion. This strength of the solution is not given. An interval of 24 hrs was allowed between successive applications and 48 hrs at weekends. Following a 2 weeks rest, the test material was applied to the same sites in the same way. The substance was judged not to be an irritant or a sensitiser.

Sensitisation

Guinea-pig : Ten animals were induced with 5% a.i. in propylene glycol under occlusion. A challenge with the same material did not produce sensitisation.

Man : (See also above, under irritation). In 25 m, a product containing 3% a.i. and 3% amyl p-dimethyl-aminobenzoate was tested. Following 24 hr pretreatment with 5% sodium lauryl sulphate under occlusion, the product was applied to the forearms under occlusion for 5 alternate 48-hr periods. After a 10-day rest period, new sites were pretreated with 10% sodium lauryl sulphate for 1 hr under occlusion. Challenge patches were then applied for 48 hrs under occlusion. Sensitisation was not found.

Photosensitivity and phototoxicity

Man : A mixture of 3% a.i. and 3% amyl p-dimethyl-aminobenzoate was tested in 35 volunteers. 0.2 ml of test material was applied to normal and stripped skin on the back (probably for 24 hrs with occlusion). Each site was then exposed to 6 MEDs of xenon radiation filtered through window glass. Observations were made 1, 3 and 24 hrs after irradiation. No abnormality was found.

Fifteen women and 25 men had the above mixture applied daily to the face and upper trunk, after which the subjects were exposed to 3 MEDs. Four women and 8 men complained of very mild itching around the eyes, but there were no visible signs of reaction in any subject.

Thirty-five volunteers (same as in first paragraph, above) had 0.2 ml of the test material applied to an area of 25 cm^2 of stripped skin on the back, which was then exposed to 3 MEDs from a xenon tube and occluded (times not specified). This procedure was repeated 5 times at intervals of 48 hrs. Ten days later 0.2 ml was applied both to stripped and normal skin followed by exposure to 3 MEDs of xenon tube radiation. After occlusion, probably for 24 hrs, the sites were inspected at 24, 48 and 72 hrs. No evidence of photocontact allergy was found.

Twenty patients known to be sensitive to benzocaine were patch tested with 5% benzocaine with occlusion and simultaneously with two commercial sunscreen preparations under the same conditions. It had previously been established by HPLC that one of these preparations contained 0.3% free benzocaine (arising as a contaminant in the manufacturing process) and the other 0.001%. All patients reacted positively to the benzocaine patch ; 11/20 to the preparation containing 0.3% benzocaine.

In a number of clinical studies, patients with allergic reactions to glyceryl-PABA were studied. One author found that the compound was a common sensitiser and that there was invariably cross-reaction with benzocaine. There was also cross-sensitivity with aniline and paraphenylenediamine. The a.i. was also capable of acting as a photo-sensitising agent. In another report, one patient was shown to have both photosensitisation and contact allergy to the compound.

In another study, 2 patients with sensitivity to the a.i. were identified among 23 patients complaining of skin disorders following exposure to the sun. Of these, 1 was found to have a photosensitivity and the other a contact allergy. In both cases the sensitivity was to PABA but not to benzocaine. Three patients with benzocaine sensitivity however, did not show a cross-reaction by glyceryl-PABA.

Subacute toxicity

Rat : Eight animals had applied 0.2 ml of the commercial lotion mentioned above daily to the skin for 32 days. No abnormality was found.

Rabbit : Over a period of 20 days, 20% of a.i. was applied to intact and abraded skin of rabbits (numbers not specified) at levels of 1, 2 and 4 g/kg bw daily. Control animals received vehicle only. No abnormal systemic or topical signs were found.

Subchronic toxicity

Rat : Groups of 10 m and 10 f were treated with 0, 55.5, 277 and 555 mg/kg bw/day in the form of a solution of 18.5% a.i. in propylene glycol (vehicle only for controls). These doses were selected to correspond with similar tests carried out with "Escalol" (another amino-benzoate) and also on the basis of work which showed that the average human female used about 4 mg/kg bw/day of the a.i. ; thus the doses were considerably higher (about 14, 69 and 239 times use level). Treatment was carried out daily by application to shaved skin, approximately 15% of body surface, 5 days a week for 13 weeks. Various haematological and biochemical tests were carried out in weeks 7 and 13. All survivors were subjected to necropsy. A selection of tissues was weighed, and histological examination carried out if abnormal weight or appearance was noted.

All animals survived. No abnormal changes in body weight or behaviour were noted. High dose males showed slight skin irritation, otherwise no skin abnormality was found. High dose females showed a significant increase in relative and absolute weights of the spleen, but the histological and haematological pictures were normal in these animals. There were slight changes in the blood picture in high dose animals, but no definite trend was present, and the changes were slight in absolute terms. There were no significant abnormalities in urine. Biochemical examination of the blood showed no definite abnormalities. NEL 277 mg/kg bw/day.

CONCLUSION

- The Committee felt that there were several gaps in the data provided.
- There were no reports of tests on oral toxicity, mutagenicity or carcinogenicity
- The concentrations used in testing for irritation and sensitisation were low in relation to the proposed use level.
- The purity of the compound and its freedom from benzocaine should be rigorously established.
- It was important to know definitely whether it was absorbed from the skin.
- Teratogenic activity would be necessary.

Classification : C.

SOURCES

Avon R D Laboratories (1981).

BGA (1984). Personal Communication.

Council of Europe Data Sheet.

Federal Register (1978) 43 38229.

Fisher (1976) Cutis 18 495.

Fisher (1977) Arch. Dermatol. 113 1299.

Fisher (1977) Contact Dermatits 3 170.

Goldman and Spstein (1969) Arch. Dermatol. 100 447.

Thune (1984) Photodermatology 1 5.

First Version	:	August	1983.
Revised	:	November	1983.
Revised	:	August	1987.
Revised	:	March	1988.

2. 2-ETHYLHEXYL-4-METHOXYCINNAMATE SCC/414/88

. FORMULA AND SYNONYMES

EEC N° 2.13
Colipa S28
Neo Heliopan AV
Parsol MEX

MW = 290

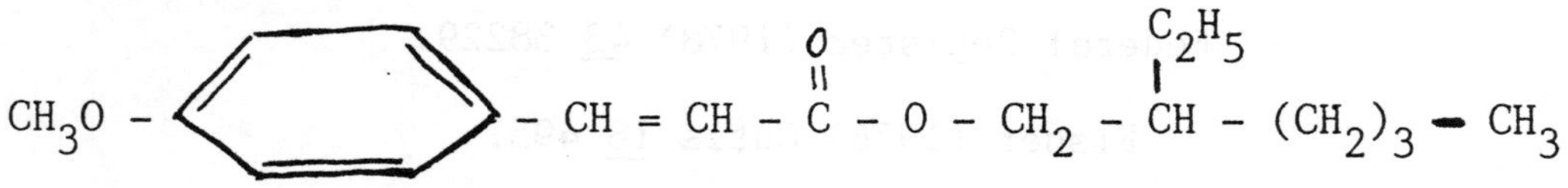

. CHARACTERISTICS

Odorless, pale yellow, slightly oily liquid.
Insoluble in water.
Miscible in alcohols, propylene glycol, mono-myristate and various oils (ref. 1).

. USE LEVELS

Maximum concentration authorized (council Directive) : 10%

. RECAPITULATION OF THE STUDIES OF TOXICITY

LD50 oral (mouse) : > 8 gr/kg (ref. 1, 2, 6)
(rat) : ≥ 20 ml/kg (ref. 3)

ORAL TOXICITY

Subacute toxicity

Rat : Three week oral study. Groups of 5 m and 5 f animals were given 0, 0.3, 0.9, and 2.7 ml/kg/bw/day by gavage for 3 weeks. All animals of the top dose groups exhibited loss of body weight and a reduced relative and absolute weight of the thymus.

Male rats showed a decrease in absolute weight of the left kidney and female rats showed a decrease in the absolute weight of the heart.
At the 2 lower doses the only significant alteration observed was an increased absolute weight of the pituitary gland in male rats receiving the lowest dose. As the number of animals was low, the investigators considered this not to be biologically relevant.
N.E.L. 0.9ml/kg bw/day.

Subchronic Toxicity

Rat : Thirteen week oral study. Four groups of 12 m and 12 f SPF rats received the compound in the diet at levels of 0, 200, 450 and 1,000 ml/kg bw/day. During the experiment the usual clinical observations were carried out, as well as extensive haematological and biochemical studies. Full gross necropsy was carried out on all survivors.
Histological investigations were carried out in half the animals of the control and top dose groups. The organs studied included the heart, lungs, liver, stomach, kidneys, spleen, thyroid and retina.
In the remaining animals histological examination of the liver only was carried out. Six control animals and 6 top dose animals were allowed to recover over 5 weeks, and then examined.

The results of the experiment showed no dose-related mortality. The kidney weights of top dose animals were increased, but were normal again in the animals allowed a recovery period ; the increase was attributed to a physiological response to an increased excretion load.
There was a dimunition of glycogen in the liver, and a slight increase in iron in the Kupfer cells, in the high dose animals. Two of these also showed minimal centrilobular necrosis of the liver with some infiltration ; similar less marked findings were attributed to infection. High dose females had increased GLDH which reversed during the recovery period. The N.E.L. was probably 450 mg/kg bw/day. (ref. 10).

CUTANEOUS AND MUCOUS TOLERANCE

DERMAL TOXICITY

Mucous membrane irritation

Rabbit : 4 animals/group.
Instillation of 0.1 ccm of a test preparation (concentration unknown) into the conjunctival sac. No further treatment in one group, subsequent washing out in another.
No signs of inflammation (ref. 5).

Rabbit : Draize test with 0.1 ml of the undiluted test liquid. The chemical was considered as "practically non-irritating to the eye". (ref. 1, 2, 6).

Skin irritation

Guinea pig, 20 animals.
Epicutaneous application of the undiluted chemical twice daily for 16 d. No signs of irritation (ref. 5).

Man, 60 subjects, 20 of them with sensitive skin.
Occlusive patches containing the undiluted test liquid were applied for 24 h. No irritation immediately after exposition nor after 24 and 48 h of observation (ref. 5).

27 men and 22 women.
Patch test using an unspecified concentration. No positive results after 24 and 48 h (ref. 1).

Man, 50 subjects, about one third of them with extremely sensitive skin. The product was very well tolerated on the skin (ref. 1, 2, 6).

Man, 54 subjects.
Draize repeated insult patch test of a 7.5% dilution in petrolatum. No irritation (ref. 1, 8).

Man, 53 subjects.
Draize repeated insult patch test at a concentration of 2%. No irritation (ref. 7).

Sensitization

Guinea pig, 20 animals.
Induction : epicutaneous application of the undiluted chemical twice daily for 16 d. Interval without treatment : 3 d. Challenge : one application daily for 3 d. No sensitization (ref. 5).

Guinea pig, 4 animals/group.
One group received 0.05 ml of the undiluted test liquid by intracutaneous injection on 5 subsequent days. In another group 0.025 ml of a 50% acetone solution were applied topically daily for 3 weeks to 2 cm^2 areas on the shaved sides.
No sensitization by either topical or intradermal route (ref. 1).

Man, 53 subjects.
Draize repeated insult patch test carried out at a concentration of 2%. No sensitization (ref. 7).

Man, 54 subjects.
Induction : a formulation containing 7.5% of test liquid diluted in petrolatum was applied 11 times for 48 h under occlusive conditions. Interval without treatment : 14 d. Challenge : application of a single dose. No reaction (ref. 8).

Phototoxicity and Photosensitization

Man, 10 subjects.
Application of patches for 24 h and subsequent UV-radiation with a suberythema dose. No signs of phototoxicity (ref. 5).
Tests which "showed that the product did not provoke photosensitization" (ref. 1, 6). No details available.

Dermal Toxicity

Rat : 13 week dermal study. Four groups of 10 m and 10 f SD rats were treated by application of various concentrations of a.i. in light mineral oil. The doses were : 0, 55.5, 277 and 555 mg/kg bw/day applied to shaved skin 5 days a week for 13 weeks. (The doses are believed to be maximally 135 times the amount used daily by the average consumer). Various laboratory and clinical tests were carried out during the experiment. All animals showed a slight scaliness at the site of application, which may have been a reaction to the vehicle. Body weight gain was greatest at the low dose. Haematology showed no significant change. SAP was elevated in high dose animals but the investigators did not believe the change to be significant. The relative liver weight in high dose animals was elevated, but microscopic appearances were normal. The n.e.l. was put at 555 mg/kg bw/day, but may be 227 mg/kg bw/day in view of the liver findings (ref. 11).

Permeation of skin

In vitro tests.

Rat : Naked rat skin : This was set up in a chamber experiment. Most of the material was found in the stripped skin ; there was less in the stratum corneum, and least in the chamber. Approximately the chamber contained : after 6 hr 1.13% ; after 16 hr, 11.4% ; and at 24 hr 17.9%. The figures for the horny layers and the stripped layers combined were, respectively, 31.4%, 44.4%, and 45.7% (percentages of applied doses). Solutions of 3% and 20% of a.i. gave similar results.

Another set of experiments was carried out in which varying amounts of "Parsol 1789" were added to the a.i. in the formulation. There seemed to be no effect on the absorption of the a.i. (ref. 14).

Pig :
A similar experiment using mini-pig skin was carried out in which "Parsol 1789" was used as well as the a.i. Using 3 sorts of formulations, about 3% of a.i. was found in the chamber in 6 hr. Using the concentrations proposed for commercial use (i.e., 7.5% of "Parsol 1789" and 2% of a.i.) about 2.2% was found in the chamber. It is calculated that the total absorption for 75 kg consumer would be about 70 mg - about 0.9 mg/kg bw. (ref. 15).

Man :
A test on human abdominal skin in a chamber has been carried out. With 7.5% a.i. about 0.03% is found in the chamber in 1 hr., 0.26% in 6 hr., and 2.0% in 18hr. Varying mixtures of a.i. and "Parsol 1789" were investigated (ref. 16).

In vivo tests :
Man : Eight healthy volunteers had small amounts of radioactive a.i. applied to the interscapular region. One group of 4 had the material applied under a watch glass ; the other 4 had it applied on gauze, with occlusion in one case. Tests for a.i. absorption were negative except for about 0.2% in the urine. The concentrations used seem not to be stated (ref. 17).

MUTAGENICITY

Salmonella mutagenesis assays were performed on strains TA-98, TA-100, TA-1535, TA-1537, and TA-1538 in the presence and absence of S9 mix. Positive results with strain TA-1538 without metabolic activation. This may have been a batch effect. The chemical also increased the frequency of sex-linked recessive lethals in Drosophila melanogaster (ref. 9).

Ames test : From another laboratory, a very weak positive was found with strain TA-1538 without activation, at 10 µl/plate ; it was not found in 2 replicates, nor in a second Ames test (ref. 18).

Chinese hamster V 79 cells. There was a very slight increase in mutant colonies with dose (ref. 19).

Test for mutagénesis and possible crossing over in S. cerevisiae : negative (ref. 20).

In vitro human lymphocyte test - negative (ref. 21).

Balb/c 3T3 cell transformation : negative (ref. 22).

Unscheduled DNA synthesis : negative (ref. 23).

Drosophila mutagenicity studies (feeding, larvae and adults) : negative (ref. 24).

Mouse micronucleus test : no effect up to 5000 mg (ref. 25).

Drosophila : somatic mutation and combination test using wing structure : negative (ref. 26).

CARCINOGENICITY

Test for inhibition of UV-induced tumours in hairless mice.

The technique involved exposure of hairless mice to repeated doses of UV simulating the solar energy spectrum. After a rest period, 3 applications a week were made to an area of skin of 12-o-tetradecanoyl phorbol-13-acetate (at first at 10 g/ml, but later at 2 g/ml as the higher concentration was irritating). Suitable controls were used. The test group was protected by 50% a.i., and 7.5% gave an effect equivalent to reducing the insolation four-fold. It had been suggested that the a.i. could itself be a promoter, but the authors found no evidence for this (ref. 25).

TERATOGENICITY

Rabbit : Groups of 20 f rabbits were mated and given 0, 80, 200 and 500 mg/kg bw/day by gavage during the period of organogenesis. Except for a slight reduction of maternal and foetal weight in the top dose, there was no evidence of embryotoxicity or of teratogenesis. The N.E.L. was put at 200 mg/kg bw/day (ref. 12).

Rat : Following a pilot study, groups at 36 rats were mated and treated with 0, 250, 500 and 1000 mg/kg bw/day of a suspension of a.i., probably by gavage, from day 6-14 of gestation. Owing to an error, the preparation of the control foetuses led to their destruction, so this part of the test was repeated under identical condition Subgroups of each dose group were allowed to litter normally and rear the offspring. No abnormalities were observed, except that the percentage of resorptions in the high dose group was elevated by comparison with the other groups. The investigator reports, however that this relatively high rate is the usual one with this strain of rat in this laboratory, and he attributes the difference to an unusually low level of resorption in the other groups. The N.E.L. may be at 1000 mg/kg bw/day (ref. 13).

CONCLUSION

The data presented suggest that the compound is not an irritant or a sensitiser ; however, the tests for sensitising capacity in man were carried out at levels below the proposed maximum use level. There is no carcinogenicity study, but tests for gene mutation and chromosomal aberration were very probably negative. The Committee would like to see a test for percutaneous absorption in man carried out under realistic conditions, and the results expressed if possible as an absorption coefficient rather than as a percentage of the applied dose. The evidence presented, however, does suggest that substantial absorption from the skin might be expected.

Classification : C.

REFERENCES

1. Avon Research and Development, Study Project ATO 187, 10.09.1982.
2. 13. Week Oral Dictary Toxicity Study in the Rat. Dr F. CAMPONOVO, 23.07.1984. HOFFMAN. LA ROCHE & CIE SA ,CH, BASLE.
3. Penetration Studies in vitro through the intact skin of nakeds Rats. Dr G.KLECAK, HOFFMANN. LA ROCHE & CIE SA , CH. BASLE 13.12.1979.
4. Penetration Studies in vitro through the intact skin of naked Rats. Dr G. KLECAK , HOFFMANN. LA ROCHE & CIE SA, CH. BASLE. 06.05.1980.
5. In vitro penetration studies on the excised mini pig skin with chemical UV - filters PARSOL 1789 and PARSOL MCX in 3 Cosmetic vehicles; Dr G.KLECAK, Th. WEBER, W. HERING, HOFFMAN. LA ROCHE & CIE SA, CH. BASLE. 19.08.1982.
6. Percutaneous absorption and excretion of radioactivity following topical administration of labelled Parsol MCX to healthy volunteers; Dr A. DARRAGH, Biological and Medical Research Insitute, St. James' Hospital, Dublin, 29.07.1980
7. Absorption and excretion of radioactivity following topical administration of radioactive Parsol MCX and Parsol 1789 to healthy volunteers. Dr A. DARRAGH, Clinical Pharmacology Unit, Biological and Medical Research Institute, St. James' Hospital, Dublin 08.08.1980.
8. Penetrationsstudie mit den Sonnenschutzmitteln Parsol 1789 und Parsol MCX - Prof. Dr G. STUETTGEN Universitätsklinikum CHARLOTTENBURG - Hautklinik und Poliklinik im Rudolf - Virchow - Krankenhaus 24.09.1982.
9. Mutagenicity evaluation of Parsol MCX in Salmonella typhimurium strains by means of different testing procedures. Dr M. SCHUEPBACH, 12.06.1980. HOFMANN LA ROCHE, CH. BASLE.
10. Mutagenicity evaluation of Parsol MCX in the Salmonella/mammalian microsome plate incorporation assay (Ames test) Dr M. Schüpbach, HLR, Basle, CH, 19.12.1985.
11. Mutagenecity evaluation of the UV-B sunscreen PARSOL MCX in chinese hamster V 79 cells in presence and absence of rat liver activation system. Dr R. STROBEL, HOFFMANN LA ROCHE, CH. BASLE.

12. Mutagenecity evaluation of the UV-B filter Parsol MCX with Saccharomyces ecrevisiae in a "treat and plate test"; A. CHETELAT, 04.07.1983. HOFFMANN LA ROCHE, CH. BASLE.

13. Chromosome analysis of human peripheral blood lymphocytes after in vitro exposition to thesunscreen Parsol MCX in the presence and in the absence of mtabolic activation. Dr. HUGENTOBLER, HOFMANN LA ROCHE, CH. BASLE. 30.01.1984.

14. Evaluation of a potential in vitro transformation of Balb/C 3T3 cells after a 3 -day -exposure period with the UV-B sunscreen Parsol MCX. Dr R. STROBEL, HLR, CH. BASLE. 23.04.1985.

15. Unscheduled DNA synthesis (UDS) essay with the UV-B sunscreen Parsol MCX using primary cultures of rat hepatocytes. Dr R. STROBEL, HLR, CH. BASLE. 26.11.1985.

16. Drosophila mutagenecity evaluation of the sunscreen Parsol MCX, Dr M. SCHUEPBACH, H.L.R. & CIE. SA - CH. BASLE. 23.03.1983.

17. In vivo mutagenicity evaluation of the sunscreen PARSOL MCX with the micronucleus test in the mouse. Dr A. L. HUGENTOBLER, HOFFMANN LA ROCHE, CH. BASLE. 02.05.1983.

18. Somatic mutation and recombination testin Drosophila melanogaster with PARSOL MCX ; Dr SCHUEPBACH, HOFFMANN LA ROCHE, CH. BASLE. 24.04.1984.

19. Embryotoxicity study in rabbits with oral administration of PARSOL MCX; Dr BUERGIN, HOFFMANN LA ROCHE. CH. BASLE. 14.10.1983.

20. Embryotoxicity and teratogenecity study in rats with oral administration of PARSOL MCX, Dr A. KISTLER, HOFFMANN LA ROCHE, CH. BASLE. 62.09.1983

21. Inhibition of Solar UVR- induced Skin Tumors in Hairless Mice by application of 2 -ETHYLHEXYL -P- METHOXYCINNAMATE. F.D. Forbes, R.E. Davies, C.P. Sambuco, C.A. Cole and F. Urbach. The Center for Photobiology / Temple University Philadelphia; 07.1986.

3. 4-TERT-BUTYL-4'-METHOXYDIBENZOYLMETHANE SCC/414/88

. FORMULA AND SYNONYMES

EEC N° 2.31

Colipa S66

Parsol 1789

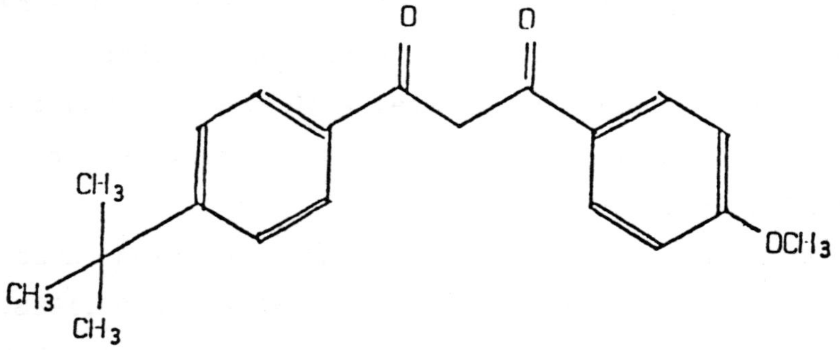

$C_{20}H_{22}O_3$

MW 310.4

. CHARACTERISTICS

Crystalline substance.

Insoluble in water.

Soluble in carbitol 15%, chloroform 15%, acetone 5%, ethanol 2%.

. USE

Use level up to 5%.

. RECAPITULATION OF THE STUDIES OF TOXICITY

LD50	oral	(rat)	:	16 gr/kg bw
	oral and ip	(mouse)	:	8 gr/kg
	dermal	(rat)	:	1 gr/kg bw ref.(1-2-3).

ORAL TOXICITY

Suchronic Toxicity

Rat : Thirteen-week study. Four groups of 12 m and 12 f animals were given the a.i. in the diet in amounts equivalent to 0, 200, 450 and 1000 mg/kg bw/day.

There were no treatment-related deaths. Food consumption was reduced in the intermediate and top dose groups. There was a fall in the red cell count in females at the intermediate and top dose levels. The plasma protein levels were somewhat higher in all dosed animals, but this did not seem to be dose-related.

At necropsy, the relative liver weights of males were increased at the top dose, and the absolute and relative liver weights of females were increased in both the intermediate and top dose animals. One top dose animal showed damage to the germinal epithelium of the testis.

Supplementary groups of 6 rats were treated with the top dose and then allowed a 4-week recovery period. At sacrifice, the liver weights of these rats were similar to those of control rats.

If the increase in liver weight is accepted as an effect, the NEL is 200 mg/kg bw/day ; otherwise it may be 450 mg/kg bw/day. (4).

CUTANEOUS AND MUCOUS TOLERANCE

DERMAL TOXICITY

Irritation of mucous membranes

Rabbit : A standard Draize test was carried out, using a.i. dissolved in diethyl phthalate, without rinsing. There was no adverse effect up to the limit of solubility, 20% (6).

Irritation of the skin

Rabbit : Five groups, each comprising 10 m and 10 f, were studied. In each group, 5 animals of each sex had the skin of the application site abraded. The amounts applied were 30, 60 and 360 mg/kg bw/day, with occlusion for 6 hrs, for 21 consecutive days. Some irritation was found in the vehicle control animals. There was a dose-related erythema in treated animals. Abrasion of the skin did not affect the findings. No treatment-related changes were found in body weight, food or water consumption, or in haematological or biochemical tests. No abnormality was found on histoligical examination, except at the sites of application (7).

Rabbit : [Two ?] groups of 6 animals were used. The a.i. was dissolved in ethanol/2-phenylethanol (50/50) in a concentration of 10% ; 0.5 ml was applied to an abraded and a non-abraded site on the skin of each animal for 4 hrs with occlusion. A very slight irritant effect was found, and was thought to be due to the vehicle. No individual data are given, and the report is somewhat difficult to interpret (8).

Tests for sensitisation

Man : Repeated insult patch test.
Eleven male and 49 female subjects were recruited ; 8 failed to complete the study. About 0.2 ml of a 10% solution was applied under occlusion for 24 hrs on 10 occasions, with rest intervals of 24 or 48 hrs. On completion of this course, a 10-day rest period was allowed and then challenge applications were made to the original site and to a new site. No adverse reaction was observed (9).

Tests for capacity to produce sensitisation

Guinea-pig : Freund's complete adjuvant test.
Two groups of 8-10 animals were used. Induction was by 3 intradermal injections, on days 0, 4 and 9, of 50% [?] emulsion of the a.i. in Freund's complete adjuvant.

Animals of the control were injected with Freund's adjuvant only. On days 21 and 35 a challenge was made by epicutaneous application of 0.025 ml of a.i. (solvent not specified) at the minimal irritant concentration and 3 lower concentrations (not specified in the report). There was no evidence of sensitisation (10).

Guinea-pig : Magnusson-Kligman maximisation test. Groups of 20-25 animals were used as test and control groups respectively. Induction was by intracutaneous injection of 0.1 ml of 5% a.i. in Freund's complete adjuvant, 5% in saline and adjuvant alone. This was followed 7 days later, by an epicutaneous application of a 20% suspension of a.i. with occlusion for 2 days. The challenge was carried out on day 21 ; 20% and 6% were applied for 24 hours. There was no evidence of sensitisation (11).

Guinea-pig : Open epicutaneous test. The experimental group consisted of 20 animals and the control group of 10. The account given is obscure, but it would appear that the maximum non-irritating concentration and the minimum irritating concentration were first determined. Solutions of 20% and 6% were applied to one flank for 21 days. Challenge was made at 21 days and 35 days by application of a 10% solution to the opposite flank. There was no evidence of sensitisation (12).

Test for potential to cause phototoxicity

Guinea-pig : Ten animals were used. Five sites were delineated on each flank. To each of 4 sites, 20% of a.i. in acetone was applied with 2% DMSO as a maximising agent. An alcoholic solution of 0.1% 8-methoxypsoralen was used on the fifth site as a positive control. Thirty minutes later, one flank was exposed to non-erythemogenic UV-A radiation at 20 J/cm^2. Sites were scored 4, 24 and 48 hrs after application. There was no evidence of phototoxicity (13).

Test for potential to cause photo-allergenic effects

Guinea-pig : Groups of 10 animals were used. In the 2 test groups, induction was carried out as follows :

(a) 4 injections of 0.1 ml of Freund's complete adjuvant were made in the neck to delineate a square.

(b) applications of 0.1 ml of a.i. in acetone were made over 8 cm^2 in this area, and 30 minutes later UVA irradiation at 10 J/cm^2 was applied.

(c) Procedure (b) was repeated 5 times over the subsequent 2 weeks.

(d) On days 21 and 35, a challenge was applied to both flanks using 0.025 ml of solution over 2 cm^2.

The left flank of each animal was then irradiated as above ; reading was at 24 and 48 hrs. In the first experimental group, the induction concentration was 10% and the challenge concentration was also 10% in the second experimental group, the concentrations were 1% and 10% respectively. A negative control group received no a.i. by way of induction, and for the positive control 3% tetrachlorosalicylanide replaced the a.i. There was no evidence of photo-allergenicity (14).

Test for potential to cause photosensitisation

Man : Twenty-five volunteers were used. The a.i. was incorporated into petrolatum at 2%, to which 2% DMSO was added as a maximising agent. A minimum erythema dose for each subject was determined by finding the time taken to produce erythema using UVA + UVB, 285-400 nm.

Induction was then carried out by application of the a.i. to two areas of the skin of the back, with occlusion. One of these was a test area, the other an irritancy control. The patches were removed after 24 hrs. After a further 24 hrs, the test area was exposed to 3 minimal erythema doses. This entire procedure was carried out 6 times, beginning on days 1, 4, 8, 11, 15 and 18. Challenge was carried out about 10 days after the completion of induction. The test material was applied to two fresh sites and occluded for 24 hrs. Radiation was then applied using 10 J/cm^2. UVA, 320-400 nm. A radiation-only site was used as a control. There was no evidence of any adverse reaction (15).

Subacute Toxicity

Rat : Dermal, Four-week study. Four groups, each of 5 m and 5 f animals were used. Dose levels and conditions were : control (abraded skin) ; 120 mg/kg bw/day (abraded skin) ; 200 mg/kg bw/day (abraded skin) ; 230 mg/kg bw/day (intact skin). Exposure was for 5 hrs a day, with occlusion, followed by rinsing. There was some skin irritation in all groups. Some changes in haematological findings occurred, but they do not appear to be significant. No significant changes were found in biochemical values. Necropsy was grossly normal, and histological examinations revealed no abnormality, either in internal organs or the skin (5).

Tests for percutaneous absorption

Rat : Isolated naked rat skin in vitro. A test was carried out using a chamber in the usual way. The a.i. was labelled with ^{14}C, and 180 $\mu g/cm^2$ was applied to the skin as a 1.5% solution in acetone or deltyl (proposed use level 5%). Radioactivity was measured in the residue on the skin, in the stratum corneum prepared by stripping, in the remainder of the skin and in the chamber fluid. The durations of the experiments were 1 and 6 hrs. Levels of radioactivity were expressed as percentages of the applied radioactivity. The results were as follows :

(a) Acetone solution : 1 hr, stratum corneum 11.5%, remaining skin 4.2%, chamber fluid, nil (total 15.7%). 6 hr : 10.1%, 10.8% and nil (total, 20.9%).

(b) Detyl solution : 1 hr, 4.1%, 5.0% and nil (total 9.1%) ; 6 hr: 4.0%, 7.3% and nil (total 11.3%).
Thus there was somewhat less absorption from the deltyl solution. The number of experiments carried out is not clear, but probably it was one at each dose and time interval (22).

In another similar experiment, the a.i. was dissolved in carbitol at a concentration of 1.5%, and amounts of 120, 360 and 1200 ug/cm^2 were applied to the skin with exposures for 1, 6, 16 and 24 hrs.

There was only trivial penetration, at most, to the chamber.
The totals found in the strippings + skin were :

At 120 µg/cm ^{2}:	1 hr 6.7% ;	6hr 16.7% ;	16 hr 27.7% ;	24 hr 44.3%.
At 360 µg/cm ^{2}:	7.8% ;	17.5% ;	39.8% ;	41.7%.
At 1200 µg/cm ^{2}:	4.6% ;	13.8% ;	25.8% ;	33.9%.

Thus, although the amounts in the skin increase with time, they do not seem to be influenced by concentration in the range studied (23).

A similar set of experiments was carried out, except that in this case they were designed to see whether the incorporation of both the a.i. and "Parsol MCX" in the vehicle made any difference to the absorption of either substance. The solvent was carbitol. Readings were at 1 and 6 hrs. The results, as far as the a.i. was concerned, were broadly similar to those of the preceding experiments (24).

Pig : Minipig skin in vitro.
It seems probable that the method used was similar to that for the rat skin, but this is not stated. The contact time was 6 hrs. The aim was to study the effects of various formulations, and to see whethe "Parsol MCX" affected absorption of the a.i., or vice versa. It is possible to extract those data which refer to the permeation of the a.i. only. A single concentration (2%) was used, in 3 different vehicle - o/w lotion, - o/w cream, and - w/o cream. The application was of 120 µg/cm ^{2} . The results showed that the total amounts found in the skin (stratum corneum + remaining skin) were as follows, for the respective formulations : 2.6%, 3.7% and 2.9%. (These were similar in the presence of 7.5% of "Parsol MCX"). The authors calculate that these values, extrapolated to man under reasonable use conditions, would suggest that 129 µg/kg bw/day would be absorbed into the skin (25).

Man : Skin permeation in vitro.
Abdominal cadaver skin was used in a chamber experiment. The a.i. was labelled with ^{14}C and formulated as 2% in a w/o cream. The cream was applied at a dose of 2.5 mg/cm ^{2} (= 50 µg a.i./cm) and left for 1, 6 and 18 hrs. After stripping, the remainder of the skin was cut horizontally to give layers which consisted of epidermis, upper corium, lower corium, and subcutaneous fat respectively.

Radioactivity was measured in the strippings, the various skin layers, and the chamber fluid. The total amounts in the skin, with the preparation containing the a.i. only, are given for comparison with the animal data summarised above. They were : 1 hr, 4.51%; 6 hr, 7.03% ; 16 hr, 15.96%. Activity was found in the fat after the longest exposure only, and then in an amount of 0.10%. No activity was found in the chamber fluid at any time. In general, activity decreased as the distance form the surface of the skin increased ; and activity increased in all layers as exposure became longer. The amounts reaching the deepest layers were very small - e.g., even at 18 hrs the percentage of radioactivity found in the lower corium was 0.34% (26).

Rat : in vivo.
A 1% solution of ^{14}C labelled a.i. dissolved in carbitol was used. This was applied to the rat skin at a dose of 120 µg/cm^2 . Whether occlusion was used is not mentioned. After 6 hrs, the skin was cleaned and the radioactivity in the washings estimated. The animals were killed, and the treated area of skin excised. The amounts of radioactivity found in the strateum corneum and in the deeper layers were estimated, and were found to be 1.4% and 2.3% respectively.
(Summary report only ; no experimental details. Origin of report not stated).

Man : in vivo.
This study was designed to test the effect of "Parsol MCX" on the absorption of a.i. In a first experiment, ^{14}C-labelled a.i. was applied to the backs of 4 healthy volunteers, 1 with occlusion and 3 without. The application was of 200 µl of a 10% solution in carbitol, spread over 10 cm^2 (= 2 mg/cm^2 a.i.). Exposure was for 8 hrs. From the subject with occlusion, the percentage of radioactivity recovered from the strippings was 0.48, and from the urine 0.08. In the absence of occlusion, the figures were 0.17 and 0.013, respectively (mean of 3 subjects). In a second experiment, the same procedure was repeated, except that unlabelled "Parsol MCX" was added to the a.i.

In this case, the respective recoveries were 0.32 and 0.04 (occluded) and 0.56 and 0.03 (mean of 3 experiments without occlusion). No radioactivity was found in the blood or faeces in any case (27).

MUTAGENICITY

An Ames test was carried out with up to 500 µg a.i. dissolved in DMSO. The test was negative with and without activation (16).

A test was carried out using Saccharomyces cerevisiae with up to 125 µg of a.i./ml. There was no evidence of mutagenic effects (17).

Tests were carried out on V79 chinese hamster lung cells to see whether the a.i. induced mutation at the HGPRT locus. Concentrations higher than 20 µg/ml difficulties because of solvent toxicity (methanol). In addition, the a.i. was cytotoxic in the absence of activation. For this reason, 20 µg/ml was the highest concentration tested. There was no evidence of mutagenic effect (18).

Mouse : Micronucleus test.
Doses of 1000, 2000 and 5000 mg/kg bw were given 30 and 6 hrs before sacrifice. The test was negative (19).

TERATOGENICITY

Rat : In a preliminary test, it was found that 1000 mg a.i./kg bw/day did not induce any maternal abnormalities, although there was some increase in resorptions. This was chosen as the top dose. In all, 4 groups of 36 animals were used. The dosing rates were 0, 250, 500 and 1000 mg/kg bw/day by gavage from days 7 to 16 of pregnancy. The numbers pregnant in each group were, respectively, 33, 35, 31 and 34. At day 21, the animals were divided into approximate half-groups ; one half was selected for sacrifice and the other to continue to delivery and rearing.

The only suspicious finding was a litter with a few [3 ?] unilateral eye defects at the top dose. Since this was confined to one litter it was not considered to be compound-related. All other groups showed no important changes. The rearing subgroup was normal. There was no compound-related maternal toxicity (20).

Rabbit : Four groups of animals were used - control, 80, 200 and 500 mg/kg bw/day (owing to an error, these target doses were not quite achieved). The a.i. was given by gavage from day 7-19 of pregnancy. The numbers pregnant in each group were, respectively, 17, 19, 17 and 19. There were some maternal deaths, probably not compound-related. In the low dose group, there was a high incidence of resorptions ; the reason for this is not clear. At sacrifice, there appeared to be a high incidence of abnormalities in the medulla oblongata region, the nature of which is not very obvious from the report. Probably however, the changes are not significant in that they do not appear to be compound-related. The report, in general, is not easy to interpret (21).

CONCLUSION

The Committee felt that the results of the mutagenicity studies might render a long-term carcinogenicity study unnecessary. It would be preferable if the data on absorption in man could be expressed in terms of an absorption coefficient rather than as a percentage of the dose applied.

The teratological study in the rabbit has many anomalous findings, and is difficult to evaluate.

Details of the absorption spectrum of the compound should be supplied.

As a matter of general principle, the Committee queried whether photo-mutagenicity studies should be carried out on this and other ultra-violet absorbing compounds.

Classification : C.

REFERENCES

1 . Korte (1980) Huntingdon Research Centre, Germany.

2 . Bonettie E.P. (1978) HLR Biological and Pharmaceutical Research Centre, Basle.

3 . Korte (1980) Huntingdon Research Centre, Germany.

4 . Bremer K.D. (1983) HLR Biological and Pharmaceutical Research Centre, Basle.

5 . Buser S. (1979) HLR Biological and Pharmaceutical Research Centre, Basle.

6 . Klecak (1978) HLR Biological and Pharmaceutical Research Centre, Basle.

7 . Kynoch S.R. et al (1980) Huntingdon Research Centre.

8 . Klecak (1982) HLR Biological and Pharmaceutical Research Centre, Basle.

9 . Koehler and Konstant (1979) Food and Drug Research Laboratories, East Orange, N.J., U.S.A.

10 . Klecak (1982) HLR Biological and Pharmaceutical Research Centre, Basle.

11 . Klecak (1982) HLR Biological and Pharmaceutical Research Centre, Basle.

12 . Klecak (1982) HLR Biological and Pharmaceutical Research Centre, Basle.

13 . Klecak (1982) HLR Biological and Pharmaceutical Research Centre, Basle.

14 . Klecak (1982) HLR Biological and Pharmaceutical Research Centre, Basle.

15 . Cuthber and D'Arcy-Burt (1985) Inveresk Research International.

16 . Schupback (1979) HLR Biological and Pharmaceutical Research Centre, Basle.

17 . Chatelet (1983) HLR Biological and Pharmaceutical Research Centre, Basle.

18 . Miltenburger H.G. (1984) Laboratorium fur Mutagenitatsprufung, Darmstadt.

19 . Hugentobler A.L. (1981) HLR Biological and Pharmaceutical Research Centre, Basle.

20 . Eckhardt K. (1984) HLR Biological and Pharmaceutical Research Centre, Basle.

21 . Hummler (1983) HLR Biological and Pharmaceutical Research Centre, Basle.

22 . Anon (1980) HLR Biological and Pharmaceutical Research Centre, Basle.

23 . Anon (1979) HLR Biological and Pharmaceutical Research Centre, Basle.

24 . Anon (1980) HLR Biological and Pharmaceutical Research Centre, Basle.

25 . Klecak et al (1982) HLR Biological and Pharmaceutical Research Centre, Basle.

26 . Stuttgen G. (1982) Freie Universitat Berlin.

27 . Darragh and Lambe (1980) Biological and Medical Research Institute, Dublin.

28 . Doskoczil et al (1979) Parfum und Kosmetik 60, 407.

First	Version	:	February	1984
Revised	Version	:	December	1987
Revised	Version	:	March	1988

CONTENTS

38th meeting of the SCC - October 10-11, 1988

ATTENDANCE LIST

SCIENTIFIC COMMITTEE ON COSMETOLOGY

38th MEETING - 10/11 OCTOBER 1988

Present

Mr AGACHE

Mr DE GROOT

Mrs DONY

Mrs ENJOLRAS

Mr FIELDER

Mr GOULDING

Mr KAPOULAS

Mrs KNAAP

Mr LOPRIENO

Mr O'MAHONY

Mr RAMOS MORGADO

Mr SCHOU

Excused

Mr GRANADOS JARQUE

Mr MUSCARDIN

Mr STÜTTGEN

COMMISSION

Mrs MASSE

Mr GONTIER (DG XI/C/2)

Mr COLLIN (expert)

SUMMARY

Opinions expressed on October 10-11, 1988 concerning the use of :

- HAIR DYES (X1/663 - 88 ; X1/716/88)

1.	RESORCINOL	(A11)
2.	4 - CHLORORESORCINOL	(A12)
3.	2 - HYDROXY - 4 - AMINOTOLUENE	(A27)
4.	2 - (2'-4'-DIAMINOPHENOXY) ETHANOL HYDROCHLORIDE	(A42)
5.	METHYL RESORCINOL	(A44)

- PRESERVATIVE AGENTS

1. OXADINE A (EEC n° 60 ; P75)
2. GLUTARALDEHYDE (EEC n° 26 ; P76)

- UV FILTERS (X1/823/88)

2, 4, 6 TRIANILINO - (p-CARBO-2'-ETHYLHEXYL-1'OXY) - 1, 3, 5 TRIAZINE (S69).

- STABILIZATING AGENTS

1 - 1 - 1 - TRICHLOROETHANE

REPORTS BY THE SCIENTIFIC COMMITTEE ON COSMETOLOGY ON THE USE OF CERTAIN HAIR DYES

. THE COMMITTEE'S MANDATE

To give its opinion on the question whether the use of certain hair dyes :
- Resorcinol
- 4 - chlororesorcinol
- 2 - hydroxy - 4 - aminotoluene
- 2 - (2' - 4' - diaminophenoxy) ethanol hydrochloride
- methyl resorcinol

is admissible from the health point of view.

1. RESORCINOL

. FORMULA AND SYNONYMES

Cl.: 76505
CAS: 108-46-3
Colipa A 11

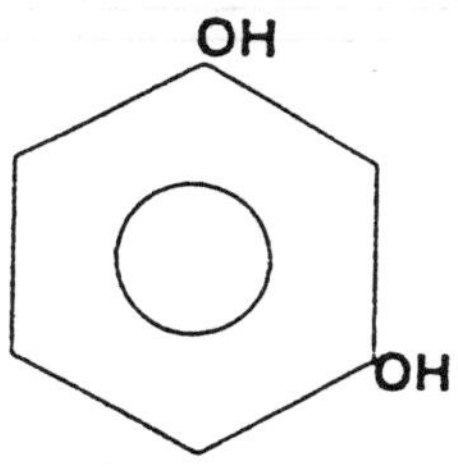

Mol.w.110.11

RESORCINOL
1,3-DIHYDROXYBENZENE
1,3-BENZENEDIOL
m-BENZEDIOL
m-DIHYDROXYBENZENE
m-DIOXYBENZENE
3-HYDROXYCYCLOHEXADIEN-1-ONE
m-HYDROXYQUINONE
3-HYDROXYPHENOL

. USE

Oxidative hair dye

max. use : 2.5%

1.25% in combination with H_2O_2

0.5% in hair lotions and shampoos

. RECAPITULATION OF THE STUDIES OF TOXICITY

LD50	oral (rat)	: 370-980	mg/kg
	dermal (rabbit)	: 3360	mg/kg

ORAL TOXICITY

Short term oral : Rat, 20 mg/kg/d., 5d./wk 12 wks.: no-effects.

CUTANEOUS AND MUCOUS TOLERANCE

DERMAL TOXICITY

Eye irritation

Rabbit : Mild conjunctival redness (1 h, 2.5 w/v in water with 0.5 sodium sulphate into 1 eye. Rabbit (24, 0.1 g into 1 eye) : severe conjunctivities, iritis, corneal opacities occluding most of the iris, corneal ulcerations extending over the entire corneal surface, keratoconus and pannus formation with no improvement during 14 days of observation.

Dermal irritation

Rabbits : 2.5% w/v for 24 h, not irritating on shaven skin (reading at 72 h) ; 0.5 g in saline solutions : negative ; 1000-7950 mg/kg or 2500-19000 mg/kg dermal : irritation and necrosis dose-related ; 3% in ointment : canthose, hypergranulosis, oedema (ear and shaven flank daily for 14 days).

Sensitization

Guinea pig : 3% in water with 2% Natrosol 250, 2% Tween 80 and 0.05% sodium sulphite : no inflammation and no allergic reactions (3 weeks treatment and 2 wks challenge reaction).

Human-skin sensitization

Resorcinol : (1% in paraffin solution) showed positive reaction in 2.1% (1.9% males and 2.2% females) of 877 persons suffering from primary contact dermatitis. Resorcinol causes primary irritation of skin and eye in doses above 10%. Resorcinol (0.2% in salicyl-alcohol) given positive response in a patch test in eczematous subject (erythema) and acute dermatitis with salicyl-resorcinol solution in alcohol applied dermally. Resorcinol (5% in water) showed allergic reactions in 7.9% of 340 patients (eczematous) tested. Resorcinol given positive reaction for cross-sensitivity/patch test with resorcinol monoacetate ; hydroquinone ; pyrocatecol ; phenol ; pyrogallol ; hydroxy-quinone ; phoroloroglucinol ; hexylresorcinol ; orcinol ; cathecol ; 4-phenyl cathecol ; pyrogallic acid ; 3,5-dihydroxybenzene ; resorcinol mono methyl ether ; resorcinol dimethyl ether ; and floroglucinol. Resor - cinol caused ochronosis and myxoedema in a patient with received 12% in ointment on the leg ulcers for 13 years before died. In a case of resorcinol poisoning in a young children (7 weeks) the compound given severe haemolytic anaemia with haemoglobinuria and a generalized papulo-squamous eruption. Resorcinol (3-10-35% in vaseline) caused urticaria in 4 patients treated dermally : 3/4 fever and headache ; 2/4 showed articular pain. One person died after cutaneous applications of an ointment with 20% resorcinol.

Dermal asorption

Rhesus monkey : 0.2% ; guinea pig : 0.6%. Rat (back skin, 30 min.) Resorcinol (1.5% and 3.0%) + p-TDA (1.5% and 3.0%) 1:1 H_2O_2 : 2.77% (ie. 75.5 nM/cm ^{2}, 332.5 µg/kg) and 4.58% (i.e. 62.46 nM/cm ^{2}, 270 µg/kg) after 96 h.

Short term dermal

Rat : 2 50 mg/kg s.c. x 14 or 30 days : no-effects. Rat (154 mg/kg/d. in arachis oil x 47 days s.c.) : goitrogenic action after 69 day (irregular absorptions could have been involved. Guinea-pig (1% and 3% in unguentum on unshaven ears and flank of male for 14 d.) : acanthosis (max. 8 d.), hypergranulosis and orthohyperkeratosis, oedema on ears and flank, and papillomatosis only ears (max. 8 d.). Female rat, Resorcinol diacetate (800 mg/kg s.c. 2 times/d. x 12-19 d. or 22-25 d.) : goitrogenic action.

Human-skin absorption

In commercial hair dye (1.225% of resorcinol) applied (dose not reported) for 25-28 min penetrated 0.076% (as % of total dose excretion). Resorcinol 2% in hydroalcholic vehicle, dermal topical applications (2 time/d., 6 d/wk x 4 wks, i.e. 48 appl.) on 3 human volunteers (0.30 mg/cm2, 12 mg/kg/day) : 2.87% (i.e. 0.34 mg/kg/day), flux rate 0.37 µg/cm /h. It has been reported a skin absorption of 0.7% In vitro : 0.86 µg/cm2/h. (46).

MUTAGENICITY AND GENOTOXICITY

The compound has been tested for the induction of gene mutations in vitro and found negative in Salmonella spot test and in two studies plate test and in E.coli. The compound did not induce micronuclea on mice treated with a total dosage of 600 mg/kg in two equal oral doses separated by an interval of 24 hours.

CARCINOGENICITY

Mice : Skin painting throughout life span (50 animal/group) : 1 squamous cell papilloma on dorsal skin (dose : 5%) ; 1 squamous cell carcinoma on ear (dose : 25%).

Rabbit : 0.02 ml of 5-10-50% 2 time/week during life-time : no adverse signs or tumours were observed.

Mice, rat : dermal topical applications (0.2% to 2.0% containing formulation) : No adverse effects.
Mice, rats : feeding study in progress at NC1.

TERATOGENICITY AND REPRODUCTION

Teratogenicity

Rat : (until to 500 mg/kg oral : no-effects. Rat (40-80-250 mg/kg, 6-15 days gestation) : slightly lower maternal b.w. gain (250 mg/kg) ; no-effects : 80 mg/kg. Rabbit (25-50-100 mg/kg oral, 6-18 days gestation) slightly lower maternal b.w. gain (100 mg/kg) ; no effects : 50 mg/kg. Mice (28-35 mg/kg s.c. appli., 5-7 d, 8-10 d. and 11-14 d. gestation) : no-effects. Mouse, rat and rabbit : dermal topical apllications with formulations containing resorcinol (from 0.2% to 2.0%).

Mouse (1.7%) : significant decrease in b.w. and increase number of unossified caudal and vetebral centra and an unossified bones in feet ; (1.7%). Rat (1.7%) : significant increase of skeletal anomalies (notched and short ribs). Rabbit : no-sign of maternal toxicity only focal alopecia.

Embryotoxicity

HET test : It had no-effect-level, including systemic effects at doses ca. 1-5 ppm. Chicken embryons (3 days) : ED50 2.4 µmol/egg ; LD50 2.9 µmol/egg ; and malformed fetuses vs. 3% control (0.9-7.3 µmol/egg).

Multireproduction

Rat : 3-generations (0.2% to 2.0% in formulations) : negative.

MISCELLANEOUS INFORMATIONS

Metabolism

Rat : (10-50-100 mg/kg s.c.) : peak in plasma after 15 min. ; ca. 90% compound equivalent eliminated after 2 h : half-lives : 23 min-8.6 h (50 mg/kg) ; 18 min-10.5 h (100 mg/kg) ; peak in liver and kidney at 1 h (0.2%-0.3%). 10 mg/kg : 7% in gastrointestinal tract (1 h), 1.4% gastrointestinal tract and feces (24 h), 93.6% in urine (24 h). Rat (2 x 50 mg/kg, 6 h interval, for 14-30 d. and 50 mg/kg with trace 14-C-Resorcinol) : after 2 h plasma level declined to ca. 90% at 15-30 d. ; half-lives : 32 min (14-30 day, fast phase) - 5.0h (14 days) and 7.0 h (30 day) slow phase.

Thyroid and Liver fixation

Rat : dermal topical applications (30 min, intervals of 30-40 d.) of hair dye solution (1.5% (=136 nM) radiolabelled compound, mixed 1:1 with H_2O_2) : trace (2.2 µg) in liver and any fixation has been observed in thyroid 4 day treatment.

Excretion

14% as sulphate, 52% as glucuronide and 11% free resorcinol, in urine of rabbit after oral dose of 100 mg/kg.

Immunosuppressive effect

Mixed lymphocytes cultures, suppression of humural responses in vitro and inhibition of the production of circulating antibodies against red blood cells in vivo : negative.

Thyroid

Antihyroid effects might be ascribed to inhibition of thyroid peroxidase.

CONCLUSION

The SCC does not consider the use of Resorcinol in the hair dyes to be linked to any particular toxic risk for consumers.
National Toxicology Program's results might confirm the absence of any type of toxic risk in the use of Resorcinol.

Classification : B.

ACUTE TOXICITY
MUCOUS MEMBRANE IRRITATION
SKIN IRRITATION
SENSITIZATION

1. Kynoch et al, (1975). "Acute Oral Toxicity to Rats of Resorcinol" Huntindon Research Centre, Cambridshire, - ENGLAND Report 4802/D1/75; 7th October 1975

2. Huntingdon research Centre, Report 4996/2D/75, 12th November 1975

3. Huntingdon Research Centre, Report 4984/3D/75, 12th November 1975

4. Prof. Schultz, (1976) : "Vergleichende Untersuchungen über das Sensibilisertungs-vermôgen verschiedener Haarfarbstoffe" Universitâts-Hautklinik, Hamburg Report24th may, 1976

5. Rudkiki et al, (1970) "The Epidemiology of Contact Dermatitis in Poland" Br. J. Derm. 83, p.543-545

SHORT TERM TOXICITY

6. Doniach et al., (1953) "The Goitrogenic action of Resorcinol in Rats". Brit. J. Exp. Pathol. 34, p.146-151

7. Burnett et al., (1976) "Teratology and Percutaneous Toxicity Studies on Hair Dyes". J. of Tox. and Environ. Health, 1, p. 1027-1040

7a. Potokar et al., (1980) "Resorcin : Subakute orale Toxizität an Ratten bei einer Versuchsdauer von 12 Wochen". HENKEL Kommanditgesellschaft auf Aktion Hauptabteilung Toxicologie - Düsseldorf Report of 29 January 1980.

LONG TERM TOXICITY

8. Stenbäck et al., (1974)."Lack of Toxicity and Carcinogenicity of some Commonly used Cutaneous Agents" Tox. and Appl. Pharmacol. 30, p. 7-13

9. Stenbäck, (1977), "Local and Systemic Effects of Commonly used Cutaneous agents : Lifetime studies of 16 Compounds in Mice and Rabbits". Acta Phamrmacol. et Toxicol., 41, p. 417_431

10. NCI Bioassays (1978). Testimony to Moss Hearings on Safety of Hair Dyes. Serial 95-98, Waschington 1978 (Results for Resorcinol published in Toxicol. Appli. Pharmacol. 1974

11. Kinkel et al., (1973). "Study of Long Term Percutaneous Toxicity and Carcinogenicity of Hair Dyes (Oxidizing Dyes) in Rats. Fd Cosmet. Toxicol., 13, p. 641-648

12. Burnett et al., (1975) "Long Term Toxicity Studies on Oxidation Hair Dyes" Fd Cosmet. Toxicol., 13, p.353-357

13. Giles et al., (1976). "Dermal Carcinogenicity by Mouse Skin painting with 2,4-Toluene-diamine alone or in representative Hair Dye Formulations". J. of Toxicol. and Eviron. Health 1, p. 433-440

14. Burnett et al., (1980) "Evaluation of The Toxicity and Carcinogenicity of Hair Dyes. J. Toxicol. Environ. Health, 6, 247-257

MUTAGENICITY STUDIES - IN VITRO STUDIES

15. Ames et al. (1975). "Hair Dyes are mutagenic : Identification of a Variety of Mutagenic Ingredients" Proc. Nat. Acad. Sci., 72, n°6, p. 2423-2427

16. Mac Cann et al.,(1975). (Detection of Carcinogens as Mutagens in the Salmonnella/microsome test : Assay of 300 Chemicals" Proct. Nat. Acad. Sci., 72, n°12, p. 5135-5139

17. Yoshikawa et al., (1976) "Studies on the Mutagenicity of Hair-Dyes." Bull. of Nation. Institute of Hygienic Sciences, 94, p. 28-32

18. Gericke, D. (1976) "Bericht über Prüfung von Substanzen auf Mutagene Wirkung". HOECHST AG, Labor für Krebsforschung, Frankfürt/main, 30.11.1976.

19a. ShahinM.M. et al., (1980) "Studies on the Mutagenicity of Resorcinol and Hydroxy-3-(p-alino)-anilino-6,N((p-amino) phenol) benzoquinone-mono-imine-1,4 in Salmonnella typhimurium. Mut. Research, 78, 213-218, 1980

19b. Lüpke et al., (1979) "Resorcin - Wirkstoff in Kosmetischen Mitteln?" Arztl. Kosmetologie 9 p 341-346

19c. Hastwell and Mc Gregor, (1977) "Testing for Mutagenic Activity of 33 Hair Colorant Components" Inveresk Research INternational (IRI) IRI Project N° 407485, August 1977

MUTAGENICITY STUDIES - IN VIVO STUDIES

20. Hossaack et al., (1977)
"Examination of the Potential Mutagenicity of Hair-Dye Constituants Using the Micronucleus Test"
Experimentia, 33, p. 377-378

20a. M. Bracher et al., (1981)
"Studies on the Potential in vivo induction of sister-chromatid excahges in rat bone marrow by resorcinol"
Cosmital Research Laboratories, Cosmital SA, Marly, Suisse, 1981

TERATOLOGY STUDIES

- cfr ref 7

21. Bio-Dynamics Inc., (1977) "A Modified segment II Teratology Study of Hair-Dyes in Mice".
Submitted to : Clairol Inc. - Stamford, Connecticut
Report Project n° 76-1677, September 7th, 1977

IMMUNOSUPPRESSIVE ACTION

22. Hossack et al., (1977)
"An in vivo and in vitro Evaluation of the Immunosuppressive Action of Resorcinol, 4-amino-diphenylamine, 2,5-diamino-1-methylbenzene and 4-amino-2-hydroxy-1-methylbenzene"
Huntingdon research Center, Report n°16/77632, Septembre 27, 1977

23. E.I. GOLDENTHAL et al., (1979)
"Lifetime Cchronic Toxicity/carcinogenesis study in rats".
International Research and Development Corporation, Michigan, USA

Report February/March 1979
(cfr file A15 - reference 21a & 21b)

24. E. GOCKE et al., (1983)
"Mutagenicity of Cosmetics Ingredients Licensed by the European Communities"

25. S. HAWORTH et al, (1983)
"Salmonella Mutagenicity test Results for 250 Chemicals"
Environ. Mutagenesis Suppl., 1, (1983), p. 3-142

26. R. CREBELLI et al, (1981)
"Mutagenicity of Commercial p-Phenylenediamine and of an oxidation Mixture of p-Phenylenediamine and Resorcinol in Salmonella typhimurium TA 98"
Fd. Cosmetol. Toxicol., Vol. 19,(1981), p.79-84

27. G.S.PROBST et al, (1982)
"Chemically induced Unscheduled DNA Synthesis in Pramary Rat Hepatocyte Cultures : A Comparison with Bacterial Mutagenicity Using 218 Compounds".
Environmental Mutagenesis, 3, (1981)p.11-32

28. H.F. STITCH et al, (1981)
"The Action of Transition Metals on the Genotoxicity of Simple Phenols, Phenolic Acids and Cinnamic Acids"
Cancer letters, 14, (1981), p. 251-260

29. Von R. SCHULZ et al, (1982)
"Untersuchungen zur mutagenen und Klastogenen Wirkung von Resorcin. Zygotgenetische Befunde an Verschiedenen Menschlichen Zelltypen"
Arnzeim.-Forsch./Drug Res., 32,(I), n°5, (1982), p. 533-536

30. F. DARROUDI & A.T. NATARAJAN, (1983)
"Cytogenetic Analysis of Human Peripheral Blood Lymphocytes (In Vitro) Treated with Resorcinol
Mutation Research, 124, (1983), p. 179-189

32. H. HOFER et al, (1983)
"Analysis of Human Chromosomes Afer Repeated Hair Dyeing"
Fd Chem. Toxic., Vol. 21, n°6, (1983), p.785-789

31. D. WILD et al, (1981)
"Mutagenic Activity of Aminophenols and Diphenols, and Relations with Chemical Structure"
Abstract presented at GUM meeting (Darmstadt - 1981)

34. J.P. SEILER, (1977)
"INhibition of Testicular DNA synthesis by Chemical Mutagens andCarcinogens, Preliminary Results in the Validation of a Novel Short Term System'
Mutation Research, 46, (1977), p. 305-310

33. K. KURANITZ et al, (1983)
"Investigation on the Effect of Repeated Hair Dyeing on Sister Chromatid Exchanges "
Fd Chem. Toxic., Vol. 21, n°6, (1983), p. 791-793

(35) C. BURNETT et al, (1981)
"Heritable Translocation Study on Two Hair Dye Formulations"
Fundam. and Applied Toxicol., 1, (1981), p. 325-328

(36) A N.P. LUEPKE, (1982)
"Embryotoxicity-Testing of Cosmetic Ingredients by HET"
IFSSC 12th International Congress - Paris, 13/17 September 1982
(Volume II, p. 231-248)

(36) B "Toxicity-Testing of Hair Dye Ingredients and Antidandruff Agents by HET (Hen's Egg Test)"
Hair and Aesthetic Medicine - Rome 24/26 March 1983 - p. 296-297

(37) A. KORHONEN et al, (1983)
"Embryotoxic Effects of Acrolein, Methacrylates, Guanidines and Resorcinol on Three Day Chicken Embryos"
Acta Pharmacol. and Toxicol., 52, (1983), p. 95-99

(38) BIO-DYNAMICS, (1982)
"A Modified Segment II Teratology Study on Hair Dyes in Rabbits"
Project n° 76-1666

(39) REPROTOX, (1978)
"Prüfung auf Embryotoxische Wirkung an der Maus - Resorzin"
REPROTOX - Münster - DEUTSCHLAND
Report 16 November 1978

(40) I. OSTERBURG, (1982)
"Resorcin : Oral (Gavage) Teratology Study in the Rat"
HAZLETON Laboratories Deutschland GmbH - Münster - DEUTSCHLAND
Report n° 64-213/5 - April 1982

(41) I. OSTERBURG, (1982)
"Resorcin : Oral (Gavage) Teratology Study in the New Zealand White Rabbit"
HAZLETON Laboratories Deutschland GmbH - Münster - DEUTSCHLAND
Report n° 60-213/1 - April 1982

(42) J.C. DINARDO et al, (1985)
"Teratological Assessment of Five Oxidative Hair Dyes in the Rat"
Toxicol. and Applied Pharmacol., 78, (1985), p. 163-166

44 A. ROUGIER & D. DUPUIS, (1981)
"Study of the In Vivo Percutaneous Absorption of Resorcinol in the rat
Influence of the Content in Resorcinol of the Applied Formulations"
Laboratoire d'Absorption Percutanée et de Métabolisme - Division
Recherche Fondamentale - Aulnay-sous-bois - FRANCE - Report 17.12.1981

45 L.J. WOLFRAM & H.I. MAIBACH, (1985)
"Percutaneous Penetration of Hair Dyes"
Arch. Dermatol. Res., 277, (1985), p. 235-241

46 D. YEUNG et al, (1983)
"Percutaneous Absorption, Blood Levels, and Urinary Excretion of
Resorcinol Applied Topically in Humans"
International Journal of Dermatology, Vol. 22, (June 1983),
p. 321-324

47 P.C. MERKER et al, (1982)
"Pharmacokinetics of Resorcinol in the Rat"
Research Communications in Chemical Pathology and Pharmacology,
Vol. 38, n° 3, (December 1982), p. 367-388

48. V. TSOMI & G. KALOPISSIS,(1982)
"Cutaneous Penetration of Same Hair Dyes in the Hairless Rat." Toxicol. European Research, Vol IV n°13, (May 1982) p. 119-126

49. Y.V. PASCHIN et al., (1976)
"Increased Antimutagenic Activity on Simple Substitues Phenols Mixed with the Hindered Phenolic Antioxidant Dibund"
Fd. Chem. Toxic. Vol; 24, n°8, (1976) p. 881-883

2. 4-CHLORORESORCINOL

. FORMULA AND SYNONYMES

C.1.: 76510
CAS : 95-88-5
Colipa A12

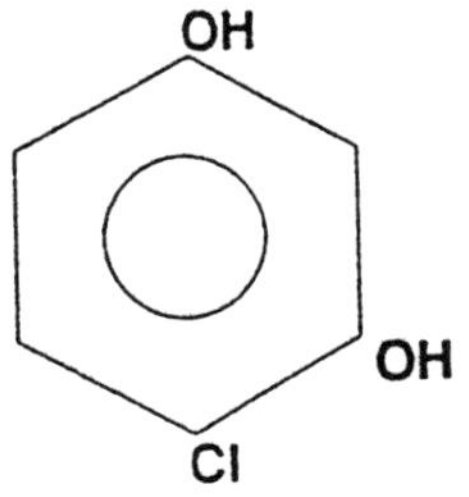

4-CHLORORESORCINOL
1,3-DIHYDROXY-4-CHLOROBENZENE

Mol.w. 144.56

. USE

Oxidation hair dye
Max. use : 3%
1.5% upon application

. RECAPITULATION OF THE STUDIES OF TOXICITY

LD50 oral (rat) : 369 mg/kg

ORAL TOXICITY

Short term oral

Rat : 20 mg/kg bw/day per os for 12 wks. (daily/5 times weekly) : no effects. Rat : 40 mg/kg bw/day for 3 wks (5 day/wk) oral gavage : slight activation of thyroid epithelium ; slight decrease in the triiodothyronin in the serum.

CUTANEOUS AND MUCOUS TOLERANCE

DERMAL TOXICITY

Eye irritation

Rabbit : 2.5% w/v into one eye : negative, mild conjunctival irritation only (2/3 animals).

Dermal irritation

Rabbit : 2.5% w/v for 24 h : negative (reading at 72 h).

Sensitization

Guinea pig : 3% epicutaneously, daily (6 d/wk) for 3 weeks : no reaction.

Dermal absorption

Rat : 1%, 2% and 3% in formulations (1:1 H_2O_2) for 30 min (clipped skin) : 0.216%, 0.217% and 0.367 in 72 h. Rat : 2% in formulation (1:1 H_2O_2) : 0.006% as $^{14}CO_2$ in expired air after 8 h. Rat : 1% in formulation (1:1 H_2O_2) for 30 min. (unclipped skin) : 0.088% in 72 h. Rat : 300 mg in acqueous sol. for 30 min. (clipped skin) : 5.47% (0.11 mg/cm^2) in 72 h.

Short term dermal

Rabbit : 2% in formulation with 6% H_2O_2 for 13 wks : no-effects.

Chronic toxicity

Mice : skin painting, 2% in formulation, 21 months : no toxicity or carcinogenicity effect.

MUTAGENICITY AND GENOTOXICITY

The compound has been tested for the induction of gene mutations in vitro and found negative in Salmonella at spot test and in several studies at plate test and in E.coli. The compound did not induce micronuclea on mice treated with a total dosage of 600 mg/kg in two equal oral doses separated by an interval of 24 hours ; the compound was negative for the induction of chromosome aberrations on human lymphocytes in vitro.

CARCINOGENICITY

A long term study is in progress at National Toxicology Programm.

TERATOGENICITY

Rat : 2 ml/kg bw in formulation on day 1-4-7-10-13-16-19 of gestation.
Rat : 50, 100, 200 mg/kg oral gavage (6 to 15 of gestation), 200 mg/kg : significant decrease in maternal weight gains, embryolethal (increase in resorptions), slight increase in fetal anomalies (not statistically significant). Minor anomalies (wavy ribs) and variations (incomplete ossification of the sternebrae, rudimentary and bilateral 14th ribs) of skeletal ; 100 mg/kg : no-effects.

MISCELLANEOUS INFORMATIONS

Organ distribution

Rat : 2% in formulation, cutaneous appl. : No marked affinity for any particular tissue (highest conc. found after 35 min 1h). Rat: oral 50 mg/kg in acqueous sol. : No special affinity to any organ.

Excretion

Rat : 50 mg/kg bw s.c. : > 96% (urine and feces) in 24 h. Rat : 50 mg/kg oral : predominantly in urine in 24 h. ; 19.3% of oral dose in bile within 3 h.

Immunosuppressive effects

4-Chlororesocinol was found negative for the induction of immuno-suppressive action evaluated by plate test and hemagglutination test when s.c. administered to 6 NMRI mice four time at the maximal tolerated dose of 1.5 mg.

CONCLUSION

The Committee requires :
- a study on skin penetration
- information on the (90-day) oral study in rats (Henkel) to establish the NEL on the thyroid
- the results of the aberration test on mammalian cells.

The Committee is waiting for the ongoing long-term study for carcinogenicity of NTP - USA.

Classification : B.

BIBLIOGRAPHY

ACUTE TOXICITY

1. KYNOCH,S.R.and LLOYD,G.K.
   "Acute oral Toxicity to Rats of 4-Chlororesorcinol,"
   Huntingdon Research Centre, 15.10.1975.

MUCOUS MEMBRANE IRRITATION

2. KYNOCH,S.R. and LIGGETT,M.P.
   "Irritant Effects of 4-Chlororesorcinol on Rabbit Eye Mucosa."
   Huntingdon Research Centre, 12.11.1975.

SKIN IRRITATION

3. KYNOCH,S.R. and LIGGETT,M.P.
   "Irritant Effects of 4-Chlororesorcinol on Rabbit Skin."
   Huntingdon Research Centre, 12.11.1975.

SENSITIZATION

4. SCHULZ,K.H.
   "Comparable Studies of Sensibilization of Different Hair Dye Ingredients."
   Universitätsklinik Hamburg (1976)

SHORT TERM TOXICITY

5. BURNETT C. et al.
   "Teratology and Percutaneous Toxicity Studies on Hair Dyes."
   J.Toxicol. Environ. Health,1, 1027-1040 (1976)
6. POTOKAR M. et al.
   m-Chlorresorcin,Subakute orale Toxizität an Ratten bei einer Versuchsdauer von 12 Wochen.
   Henkel KGaA, Düsseldorf, 05.06.1979.

LONG TERM TOXICITY

7. BURNETT C. et al.
   "Evaluation of the Toxicity and Carcinogenicity of Hair Dyes."
   J.Toxicol. Environ. Health, 5, 247 (1980)

MUTAGENICITY STUDIES

3. AMES et al.
   "Hair Dyes are Mutagenic,Identification of a Variety of Mutagenic Ingredients."
   Proc.Nat.Acad.Sci. 72, 2423-2427 (1975)
9. BILLHIMER
   "Summary of Salmonella Microsomal Assay for bacterial Mutagenesis."
   Hill Top Research, 26.04.1977.

10.YOSHIKAWA K et al.
"Sudies on the Mutagenicity of Hair Dyes."
Report of the National Institute of Hygiene, 94, 28 (1976)

11.MUELLER,U.,
"Prüfung des Farbstoffes 1,3-Dihydroxy-4-chlorobenzol auf Mutagenität im Bakterientest."
Battelle-Institut,Frankfurt (1976)

13.HOSSACK, David J.N.
"Micronucleus Test on 4-Amino-2-hydroxytoluene, 4-Chlororesorcinol and 1-Naphtol.
Huntingdon Research Centre, 26.03.1976 SWF6/76175

IMMUNOSUPPRESSIVE ACTION

12.GERICKE,D.
"Bericht über die Prüfung auf immunosuppressive Wirkung von 11 Substanzen der Hans Schwarzkopf GmbH,Hamburg."
Krebsforschunglabor der Hoechst AG, Frankfurt/M., 04.01.1977.

14.G.K. LLOYD, M.P. LIGGETT, S.R. KYNOCH and R.E. DAVIES
"Assessment of the acute toxicity and potencial irritancy of hair-dye constituents."
Fd Cosmet. Toxicol. 15, 607-610 (1977)

15J. BUESCHING
"Die Wirkung von 4-Chlorresorcin, m-Aminophenol und 1-Hydroxy-3,4-Methylendioxybenzol auf die Schilddrüse der Ratten nach oraler Appikation." Wella AG 05.02.1982.

16. H. HOFER et al.
"Studie zur Resorption und Toxikokinetik von 4-Chlorresorcin."
österreichisches Forschungszentrum Seiberdorf ,Januar 1986.

17. H. BELLINGER
"Amestest mit Chlorresorcin." Henkel KGaA. 14.09.1979.

18. J.C. PICCIANO et al.
"Evaluation of the Teratogenic and Mutagenic Potential of the Oxidative Dyes : 4-Chlororesorcinol, m-Phenylenediamine, and Pyrogallol."
J.Am.Coll.Toxicol. 2, 325-333 (1983)

3. 2-HYDROXY-4-AMINOTOLUENE

. FORMULA AND SYNONYMES

CAS: 2835-95-2

Colipa A27

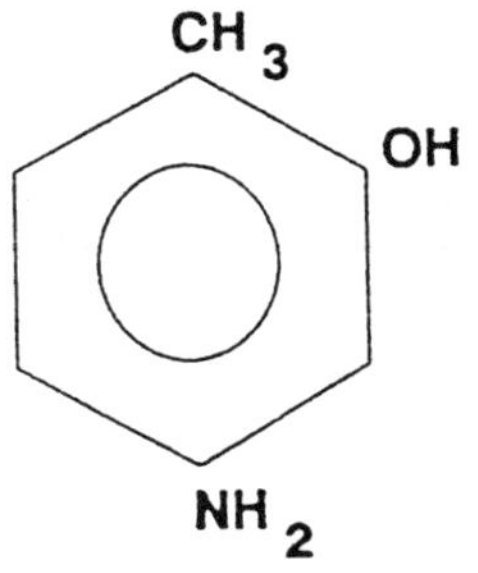

p-AMINO-o-CRESOL
1-METHYL-2-HYDROXY-4-AMINOBENZENE
2-HYDROXY-4-AMINOTOLUENE
4-AMINO-2-HYDROXYTOLUENE

Mol.w. 123.1

. USE

Oxidative hair dye
Max. use : 3%
1.5% with H_2O_2

. RECAPITULATION OF THE STUDIES OF TOXICITY

LD50 oral (rat) : 3600 -> 10000 mg/kg

ORAL TOXICITY

Short term oral

Rat : up to 2700 mg/kg oral intubation for 2 weeks (5 days a wk): disseminated white lesions in the pancreas (dose-related), no other dose-related toxic effects.
Rat : oral intubation, 300 to 2700 mg/kg bw/day for 13 weeks (5 day wk) : dose-related adverse effects at all doses (nervous system symptoms, decrease in the hematological values, increase in the relative weight

of the liver) ; at high doses (≥300 mg/kg) dose-related but not significant : increase or decrease in organ weights, histopathological changes in liver, stomach, pancreas, kidney and spleen ; liver, renal and adrenals enlargement).

Rat : oral up to 180 mg/kg bw/day for 13 weeks (5 days a week, by gavage) : no-effects.

CUTANEOUS AND MUCOUS TOLERANCE

DERMAL TOXICITY

Eye irritation

Rabbit : 2.5% w/v for 10 sec. : negative, mild conjunctival irritation only.

Dermal irritation

Rabbit : 2.5% for 24 hours : negative. Another study the compound resulted mild irritant.

Sensitization

Guinea pig : 3% in CMC and Tween 80, epicutaneously daily for 13 weeks (6 day a week) and other 2 weeks : weak sensitization in 4/19 animals.

Dermal absorption

Rat : 1% in formulation (1:1 with H_2O_2) for 30 min : 3.62 $\mu g/cm^2$ (male) and 5.64 $\mu g/cm^2$ (female).

Human absorption

1% in formulation (1:1 with H_2O_2) to dry hair at a lotion/hair ratio of 1.5-2, worked into hair of 3 human volunteers for 5-8 min. and left on for other 20 min., total urinary excretion : 0.2% applied dose.

MUTAGENICITY AND GENOTOXICITY

The compound was tested for gene mutation and found positive on Salmonella and negative in E.coli. The compound did not induce micronuclea on mice treated orally by gastric intubation.
Negative results obtained in several studies on Salmonella were performed at lowest doses respect to doses used in positive studies.
A formulation containing 2-hydroxy-4-aminotoluene have been tested for SCE in vivo in lymphocytes of human volunteers exposed for different times produced negative results.

TERATOGENICITY

Rat : oral up to 180 mg/kg bw/day (0.5% in CMC, 10 ml/kg) from day 6 to 15 of gestation : no embriotoxicity, embryoletalithy or teratogenicity effects.

MISCELLANEOUS INFORMATIONS

Immunosuppressive effects

Mouse : 4 x 7.75 mg s.c. injection, by plate-test (cell-mediated immunity) and haemagglutination test (humorally mediated immunity): negative.

CONCLUSION

The Committee requires a study on chromosomal aberration test in mammalian cells in vitro.

It appears to be judicious to look forward to a warning caution on the package following the high rates of sensitization (19%) which are observed in the animal studies.

Classification : B.

GENERAL DATA - BIBLIOGRAPHY

1. Beilsteins - Handbuch der organischen Chemie
13, 574 (1930)

ACUTE TOXICITY - BIBLIOGRAPHY

2. KYNOCH, S.R. & LLOYD, G.K.
"Acute oral toxicity to rats of 4-amino-2-hydroxy-toluene"
Huntingdon Research Centre, Report n° 4852/D 1/75,
15 October 1975

5. LLOYD G.K. et al
"Assessment of the acute toxicity and potential irritancy of hair dye constituents."
Fd. Cosmet. Toxicol. 15, 607 (1977)

MUCOUS MEMBRANE IRRITATION - BIBLIOGRAPHY

3. KYNOCH S.R. & LIGGETT M.P.
"Irritant effects of 4-Amino-2-hydroxy-toluene on rabbit eye mucosa"
Huntingdon Research Centre, Report n° 4995/2D/75
(12.11.1975)

5. LLOYD G.K. et al
"Assessment of the acute toxicity and potential irritancy of hair dye constituents"
Fd. Cosmet. Toxicol. 15, 607 (1977)

SKIN IRRITATION

4. KYNOCH S.R. & LIGGEST M.P.
"Irritant effects of 4-amino-2-hydroxy-toluene on rabbit skin"
Huntingdon Research Centre, Report n°. 4989/3D/75
(12.11.1975)

SENSITIZATION

6. SCHULZ K.H.
"Vergleichende Untersuchungen über das Sensibilisierungs-vermögen verschiedener Haarfarbstoffe"
Universitäts-Hautklinik Hamburg
(24.5.1976)

IMMUNOSUPPRESSIVE ACTION

7. GERICKE D.
"Bericht über die Prüfung auf immunsuppressive Wirkung von 11 Substanzen der Hans Schwarzkopf GmbH, Hamburg"
Krebsforschungslabor Hoechst AG
(4.1.1977)

8. HOFER H. et al
"Dosisfindungsstudie mit 4-amino-2-hydroxy-toluene"
Österreichische Studiengesellschaft für Atomenergie GmbH, Forschungszentrum Seibersdorf, SGAE-Ber. n° 2998, BL-248
(November 1978)

9. HOFER H. et al
"13-Wochen-Toxizitätsversuch mit 4-Amino-2-hydroxy-toluol an Ratten"
Österreichische Studiengesellschaft für Atomenergie GmbH, Forschungszentrum Seibersdorf, SGAE-Ber N° A 0065, BL-276/79 (August 1979)

10. HOFER H. et al
"13-Wochen-Toxizitätsversuch mit 4-amino-2-hydroxy-toluol an Ratten, ergänzende Versuch"
Österreichische Studiengesellschaft für Atomenergie GmbH, Forschungszentrum Seibersdorf, SGAE-Ber N° A 0076, BL-282/79 (September 1979)

MUTAGENICITY STUDIES IN VITRO

11. YOSHIKAWA K. & UCHINO H. & KURATA H.
"Studies on the Mutagenicity of hair dye"
Report of the Japan National; Hygien. Lab. n° 94,28
(May 1976)

12. BILLHIMER W.
"The Salmonella/microsomal assay for bacterial mutagenic activity of 4-amino-2-hydroxy-toluene"
Microbiological Associates for Hill Top Research
(21.3.1977)

13. WALLAT S.
"1-Methyl-2-hydroxy-4-amino-benzene : Prüfung auf Mutagenität im Ames-Test"
Henkel KGaA (AbschluBericht vom 30.6.1982)

14. HOSSACK D.J.N. et al
"Ames metabolic activiation test to assess the potential mutagenic effect of 1-Methyl-2-hydroxy-4-amino-benzene"
Huntingdon Research Centre, Report n° WLA/5/77380
(10.6.1977)

15. HASTWELL R. & McGREGOR D.B.
"Testing for mutagenic activity of various hair colourants in Salmonella typhimurium"
Inveresk Research International - Report n° 761
(May 1977)

16. MÜLLER U.
"Prüfung des Farbstoffes 1-Methyl-2-hydroxy-4-amino-benzene auf Mutagnität im Bakterientest"
Battelle-Institut, Frankfurt/Main
(Januar 1976)

MUTAGENICITY STUDIES IN VIVO

17. HOSSACK D.J.N. & RICHARDSON J.C.
"Micronucleus test on 4-amino-2-hydroxy-toluene 4-chloro-resorcinol and 1-naphtol"
Huntingdon Research Centre - Report SWF 6/76175
(26.3.1976)

SKIN ABSORPTION
STUDY IN ANIMALS

18. BÄRTNIK F. et al
"1-Methyl-2-hydroxy-4-aminobenzol
Ermittlung der cutanen Resorption aus einer Haarfarbstoff-formulierung"
Institute for Toxicology of Messrs. Henkel
(10.7.1984)

19. BARTNIK F et al
"1-Methyl-2-hydroxy-4-aminobenzol
Ermittlung der Organverteilung nach cutaner Applikation
aus einer Haarfarbstofformulierung"
Institute for Toxicology of Messrs. Henkel
(17.9.1984)

STUDY IN MAN

20. WOLRAM L.J., MAIBACH H.I.
"Percutaneous penetration of hair dyes"
Archives of Dermatological Research 227, 235-241
(1985)

MUTAGENICITY IN VIVO

21. TURANITZ K et al
"Investigations on the effect of repeated hair dyeing on sister
chromatid exchanges"
Food and Chemical Toxicology Vol. 21, n° 6, 791-793
(1983)

TERATOGENICITY

22. OSTERBUR I.
"4-amino-2-hydroxytoluol embryotoxicity study in the rat"
Hazleton Laboratories Deutschland GmbH
(16.2.1984)

4. 2(2'-4'-DIAMINOPENOXY) ETHANOL HYDROCHLORIDE

. FORMULA AND SYNONYMES

CAS: 66422-95-5

Colipa A42

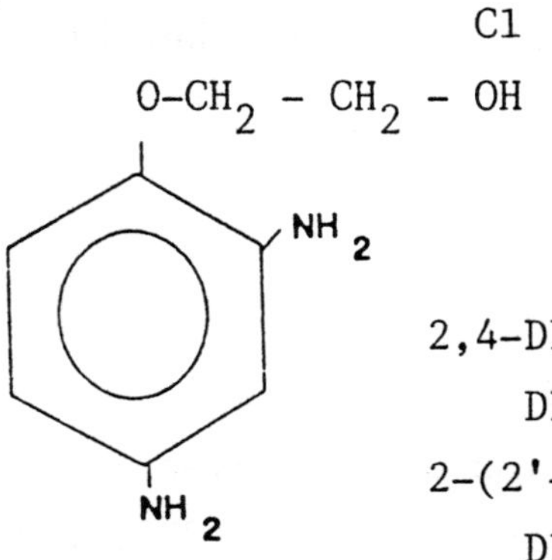

2,4-DIAMINOPHENOXYETHANOL DIHYDROCHLORIDE

2-(2'-4'-DIAMINOPHENOXY) ETHANOL DIHYDROCHLORIDE

(DIAMINO-2',4'-PHENOXY)-2-ETHANOL DICHLORIDRATE

Mol.w. 241

1-β-HYDROXYETHYLOXY-2,4-DIAMINOBENZENE

. USE

Oxidative hair dye

Max. use 4%

2% in combination with H_2O_2

. RECAPITULATION OF THE STUDIES OF TOXICITY

LD50	oral	(rats)	:	1113 - 1160	mg/kg
		(mice)	:	1745	mg/kg

ORAL TOXICITY

Short term oral

Rat : 56 mg/kg bw/day oral for 3 months : no-effects. Mouse and rat: up to 0.2% oral for 12 weeks, toxic effects in mice (0.1%) and rats (0.2%). No effects level : 0.05% (mice) and 0.1% (rats).

CUTANEOUS AND MUCOUS TOLERANCE

DERMAL TOXICITY

Eye irritation

Rabbit : 4% aqueous solution (pH 2,5), without rinsed : pratically not irritating (reading at 24, 48, 72 h. and days 4,7).

Skin irritation

Rabbit : 4% in proplene glycol for 24 h : mildy irritating.

Sensitization

Guinea pig : compound with 50% Freud's complete adjuvant, challenge reaction (25% in petrolatum for 24 h) : 3/10 sensibilized (middle potential of allergenicity).

Dermal absorption

Hairless rat : 0.40% (= 23.65 nM) in commercial vehicle + 20 vol H_2O_2 for 40 min.: 5.05 nM (= 0.84 µg/cm^2) ; 0.33% (= 23.65 nM) in complete formulation : 2.83 nM (= 0.47 µg/cm^2) ;

(a) 0.40% (23.65 nM)

(b) 0.80%

(c) 1.20% in commercial vehicle (1:1 with 20 ml H_2O_2)

for 40 min : (a) 0.84 µg/cm^2 (5.03 nM)

(b) 1.34 µg/cm^2 (7.95 nM)

(c) 1.58 µg/cm^2 (9.42 nM) after 4 days.

MUTAGENICITY AND GENOTOXICITY

The compound has been tested for gene mutations in vitro in several experiments on Salmonella typhimurium and found negative on 5 studies with 1 only positive study.

Negative results were also obtained in Salmonella with a urinary assays on rat treated oral, i.p. and by topical applications, and on mice treated dermal. Negative data for gene mutations in vitro were obtained on reversions systems in E.coli (2 studies) and S. cerevisiae XV185-14C, forward mutation assay in S.pombe P1 and V79 hamster cell line (HPRT). Negative data were also obtained in vivo on D.melanogaster and in mouse spot-test with up to 1500 mg/kg.
The compound was anable to induce chromosome aberrations in vitro on CHO cell line and in human lymphocytes, micronuclea and dominant lethals on mice in vivo. The compound did not induce gene conversion on S. cerevisiae D4 (2 studies), UDS on HeLa human cell line.

CARCINOGENICITY

Mouse (0.07% and 0.04% in water) and rat (0.1% and 0.05% in water) for 104 weeks : no carinogenic effects.

TERATOGENICITY

Rat : 50-100-200 mg/kg/d. gastric intubation, 6 to 15 of pregnancy : increase in the incidence of skeletal anomalies and the proportion of litters containing such foetuses (dose-related, statistically significant at all doses), an increase in skeletal variants (extra rib and variant sternebrae) and lower values, not statistically significant, for litter and mean foetal weight at 200 mg/kg. Mouse up to 1500 mg/kg topically on shaven skin (6 to 15 post-fertilization): negative.

CONCLUSION

The SCC does not consider the use of 2,4-Diaminophenoxyethanol in the hair dyes to be linked to any particular toxic risk for consumers.

It appears to be judious to look forward to a warning caution on the package following the high rates of sensitization (30%) which are observed in the animal studies.

Classification : A.

BIBLIOGRAPHY - ACUTE TOXICITY
MUCOUS MEMBRANES TOLERANCE
SKIN IRRITATION
SENSITIZATION

1. PROF. SEGRE (1976)
"Substance : 7038 V - Solution aquese - Toxicité aiguë"
Lab. of "Universita di Siena" - Instituto di Farmacoligca
ITALY Report from 13.10.1976

2. K.G. DOSSOU (1977)
"Toxicité aiguë par voie orale - IMEXINE OAJ (sous forme de chlorhydrate)"
Lab. of "Recherche Fondamentale" - Société L'OREAL
Aulnay/s/Bois - France
Report from 1.6.1977

3. K.G. DOSSOU, (1979)
"Détermination de l'indice d'irritation oculaire - IMEXINE OAJ"
Lab. of "Recherche Fondamentale" - Société L'OREAL
Aulnay/s/Bois - FRANCE
Report form 2.2.1979

4. K.G. DOSSOU, (1979)
"Indice d'irritation primaire - IMEXINE OAJ"
Lab of "Recherche Fondamentale" - Société L'OREAL
Aulnay/s/Bois - FRANCE
Report from 31.1.1979

5. PROF. P.E. FOURNIER (1978)
"Test de sensibilisation (test de Magnusson) sur cobayes d'un produit dénommé Colorant IMEXINE OAJ - société L'OREAL"
Lab. of "Unité de Pharmacologie Clinique, Hôpital Fernand Widal" - Paris - FRANCE
Report from 10.5.1978

90-DAY ORAL STUDY

6. PROF. P.E. FOURNIER, (1978)
"Etude de toxicité chronique chez le rat - Colorant capillaire "D"
(IMEXINE OAJ) - Laboratoires de L'OREAL"
Lab. of "Société de Recherches Biologiques" - Asnières
FRANCE
Report from 10.10.1978

MUTAGENICITY STUDIES - IN-VITRO STUDIES

7. M.M. SHAHIN et al, (1980)
"Structure-Activity Relationships within a Series of 2,4-Diaminoalkoxybenzene compounds"
Mutation Research, 79, (1980), p. 289-306

8. G. MOHN et al, (1982)
"Mutagenic activity of 2-(2',4'-diaminophenoxy)ethanol in Strains TA 1538 and TA 98 of Salmonella Typhimurium"
Mutation Research, 102, (1982), p. 313-318

9. M.M. SHAHIN et al, (1982)
"Mutagenic evaluation of the Hair-Dye Component 2,4-diaminophenoxyethanol in Salmonella Typhimurium/microsome plate test and Saccharomyces Cerevisiae Strains D4 and XV185-14C"
Mutation Research, 102, (1982), p. 319-329

10. N. LOPRIENO et al, (1982)
"Mutagenic Studies on the Hair Dye 2-(2',4'-diaminophenoxy)ethanol with Different Genetic Systems"
Mutation Research, 102, (1982), p. 331-346

11. S. VENITT et al, (1984)
"The Hair-Dye Reagent 2-(2',4'-diaminophenoxy)ethanol is Mutagenic to Salmonella Typhimurium"
Mutation Research, 135, (1984), p. 31-47

12. BARALE et al, (1984)
"A battery approach Versus a single test for evaluating the Genotoxic activity of 2-(2',4'-diaminophenoxy)ethanol - A comment on the preceding Paper by VENITT, CROFFTON-SLEIGH and OSBORNE"
Mutation Research, 135, (1984), p. 49

13. R.M. HASTWELL et al, (1984)
"Lack of Genotoxic Activity of 2-(2',4'-diaminophenoxy)ethanol in Escherichia coli Strains WP 2, WP 2uvrA and WP2uvrA/recA"
Mutation Research, 102, (1982), p. 309-312

14. F. DARROUDI et al, (1982)
"Test for Induction of Chromosomal Aberrations in Chinese Hamster Ovary Cells (In-Vitro) by 2-(2',4'-Diaminophenoxy) ethanol"
Mutation Research, 102, (1982), p. 351-355

MUTAGENICITY STUDIES - IN-VIVO STUDIES

15. W.G.H. BLIJLEVEN, (1982)
"Mutagenicity of 2-(2',4'-diaminophenoxy)ethanol in Drosophila melanogaster"
Mutation Research, 102, 1982, p; 347-349

16. J.C. RICHARDSON et al, (1982)
"An evaluation of the Mutagenic Potential of 2-(2',4'-diaminophenoxy)ethanol Using the Micronucleus Test"
Mutation Research, 102, 1982; p. 357-360

17. D.J. BRUSICK et al, (1982)
"The acitivity of 2-(2';4'-diaminophenoxy)ethanol in 3 Genetic Toxicity Bioassays"
Mutation Research, 102, 1982, p. 361-372

MUTAGENICITY STUDIES - MISCELLANEOUS

18. G. KALOPISSIS, (1981)
"Experimental Studies on a New Hair Dyeing Ingredient 2-(2',4'-diaminophenoxy)ethanol : Lack of Genotoxic Properties"
Toxicological European Research Volume III, n° 4, Juillet 1981, p. 191-195

TERATOGENICITY STUDIES

19. A.M. BOTTOMLEY et al, (1981)
"Effect of 2.4-Diaminophenoxyethanol, Dihydrochloride (OREX 108) on Pregnancy of the Rat"
Huntingdon Research Centre - Cambridgeshire - ENGLAND
Report LRL/7/801091 - 5 February 1981

20. R.P. BELLIES et al, (1978)
"Teratology Evaluation of OREX 108 (2,4-diaminophenoxyethanol, dihydrochoride) in Mice"
Litton Bionetics Inc. - Kensington, Maryland - USA
LBI Project n° 20807 - November 1978

PERCUTANEOUS ABSORPTION

21. V. TSOMI AND G. KALOPISSIS, (1982)
"Cutaneous Penetration of Some Hair Dyes in the Hairless Rat"
Toxicological European Research, Vol. IV, n° 3, (Mai 1982)

22. V. TSOMI AND G. KALOPISSIS, (1982)
"Percutaneous Absorption on Hairless Rat of Two Hairdyestuffs (couplers):2,4 Diaminoanisole and 2-(2',4'-Diaminophenoxy) ethanol"
Toxicological European Research, Volume IV,n°4,(July 1982) pp. 205-208.

CARCINOGENICITY STUDY

23. N. KUWAHARA et al.,(1983)
"Carcinogenicity Test of IMEXINE OAJ"
Department of Experimental Pathology - Cancer Institute -Tokyo
Report April 15,1983

5. METHYL-RESORCINOL

. FORMULA AND SYNONYMES

CAS: 608-25-3

Colipa A44

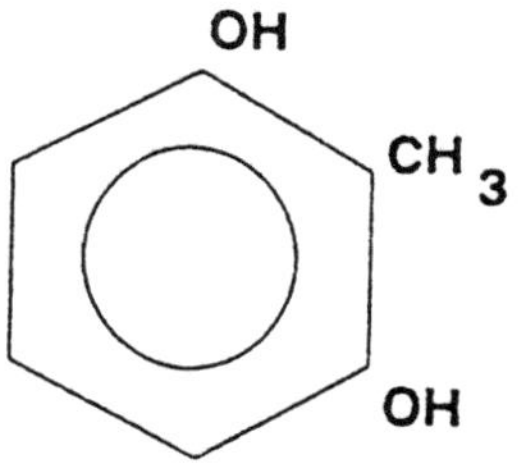

Mol.w.124

2-METHYLRESORCINOL
1,3-DIHYDROXY-2-METHYL-BENZENE
1,3-BENZENEDIOL-2-METHYL
2,6-DIHYDROXYTOLUENE

. USE

Oxidative hair dye

Max.use : 2.0%

1% in combination with H_2O_2

. RECAPITULATION OF THE STUDIES OF TOXICITY

LD50	oral	(rats)	:	> 5000 mg/kg (0.2% in hair dye basic cream)
		(mice)	:	390 mg/kg

ORAL TOXICITY

Short term oral

Rat : Up to 180 mg/kg/day oral (5 times a week) for 12 weeks : no-effects.

CUTANEOUS AND MUCOUS TOLERANCE

DERMAL TOXICITY

Eye irritation

Rabbits : 5% w/v in water : mild conjunctival irritation (after 24 h., disappeared after 7 days) ; cornea and iris : no irritation.
Rabbit : 2% in basic cream (6% H_2O_2) for 10 sec. : slightly irritating.

Dermal irritation

Rabbit : 10% for 4 h. : mild reversible dermal irritation.
Rabbit : 2% in basic cream (6% H_2O_2) for 4 h. : not irritating.
Mouse : 10% in aqueous, 2 appl./day for 5 days on same skin area : mild skin redness after 5 appl.

Human dermal irritation

5 human volunteers, 10% aqueous solution on forearm skin for 30 min. (30 sec. intervals), open method : not irritating. 5 human volunteers, 10% aqueous solution on upperarm skin for 2 h., occlusive conditions : no irritation until to 24 h.

Sensitization

Guinea pig : 5% aqueous : negative. Guinea pig : 2% in basic cream (6% H_2O_2) : negative up to 72 h. (challenge 1%, 21 and 28 day).

Dermal absorption

Rat : 16.9 mg (= 0.136 mMol) / 6.25 g basic cream + 6% H_2O_2 (1:1): 0.48% in 24 h.

Short term dermal

Rabbit : 1% in formulation (1:1, 6% H_2O_2) dermal topical application twice weekly for 13 wks. : no-affects (slight thickening in treated skin).

MUTAGENICITY AND GENOTOXICITY

The compound was tested for gene mutation in vitro and found negative on Salmonella. The compound did not induce chromosome aberrations in vitro on CHO cells and ; was inable to induce micronuclea in vivo on mice with oral doses up to 2 x 350 mg/kg.

TERATOGENICITY

Embrytoxicity

Chicken-embryo : HET test : moderately toxic, no-evidence of teratogenicity : LD50 = 6.1 mg/egg (day 1) - LD50 = 1.08 mg/egg (day 5).
Rat : 1% in formulation, 2% ml/kg topical applications on day 1-4-7-10-13-16-19 of gestation. Study inadequate (treatment every 3 days ; only one dose tested).

MISCELLANEOUS INFORMATIONS

Excretion study

Rat : 20 mg in water s.c. : > 90% as glucuronide or sulphate after 24 h. (urine and feces), no expired air. Rat : 40 mg in water, oral : completely absorbed by the intestine ; > 90% in 24 h. (urine).

CONCLUSION

Is required a teratogenicity study.

Classification : B.

BIBLIOGRAPHY

1. ACUTE ORAL TOXICITY - CHEMICAL SUBSTANCE
M. POTOKAR
2-Méthylresorcin
Prüfung der akuten oralen Toxizität an der Maus
Henkel KGaA, Abschlußbericht, July 1976

2. ACUTE ORAL TOXICITY - HAIR DYE FORMULATION
W. KÄSTNER
2-Methylresorcin als Bestandteil einer Oxidationshaarfärbecreme.
Prüfung der akuten oralen Toxizität an der Ratte.
Henkel KGaA , Abschlußbericht, 15 Oktober 1982.

3. MUCOUS MEMBRANE IRRITATION - CHEMICAL SUBSTANCE
M. POTOKAR , P. NAVRADE
2-Methylresorcin
Prüfung auf primäre Schleimhautreizung am Kaninchen.
Henkel KGaA, Bericht, 02.10.1981.

4 MUCOUS MEMBRANE IRRITATION - HAIR DYE FORMULATION
W. KÄSTNER
2-Methylresorcin als Bestandteil einer Oxidationshaarfärbecreme.
Prüfung auf primäre Schleimhautreizung am Kaninchenauge.
Henkel KGaA, Bericht , 22.11.1982.

5. SKIN IRRITATION
M. POTOKAR
2-Methylresorcin
Prüfung auf pimäre Hautreizung mittels Okklusionsmethode(Patch-Test) am Kaninchen.
Henkel KGaA, Abschlußbericht, Dezember 1981.

6. SKIN IRRITATION - HAIR DYE FORMULATION
W. KÄSTNER
2-Methylresorcin als Bestandteil einer Oxidationshaarfärbecreme. Prüfung auf primäre Hautreizung mittels Okklusivmethode (Patch-Test) am Kaninchen.
Henkel KGaA, Abschlußbericht, 29.10.1982.

7. SKIN IRRITATION - CHEMICAL SUBSTANCE
M. POTOKAR
2-Methylresorcin
Prüfung auf Hautreizung nach wiederholter Applikation an haarlosen Mäusen
Henkel KGaA, Abschlußbericht, Juli 1976.

8. HUMAN SKIN IRRITATION - CHEMICAL SUBSTANCE
M. POTOKAR
2-Methylresorcin
Prüfung auf primäre Hautreizung nach wiederholter Benetzung der Haut an fünf Versuchspersonen.
Henkel KGaA, Abschlußbericht, September 1982.

9. M. POTOKAR
idem mit Okklusionsmethode

10. SENSITIZATION - CHEMICAL SUBSTANCE
M. POTOKAR
2-Methylresorcin
Prüfung auf allergieauslösende Eigenschaften am Meerschweinchen
Henkel KGaA ,Abschlußbericht, Dezember 1976.

11. SENSITIZATION - HAIR DYE FORMULATION
W. KÄSTNER
2-Methylresorcin als Bestandteil einer Oxidationshaarfärbecreme. Prüfung auf sensibilisierende Eigenschaften am Meerschweinchen.
Henkel KGaA. Abschlußbericht, 10.12.1982.

12. 90 - DAY ORAL TOXICITY: RAT

M. POTOKAR, F. BARTNIK, W. PITTERMANN AND C. BUSCH
2-Methylresorcin
Subacute orale Toxizität an Ratten bei einer Versuchsdauer von 12 Wochen.
Henkel KGaA, Abschlußbericht, November 1980.

13. 13-WEEK PERCUTANEOUS TOXICITY-HAIR DYE FORMULATION (RABBITS)

C. BURNETT,E.J. GOLDENTHAL,S.B. HARRIS,F.X. WAZETER,J. STRAUSBURG
R. KAPP AND R. VOELKER
J.Tox. Environment. Health 1 (1976) 1027 - 1040
Teratology And Percutaneous Toxicity Studies On Hair Dyes.

14. CUTANEOUS ABSORPTION

F. BARTNIK and F. WINGEN
2-Methylresorcin
Cutane Resorption sowie Ausscheidung nach subcutaner und oraler Applikation. Versuche mit Ratten
Henkel KGaA, Abschlußbericht, 11.7.1979.

15. MUTAGENICITY STUDIES:IN VITRO AMES-TEST

S. WALLAT
1,3-Dihydroxy-2!methyl benzene: Prüfung auf Mutagenität im Ames-Test.
Henkel KGaA,Abschlußbericht vom 04.07.1980.

16. IN VITRO CHROMOSOME ABERRATIONS TEST

F. DARROUDI
Evaluation to compound Methyl-2-resorcine in the chromosome aberrations test:Chinese Hamster Ovary cells(in vitro).
University of Leiden(The Netherlands),Final report,25.02.1982.

17. EMBRYOTOXICITY

F.H. KEMPER and N.P.LÜPKE
Bericht zur toxikologischen Prüfung am Hühnerembryo mit RO 261
Universität Münster,1.Dezember 1982.

18. TERATOLOGY

WIL Research Laboratories , Inc.

Project N°:WIL-29002.Combined Suchronic Feeding, Teratology and Dominant Lethal Study on 2-Methyl Resorcinol in the Sprague-Dawley Rat. Part B Teratology STudy. Final Report, December,1983

REPORTS BY THE SCIENTIFIC COMMITTEE ON COSMETOLOGY CONCERNING THE USE OF TWO PRESERVATIVE AGENTS IN COSMETIC PRODUCTS

. THE COMMITTEE'S MANDATE

To give its opinion whether the use of two preservative agents :
- 4,4-dimethyl-1,3-oxazolidine up to 0.1%
- glutaraldehyde up to 0.1%

is admissible from the health point of view.

1. 4,4-DIMETHYL-1,3-OXAZOLIDINE

. FORMULA AND SYNONYMES

EEC n° 60
Colipa P75
dimethyl oxazolidine
Oxadine-A
Oxaban-A

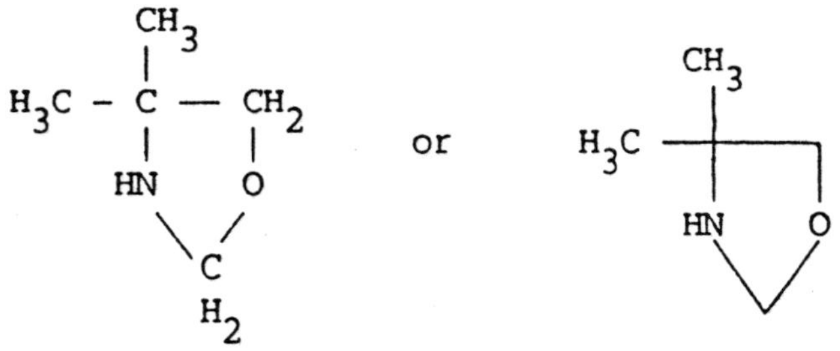

$C_5H_{11}NO$

MW 101

. CHARACTERISTICS

The material is described as an aqueous mixture of two active ingredients, viz. 4,4-dimethyl-1,3-oxazolidine or 4,4-dimethyl-1-oxa-3-azacyclopentane and (as an impurity, generated during synthesis)

3,4,4-trimethyl-1,3-oxazolidine or 3,4,4-trimethyl-1-oxa-3-azacyclopentane. The levels of the two active ingredients in Oxaban-A are c. 76% and c. 2% respectively. The commercial product also contcains c. 3% 2-amino, 2-methyl-1-propanol, traces of formaldehyde and c. 19% water. The concentration of active ingredients is c. 87.4% or c. 78%. The material is soluble in water and in oil.

. USE

Use level of the commercial mixture up to 0.1% both in rinsed-off-, and in non-rinsed off cosmetics.

. RECAPITULATION OF THE STUDIES OF TOXICITY

LD50	oral	(rat)	:	956 mg/kg
		(mouse)	:	890 mg/kg (ref.11)
	dermal	(rabbit)	:	970-2000 mg/kg

Because the oral and dermal values are comparable, dermal absorption is probably high.

LC50	(rat)	:	11,66 mg/l	(-1-hour inhal.)

ORAL TOXICITY

In a 28-day study, groups of 6 rats/sex were orally dosed with 0, 50, 200 or 600 mg Oxaban A/kg bw/day, 5 times weekly. The top dose caused clinical symptoms and growth depression, changes in several haematological characteristics (decreases in Hb and PCV, and increases in WBC, neutrophils and thrombocytes) and decreases in serum levels of total protein and albumin. At autopsy, 10/12 rats in the top-dose group showed gross changes in the stomach. The relative liver weight showed increases with 200 and 600 mg/kg, while the relative weights of the spleen, adrenals and testicles were increased with 600 mg/kg. Of the 5 organs weighed and examined microscopically only the stomach of top-dose rats showed treatment-related changes (Ref. 16).

CUTANEOUS AND MUCOUS TOLERANCE

DERMAL TOXICITY

An 87.4% solution was severely irritating to the rabbit eye and the changes produced showed no recovery over 72 hours. Dilutions of 1000 ppm and 5000 ppm of the commercial solution (equal to 0.08% and 0.4% of the active material) were not irritant to the eye (Ref. 2, 3 and 4).

In rabbits a primary skin irritation test, using 0.5 ml of the 87.4% aqueous solution on an occlusive patch for 24 hrs, caused slight to severe signs of oedema, erythema and deep tissue reaction (Ref.5). A test on rabbit skin with 0.5 ml applied for 4 hrs, produced signs of irritation but not necrosis (Ref. 14). A repeated skin application test with 3% aqueous emulsion produced irritation in 70 out of 101 human volunteers (Ref. 15).

The material showed skin sensitizing properties in the guinea pig (Ref. 6). In a well-conducted study in 101 human volunteers no evidence of sensitization was found in repeated tests with a 3% aqueous solution (Ref. 7, 15).

In a sub-chronic (13-week) dermal toxicity test in rats with Oxadine-A at dose levels of 0, 1.95, 19.5 and 195 mg a.i./kg, applied daily 5 days/week in a 1:1 ethanol/water solution, the top dose induced increased haematopoietic activity, as well as severe skin irritation. No other treatment-related abnormalities were found despite extensive clinical-chemical and histological examination (Ref. 9).

MUTAGENICITY

Results from an Ames test showed a reproducible dose-related increase in revertants in TA 98 and TA 100, but this was less than twice background and was not significant (Ref. 10 ; Ref. 18). Negative results were also obtained in a second assay using TA 98 and TA 100, that has been briefly reported (Monte et al., Fd Chem. Tox. 21(5)695-696(1983).

A fluctuation test in the S. typhimurium strains was negative (Ref.17). A chromosomal aberration test in human lymphocytes in vitro with and without S-9 mix was positive at all doses (Ref. 19). An in vitro test in CHO cells showed dose-related increases in activity, and a mouse lymphoma assay was also positive (file dated 11-5-83). A chromosomal aberration test in vivo in bone marrow cells of rats treated intraperitoneally with 20, 40 or 80 mg/kg was negative (Ref.20).

TERATOGENICITY

A teratogenicity study with 4 groups of 16 rabbits, treated dermally with 0, 30, 100 or 300 mg Oxaban-A/kg for 6 hr per day during days 7 through 19 of pregnancy, did not reveal signs of systemic embryonal or maternal toxicity or teratogenicity (Ref. 21).

CONCLUSION

The product was found to be only moderately toxic in short-term oral and dermal studies in rats. The no-effect levels were 50 and 25 mg/kg respectively (expressed as commercial product). Studies on dermal absorption are not available but data on acute oral and dermal toxicity suggest dermal absorption to be considerable. No evidence of embryotoxicity or teratogenicity was found in rabbits treated dermally with up to 300 mg/kg. Genotoxic properties were established in several in vitro tests, but one in vivo study for chromosomal aberrations in rats was negative. Further testing is necessary to confirm absence of genotoxic potential in vivo. It should be noted that the material is nitrosatable.

Classification : B.

References

1. Altech Laboratories Report n° AL-1-54 (June 20, 1973)

a. Altech Laboratories Report n° AL-1-59 (July 31, 1973)

2. Altech Laboratories Report n° AL-1-53 (June 19, 1973)

3. Altech Laboratories Report n° AL-1-118 (August 6, 1974)

4. CSC Bioassay Report 30775 (July 1, 1975)

5. Altech Laboratories Report n° AL-1-56 (July 10, 1973)

6. IMC Pharmacology Laboratory PL Report n° 60 (August 4, 1978)

7. Product Investigations Inc, Report n° PI-09-46-571C (July 24, 1979)

8. IMC Pharmacology Laboratory Report n° PLR-68 (February 23, 1979)

9. Avon Products Inc, (April 24, 1980)

10. EGEG Mason Research Institute (May 23, 1980)

11. Manufacturers Data

12. " " (July 2, 1973)

13. 'Acute Inhalation Toxicity Study (LC 50) in albino rats
International Research and Development Corporation - November 30, 1973

14. Manufacturers Data (March 3, 1975)

15. International Minerals and Chemical Corporation Report n° PLR-79
(August 22, 1979)

16. NOTOX 0542/697 - June 1987
Assessement of the subacute oral toxicity of Oxaban^{R} A in the rat :
28-day study

17. Toxicol Laboratories Ltd. Report n° : 36/8303 dated April 1983

18. " " " " n° : 35/8303 " " "

19. Toxicol Laboratories Ltd. Report SR 634 Reported 9/8302 March 1983

20. " " " " 216/8303 March 1984

21. "Dermal Teratology Study in Rabbits"
International Research and Development Corp. Feb 13, 1986

22. "Micronucleus test in Bone Marrow Cells of the Mouse with OXABAN A", RCC NOTOX b.v., Project n° 012522, July 1989, 's-Hertogenbosch, the Netherlands

23. "Detection of Single Strand Breaks in Rat Testicular DNA by Alkaline Elution", Microbiological Associates Inc, Study Number T 5562.390 June 1989, Bethesda, MD, USA

2. 1,5-PENTANEDIOL

. FORMULA AND SYNONYMES

EEC n° 26
Colipa P76
CAS Reg n° 111-30-80
GLUTARALDEHYDE

CHO - CH_2 - CH_2 - CH_2 - CHO $C_5H_8O_2$
MW: 100,13

. CHARACTERISTICS

Readily soluble in water and most organic solvents. The substance reacts with amino groups and forms cross links.

. USE

Used as a preservative in rinsed off and non-rinsed off cosmetics up to 0.1%.

. RECAPITULATION OF THE STUDIES OF TOXICITY

LD50	oral (rats)	:	60 - 650	mg/kg	
	dermal(rabbits)	:	735	mg/kg	(50% concentr)
			900		(40% concentr)
			2000 - 3200		(25% concentr)
LC50	(inhal - rats)	:	12,6	mg/1	(8h to saturated vapour)
			23,5 - 40	ppm	(4h to 50, 20 and no ppm)

ORAL TOXICITY

Short-term

Oral administration of 0.1, 0.5, 1.0 and 1.6 g/kg bw/day to rats with the diet for 7 days produced growth depression and decreased weights of liver and kidneys with 1.6 g/kg. No changes were seen with 1.0 g/kg (Colipa ref. 27). Two-week oral treatment of rats with 100 or 1000 ppm in drinking water (13 and 103 mg/kg bw/day respectively) revealed hyperplasia of mucous glands in the stomach with 103 mg/kg, while 13 mg/kg was a NEL (ref. 28). A 13-wk study in rats with up to 100 mg/kg bw in the drinking water showed growth retardation at the top-dose. There were no other relevant changes and the NEL was 25 mg/kg bw for males and 37.5 mg/kg bw for females (ref. 47).

CUTANEOUS AND MUCOUS TOLERANCE

DERMAL TOXICITY

Eye irritation

In rabbits was severe with concentrations above 1%, with concentrations between 0.25 and 1% only slight and with 0.1% absent.

Skin irritation

In rabbits with a 45% solution was severe and was accompanied by oedema and necrotic foci. With 2% only mild erythema occurred and with 1% no effects were seen. In humans a 5% solution caused severe skin reactions after 2 occluded applications. Repeated open applications of 5% or less did not induce skin reactions in humans, but in another study 0.5% was slightly irritant. Irritation seemed more likely if the applications were to areas where the stratum corneum is thin.

Sensitization

Was examined in several studies in humans. Thirteen open applications of 5% a.i. to 20 healthy subjects followed by an open challenge with 5% produced no reactions. With 7 occluded applications of 1% a.i. and 3 occluded applications of 2%, followed by a challenge with 2% also occluded produced mild reactions in 6/20.
Applications of 0.1%, 0.2% and 0.5% under occlusion for 48 hrs, 10 times to 109 subjects, followed by a challenge to the same concentration as that of the induction did not produce reactions with 0.1 and 0.2%. With 0.5% a few reactions occurred during induction, and 2/109 showed reactions to the challenge.
There are several reports of sensitization to the a.i. in medical and paramedical personnel.

Dermal treatment

Of mice with 50 µl of various aqueous dilutions on 9 days during 2 weeks caused death of all animals treated with 25 or 12.5 mg a.i. per mouse, and growth retardation with 2.5 mg a.i. per mouse. No significant changes were seen with 1.25 mg a.i./mouse, which was equivalent with 80 mg/kg bw/day.

Dermal absorption

Of 0.075%, 0.75% or 7.5% (1,5) -^{14}C-glutaraldehyde, applied in vivo under occlusion for 24 hrs, was between 10 and 53% in the rabbit and less than 9% in the rat. The highest absorption occurred with the high concentration, due to the skin damage. The percentages absorption mentioned, do not include the considerable amounts of radioactivity which remained in the stratum corneum under the area of application. This retention may explain the long $t\frac{1}{2}$ of 112 hrs in rats and of 77 hrs in rabbits observed upon dermal dosing, whereas after intravenous dosing the $t\frac{1}{2}$ was only 9-11 hrs in rats and 12-18 hrs in rabbits. Dermal dosing resulted in reasonably uniform tissue distribution. Up to 80% was excreted as CO_2 and up to 20% in the urine (ref. 33).

MUTAGENICITY

Mutagenicity

Several gene mutation test were conducted with S.typhimurium strains, but none was positive. The exposure concentrations were low, due to toxicity of the compound. No indications of mutagenicity were observed in a HGPRT test with CHO-cells treated in vitro with up to 0.1 mg/ml (ref. 40). In a SCE test with CHO-cells exposed to various concentrations up to 0.1 mg/ml, a significant increase in exchanges was seen only with 0.05 mg/ml (not with 0.1 mg/ml).
The positive result was considered a chance event (ref. 40). An UDS test with rat hepatocytes did not induce positive effects but the positive control substances were sometimes negative also (ref. 40).
In a dominant lethal assay, mice received one oral treatment with 30 or 60 mg/kg (respectively one tenth and one fifth of the LD_{50}).
No increased lethality of embryos was observed (ref. 41).

Teratogenicity

Was examined in mice, treated by gavage with a 2% commercial product in amounts providing 16, 20, 24, 40, 50 and 100 mg a.i./kg bw/day on days 6-15 of pregnancy. Increased incidences of resorptions and of malformed foetuses were seen with 40 mg/kg and more. With 24 mg/kg and less no increased incidence of malformations was seen, but signs of toxicity occurred at all dose levels (ref. 42).
In a second teratogenicity study in mice (dosed orally with 3.3, 10 or 30 mg/kg bw/day on days 7-12 of pregnancy) the percentage of resorbed and dead foetuses were relatively high in the 2 high-dose groups. The number of malformations was relatively high in all dose groups although the significance of these findings was not clear (ref. 44). Rats dosed orally with 25 or 50 mg/kg bw/day, on days 6-15 of pregnancy, did not show increases in resorptions, deaths or malformations, although the top-dose induced decreased body weight gain (IBT-study ; ref. 43).

MISCELLANEOUS INFORMATIONS

Short term inhalation

By rats of 0.2, 1.0 and 3.0 ppm in the atmosphere for 9 days, caused reduced intake of food and water, weight loss and mortality with 3.0 ppm, haematological and clinical-chemical changes with 1.0 and 3.0 ppm. A few signs of toxicity were seen with 0.2 ppm (ref. 31).
A 90-day inhalation study in rats at 0.02, 0.05 and 0.2 ppm, 5 hrs/day, 5 days/week, revealed growth depression, and clinical signs of mucosal irritation (perinasal wetness and discharge) at 0.2 and 0.05 ppm. Microscopic examination showed lesions of the heart in 3/20 males of the high-dose group (which the authors did not consider to be treatment-related). The NEL was probably 0.02 ppm, but individual data were not provided (ref. 32).

CONCLUSION

The available data showed that in short-term animal studies the substance was highly toxic by inhalation and moderately toxic when administered orally. Dermal absorption in animals was considerable. Teratogenic properties were observed at oral dose levels which produced maternal toxic effects, but foetal toxicity occurred at lower dose levels, and a no-effect level was not identified. Genotoxicity test in vitro were generally negative.
No information was available on chromosomal aberration. A dominant lethal test in mice was negative. Long-term studies were not available The substance has sensitizing properties.
A further teratogenicity study in mice is needed to identify a NEL, since this seems to be critical in assessing safety factors in use. The mutagenicity data should be supplemented with a chromosomal aberration test in vitro with mammalian cells.

Classification : B. (but not to be used in aerosol sprays)

GLUTARALDEHYDE STUDIES

1. Glutaraldehyde 50 % aqueous solution range finding toxicity studies Chemical Hygiene fellowship - Carnegie-Mellon Inst. of Research April 19 1977 (report 40-50)
2. Rang finding tests on GLutaraldehyde, 45 % aqueous (pentanedial 45 %) Mellon Institute - Pittsburgh (report 27 - 137)
3. Glutaraldehyde 25% aqueous range finding toxicity studies - Carnegie Mellon Institute of Res. - Pittsburgh 14.9.77 (report 40-120)
4. Aqucar 514 Water treatment Microbiocide - Acute toxicity and Irritancy study Bushy Run Research Center Export Pennsylvania 26.11.84 (report 47-166)
5. Glutaraldehyde dilutions, acute peroral toxicity studies - Bushy Run Research Center 2.8.82 (report 45-124)
6. Glutaraldehyde dilutions, percutaneous toxicity and eye irritation studies - Bushy Run Research Center 1.6.81 (report 44-65)
7. Glutaraldehyde dilutions, (45 %, 15 % and 10 %) acute percutaneous studies Bushy Run Research Center 10.6.85 (report 48-51)
8. Glutaraldehyde acute inhalation toxicity with chamber analysis Bushy Run Research Center 28.2.84 (report 47-21)
9. Glutaraldehyde four hour LC 50 inhalation study on rats 7.1.82 Bushy Run Research Center (report 44-96)
10. Glutaraldehyde dilutions primary skin and eye irritancy studies 14.11.84 (report 47-33)
11. Dermal sensitizing potential of Glutaraldehyde - a review and recent observations - J. Toxicol. Cut & Ocular Toxicol 3 (3)

12. Topical Glutaraldehyde - Percutaneous Penetration and skin irritation Arch. Dermatol. Res 1985 - 227:242-244
W.G. Reifenrath et al

13. Eccrine anhidrosis due to Glutaraldehyde, formaldehyde and iontophoresis
Jouranl of investigative Dermatology B. Gordon et al 1969

14. Tropical Glutaraldehyde fosr plantar hyperhydrosis
L. Juhlin et al. Arch. Derm. Vol. 97 March 1968

15. Contact dermatitis from Glutraldehyde
Arch. Derm. Vol 105 Jan. 1972 Jordan et al

16. Allergic contact dermatitis due to Cidex
T.C. Lyon Unites States Army Insitute of dental research, Walter Reed Army Medical Center
Oral Surgery - Dec. 1971

17. Glutaraldehyde and formaldehyde allergy - Neering et al.
Department of Dermatology
Vrije Universiteit Amsterdam
Contact Derm. Newsletter 16, 1974

18. Glutaraldehyde and contact dermatitis - Sanderson, Cronin - Brit. Med. J. 3 ; 802 1968

19. Glutaraldehyde Contact Dermatitis
Gordon Contact dermatitis Newsletter 15, 1974

20. Glutaraldehyde occupational dermatitis
K.S. Hansen - Dept. of Dermatology
Marselisborg Hospital Aarhus, Denmark

21. Glutaraldehyde (Pentanedial)Allergic Contact Dermatitis
Maibach et al
Arch. Dermatol. Vol. 113, Feb. 1977

22. Contact Dermatitis due to sensitivity to Cidex (Activated Glutaraldehyde) Harman et al. - Contact dermatitis Newsletter 11, 1972

23. Localised epidermolysis bullosa
Desgroseilliers et al. Arch. Dermatol. (Vol. 109, Jan. 1974)

24. The use of graded concentrations in studying skin sensitizers experimental contact sensitization in man
Marzulli et al. Fd. Cosmet. Toxicol. 12, p. 219-227, 1974

25. Dose response relationship in allergic contact dermatits : Glutaraldehyde containing liquid fabric softener
Weaver et al. Contact Dermatitis 1977, 3:65-68

26. Glutaraldehyde : cross-reactions to formaldehyde
Maibach - Contact dermatitis 1, 326-327 1975

27. Glutaraldehyde - Results of feeding in the diets of rats for 7 days.
Carnegy Mellon inst. Pittsburgh 29.8.75 (report 38-111)

28. Glutaraldehyde : two-week inclusion in drinking water of rats 19.3.85
Bushy Run Research Center Export

29. Evaluation of the subacute dermal toxicity of Glutaraldehyde in mice
Bushy Run Research Center 7.12.81 - report 44-107

30. Glutaraldehyde vapor nine day inhalation study on rats 8.11.83
Bushy Run Research Center - report 46-95

31. Glutaraldehyde vapor nine day inhalation study on rats 4.11.83
Bushy Run Research Center - Report 46-63

32. Glutaraldehyde vapor subchronic inhalation study on rats
Bushy Run Research Center 15.12.83 - report 46-IOI

33. Skin penetration and pharmacokinetics of Glutaraldehyde in rats and rabbits 15.1.85
Bushy Run Research Center - Report 47-I97

34. A new Salmonella tester strain (TA 102) with A-T base pairs at the site of mutation detects oxidative mutagens.
D.E. Levin et al.
Proc. Nath. Acad. Sci USA Vol. 79, pp. 7445-7449 Dec. 1982

35. Mutagenicity of aldehydes in Salmonella Sasaki et al.
Mutation Research 54, 251, 1978

36. Glutaraldehyde Salmonella/Microsome (Ames) Bacterial Mutagenicity Assay. Bushy Run Research Center 11.12.81
Report 44-I3I

37. Report on the mutation test for Glutaraldehyde Mizuno et al.
Research and Development
Maruishi Pharmaceutics Ltd.

REPORT BY THE SCIENTIFIC COMMITTEE ON COSMETOLOGY

CONCERNING THE USE OF

2,4,6-TRIANILINO-(p-CARBO-2'-ETHYLHEXYL-1'-OXY)-1,3,5-TRIAZINE

. THE COMMITTEE'S MANDATE

To give its opinion on the question whether the use of this compound (S69) as UV filter at the maximal concentration of 5% in sunscreen preparatives is admissible from the health point of view.

. FORMULA AND SYNONYMES

Colipa : S69
UVINUL
2,4,6-trianilino-(p-carbo-2'ethylhexyl-1'-oxy)-1,3,5-triazine.

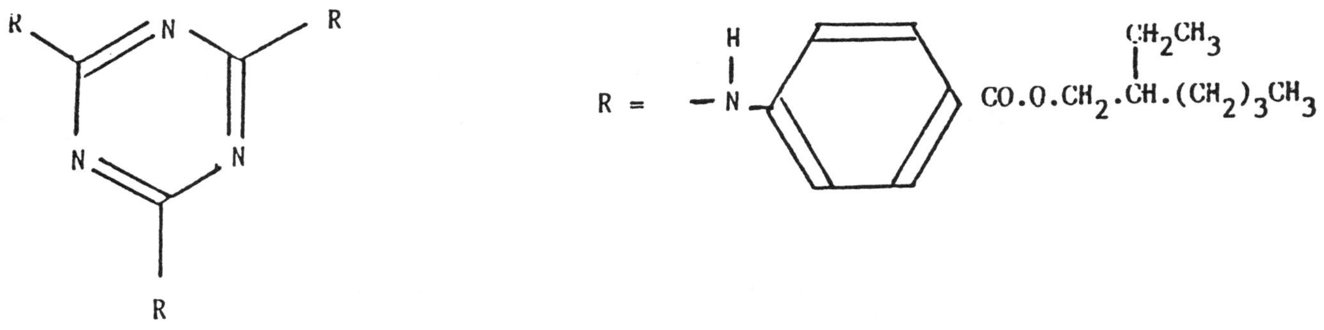

$C_{48}H_{66}N_6O_6$ MW: 823,1

. CHARACTERISTICS

Stated by manufacturer to be more than 98% pure.
Insoluble in water ; soluble in isopropyl myristate, olive oil, ethanol.
Absorption maximum : 312 nm.

. USE

Used as a stabiliser in light-sensitive plastics, dyes, etc...
Proposed use level in suncreen preparations : up to 5%.

. RECAPITULATION OF THE STUDIES OF TOXICITY

LD50	oral (rat and mouse)	: > 10 gr/kg bw	
	dermal (rat)	: > 2 gr/kg	(ref. 1 → 5)

ORAL TOXICITY

Thirteen-week study. Groups of 10 male and 10 female rats were given 0, 1000, 4000 and 16000 ppm in the diet. There was a dose related increase in the weights of the liver in female rats only ; there was no evidence of damage on histological examination, and clinical chemistry tests were normal. The authors therefore did not consider the liver findings to be significant, and the no-effect level was put at 16000 ppm, or about 1150 mg/kg bw/day (ref. 19).

CUTANEOUS AND MUCOUS TOLERANCE

DERMAL TOXICITY

Four standard Draize eye test are reported. Evaluation is uncertain in two of the experiments because of doubts about the concentrations use. Slight changes were found with a 10% solution in olive oil, with and without rinsing. Findings were normal after 48 hours.
A suspension of (probably) 50% in saline caused no abnormality.
A suspension, probably of 41%, caused slight changes only. In another study, up to 50% was used in olive oil ; no abnormalities were found.

Overall, the substance appears to be only slightly irritating, if at all (ref. 6->10).

Chick chorio-allantoic membrane

The probable concentrations of a.i. used were 0.64% and 6.4%. No abnormalities were found at either concentration (ref. 11).

Skin irritation

Two groups of 6 rabbits were used, one group with scarified skin and one group without. A 10% solution of a.i. in olive oil was applied for 24 hours with occlusion. There was definite erythema in 4/6 rabbits with sacrification, and slight erythema in 2/6 rabbits with intact skin. No abnormality was found after 7 days.

In another experiment, groups of 3 male and 3 female NZW rabbits were used. A 50% suspension of a.i. in physiological saline was applied to intact and scarified skin with occlusion for 24 hours. Vehicle controls were used. No abnormality was found.

In another experiment, groups of 6 males and 6 females were used 3 of each sex had scarification of the area of application. Concentrations of up to 2% of the a.i. were applied for 24 hours with occlusion ; the material was formulated in various o/w creams, as emulsions, and in the form of commercial preparations.

The commercial preparations had no adverse effects, but the concentrations of a.i. were only 0.9% and 1.8%. The emulsions and o/w preparations showed slight erythema and oedema in the first few days, but the maximum Draize score at any one reading was 2.

In another experiment, groups of 3 male and 3 female rabbits were tested with intact and scarified skin. The maximum concentration tested was 50% in olive oil. Applications were for 24 hours with occlusion. There was slight oedema for the first 2 or 3 days at 25% and 50% ; the highest score at any time was 2.

In another experiment, a 50% suspension in water was applied under semi-occlusive conditions for 24 hours to 3 animals. There was no evidence of irritation (ref. 9, 12, 13, 14, 16).

Guinea-pig

A commercial preparation containing 2% of a.i. was applied daily for 5 days. No abnormality was found (ref. 9).

Man

Fifty subjects were tested, 28 males and 32 females. Concentrations of 5% and 10% were applied as emulsions and as oily solutions for 24 hours with occlusion. There was one reaction to the 5% solution in oil. Otherwise no abnormality was found (ref. 15).

Sensitization

Guinea-pig

A commercial formulation containing 2% of a.i. was used. It was applied daily to the skin, 5 days a week for 3 weeks. After a 2 week rest, the same preparation was applied 3 times to a fresh site. No reaction was found (ref. 9).

A Magnusson-Kligman maximisation test was carried out in 40 animals, 20 test and 20 controls. The induction concentration of a.i. was 5% in olive oil intradermally, and 60% dermally, with occlusion for 48 hours. The challenge was made with 40% solution in olive oil. There were no significant differences between control and test groups (ref. 17).

Man

Sixty subjects were tested by applications of a commercial preparation containing 2% a.i. applied for 24 hours with occlusion. Of the 60 original subjects, 10 had the test material applied to the same sites 5 times at intervals of 48 hours with occlusion. This procedure was repeated after 10 days. No reaction was seen (ref. 9).

Photosensitization

Guinea-pig

Two groups of 9 animals were used. The concentration of a.i. was probably 0.5%. The positive control was 3,3',4,5-tetrachlorosalicyl-anilide. The applications were followed by 15 minutes of UV irradiation and the procedure was repeated 5 times. After a 10 day rest, 2 challenge applications were made, followed by irradiation as before. Later, 2 compounds were applied at a lower concentration and irradiation was again carried out. There was no evidence of photosensitization. The positive control, however, gave either very weak reactions or none ; and the maximum concentration of a.i. tested was probably only 0.5% (ref. 24).

Allergic effects

Man

A 1% solution of a.i. in olive oil was applied in a panel of 8 subjects known to be allergic to para-aminobenzoic acid derivatives. It is stated that no reaction was produced, but no details are given (ref.18).

Tolerance on repeated use

Man

In 45 subjects, of whom 14 had sensitive skin and allergic conditions a commercial formulation containing 2% a.i. was applied daily. During 3 weeks of exposure no adverse reaction was seen (ref. 9).

Percutaneous absorption

Man

A 0.5% solution of a.i. in ehtanol/hexane was applied to the forearms of 8 subjects. After 30 minutes the area was repeatedly stripped, 20 times in all. The concentration of a.i. in each stripping was estimated by HPLC.

Since the amount of percutaneous penetration can be estimated from the degree of penetration of a substance into the stratum corneum in the first 30 minutes after application, it should be possible to calculate the amount of percutaneous absorption of the a.i. from the figures given. However, there are various difficulties in doing this from the data given in the report. It is probable, however, that percutaneous absorption is likely to be considerable.
An investigation of the effect of vehicle on the permeation of a.i. into the stratum corneum, under occlusion, was made using a photo-acoustic method. It is difficult to evaluate this, since the technique is experimental, but possibly permeation is greater from a w/o emulsion than an o/w emulsion (ref. 20).

MUTAGENICITY

A standard Ames test was carried out. The maximum concentration used was 5000 microgram/plate (limit of solubility was 500 microgram). There was no increase in revertants, with or without activation (ref.21).

Mouse
Micronucleus test : The test dose was 2100 mg/kg bw. No adverse effect was found (ref. 22).

EMBRYOTOXICITY

Two series of Chick embryos were used, injections being made on day 1 and day 5 of incubation, respectively. The doses were lower on day 5 than on day 1. At the highest doses applied, nearly 50% of the embryos died, and the LD50 was about 45 mg and 25 mg respectively. Mortality in each dose group was dose related. There was a significant increase in the length of the metatarsus in the group given 10 mg on the fifth day. There were also some changes in blood chemistry values (ref. 23).

MISCELLANEOUS INFORMATIONS

Rat : doses up to 500 mg/kg bw by mouth had no effect on the blood pressure, nor on carrageenan-induced oedema of the paw (ref. 25-26).

Conclusion

CONCLUSION

The Committee felt that the concentrations of a.i. used in some of the tests for irritation of mucous membranes and for sensitization were low in relation to the proposed use level. On the whole, however, these tests were acceptable. Tests for acute oral and dermal toxicity were acceptable. Tests for mutagenicity were acceptable. The tests for phototoxicity and photosensitization were unsatisfactory.

Further testing for percutaneous penetration would be required ; if it is marked, consideration of the need for studies in pharmacokinetics and teratology/reproduction would be required.

The data provided were not such as to suggest a need for tests for inhalational toxicity or short term dermal toxicity.

Classification : B.

REFERENCES

1. Leuschner F. (1972) Laboratorium für Pharmakologie und Toxikologie, Hamburg

2. Leuschner F. (1973) Laboratorium für Pharmakologie und Toxikologie, Hamburg

3. Leuschner F. (1973) Laboratorium für Pharmakologie und Toxikologie, Hamburg

4. Hildebrandt & Kirsch (1987) BASF Aktiengesellschaft

5. Kirsch & Kiecza (1987) BASF Aktiengesellschaft

6. Leuschner F. (1972) Laboratorium für Pharmakologie und Toxikologie, Hamburg

7. Hoffberger (1972) - No institute name given

8. Leuschner F. (1975) Laboratorium für Pharmakologie und Toxikologie, Hamburg

9. Hopf G. (1973) Hamburg. Institute name not iven

10. Hildebrandt & Kirsch (1987) BASF Aktiengesellschaft. Ludwigshafen

11. Lüpke N.P. (1986) Institut für Pharmakologie und Toxicologie, Westfalische Wilhelms-Universität, Münster

12. Leuschner F. (1972) Laboratorium für Pharmakologie und Toxikologie, Hamburg

13. Leuschner F. (1973) Laboratorium für Pharmakologie und Toxikologie, Hamburg

14. Leuschner F. (1975) Laboratorium für Pharmakologie und Toxikologie, Hamburg

15. Tronnier J. (1985) Hautklinik der Stadtischen Kliniken, Dortmund

16. Hildebrandt and Kirsch (1987) BASF Aktiengesellschaft, Ludwigshafen

17. Gelbke H.-P. (1987) BASF Aktiengesellschaft, Ludwigshafen

18. Ippen H. (1987) Universitäts Hautklinik, Göttingen

19. Kirsch et al (1987) BASF Aktiengesellschaft, Ludwigshafen

20. Kolmel et al (1987) Universitäts Hautklinik, Göttingen

21. Gelbke H.-P. (1987) BASF Aktiengesellschaft, Ludwigshafen

22. Gelbke H.-P. (1987) BASF Aktiengesellschaft, Ludwigshafen

23. Kemper and Leupke (1985) Institüt für Pharmakologie und Toxikologie. Westfälische Wilhelms-Universität

24. Leuschner F. (1975) Laboratorium für Pharmakologie und Toxikologie, Hamburg

25. Behrmann (1973) Abt. Pharmakologie. Beiersdorf Aktiengesellschaft

26. Leuschner F. (1975) Laboratorium für Pharmakologie und Toxikologie, Hamburg

Evaluation of the safety of stabilizers for 1,1,1-trichloroethane as a propellant in aerosol cans

		Classification
1.	1,2-butylene oxide	D
2.	dimethoxymethane	C
3.	nitromethane	C
4.	nitroethane	A
5.	1-nitropropane,	-
6.	methoxypropane, 2 ol	A
7.	n-propanol	A
8.	isobutanol	D
9.	tertiary butyl alcohol	C
10.	tertiary pentyl alcohol	A
11.	2-methyl-3-butyn-2 ol	C

1. 1,2-butyleneoxide = 1,2-epoxybutane CAS Reg.No 106-88-7

1 ppm = 2.94 mg/m^3 1 mg/liter = 340 ppm

Maximal use level 1 % of the propellant gas

This substance is moderately anaesthetic and irritating to the skin and eyes, but skin irritation is not likely to follow vapor exposure alone.

The acute toxicity is moderate: the oral LD_{50} in rats was 500 mg/kg, an inhalation LC_{50} in rats was 4000 ppm.

In sub-acute inhalation studies in rats and mice, deleterious effects including mortality occurred with 1600 ppm, while 800 ppm was without effect. However, in reproduction studies in rabbits, mortality occurred with 250 and 1000 ppm.

In sub-chronic inhalation studies in rats and mice, 150 ppm was a no-toxic effect level (NEL), while 600 ppm caused growth retardation and changes of the nasal mucosa.

Mutagenic properties were established in the Ames test and in the mouse lymphoma test. The substance also induced chromosomal aberrations and sister chromatid exchanges in CHO-cells in vitro, and was positive in Drosophila (NTP TR329, 1988).

In long-term inhalation studies, the substance was found to be carcinogenic in rats, not in mice (NTP TR329, 1988).

Evaluation

In view of its mutagenic and carcinogenic properties, this substance should not be used as a stabiliser in aerosol cans.

Classification: D

October, 1988

2. Dimethoxymethane = dimethylformal = methylal CAS Reg.No 109-87-5

TLV: 1000 ppm or 3100 mg/m^3. STEL: 1250 ppm

Maximal use level: 2.0 % of the propellant

This substance is slightly anaesthetic and moderately irritating to the skin and to mucous membranes. No sensitisation properties have been observed.

The acute toxicity is low: the oral LD_{50} was 5.7 g/kg in rabbits. The inhalation LC_{50} in rats was 15.000 ppm, in mice and guinea pigs around 18.000 ppm.

Short-term inhalation exposure of mice and guinea pigs to 11.000 ppm during 7 h. on 15 days, induced irritation of mucous membranes, anaesthesia, fatty degeneration of the liver and kidneys, inflammation of the lungs, CNS-effects and high mortality.

The substance is well-absorbed through the skin, and metabolized to methanol, which can be detected in the blood.

Evaluation

No information is available on sub-chronic-, or chronic toxicity, or on possible effects on reproduction or teratogenicity. Results of mutagenicity tests are also lacking. An evaluation of this substance is therefore, not possible.

Classification: C.

October, 1988

3. Nitromethane

CAS Reg.No 75-52-5

TLV 100 ppm = 250 mg/m^3; STEL 150 ppm = 375 mg/m^3.

Maximal use level in the propelling agent is 1.0 %.

The acute toxicity is considerable: the oral LD_{50} in rats and mice was c. 950 mg/kg. Inhalation exposure for brief periods to levels above 10.000 ppm may induce symptoms of irritation, narcosis, weakness, and salivation. Microscopic changes were seen in the liver and kidneys.

In a 6-month inhalation study, rats and rabbits exposed to 98 or 745 ppm showed slightly depressed haemoglobin levels. The high dose induced growth depression and decreased haematocrit and haemoglobin levels in rats. In rabbits, the high dose induced decreased thyroxine levels in the blood, and increased thyroid weights, suggesting goitrogenic activity.

No mutagenic properties were observed in an Ames test, a Drosophila test and a micronucleus test.

The substance has been selected for extensive examination by the NTP. The studies will be initiated early in 1988.

Evaluation

On the basis of available toxicity data, evaluation is not possible.

Classification: C.

October, 1988.

4. Nitroethane CAS Reg.No 79-24-3

TLV 100 ppm = 310 mg/m^3; STEL 150 ppm = 465 mg/m^3

Maximal use level in the propellant is 1.0 %

The acute toxicity is considerable: the oral LD_{50} in rats was 1100 mg/kg and in mice 860 mg/kg. The LC_{50} in rabbits was between 1000 and 5000 ppm but in guinea pigs c. 10.000 ppm.

This substance is more irritating to the respiratory tract than is nitromethane. Inhaled nitroethane results in increased blood levels of nitrate. It is rapidly metabolized and is completely eliminated from the body in 30 h., mainly via the lungs.

An Ames test with and without metabolic activation was positive especially with the strain TA 100. A micronucleus test in mice was negative.

In a chronic inhalation study, rats were exposed to concentrations of 100 or 200 ppm, 7 hr/day, 5 days/week for 2-yr. There were no effects of exposure on survival,haematology, clinical chemistry, organ weights, or on non-neoplastic or neoplastic pathology. Body weights were slightly lower in exposed rats. (Griffin et al., Ecotoxicol. Environ. Saf. 16(1)11-24(1988)).

In a project on photochemical smog, a number of inhalation studies has been conducted with a mixture of diethylhydroxyl-amine (DEHA 10-20 ppm), nitroethane (c.10 ppm) and diethylamine hydrogen sulfite (at its vapour pressure). A chronic (2.5-yr) study in rats did not reveal changes of any significance (Heicklen et al., Environm. Res. 26 (1981) 258-273).
Chronic (2-yr) inhalation exposure of mice to the mixture, resulted in an increased tumour incidence of marginal statistical significance in male mice, whereas in females the total incidence of tumours was significantly decreased (Heicklen et al., Environm. Res. 27 (1982) 277-289).
A reproduction study in mice exposed to the mixture over 3 successive generations did not reveal any deleterious effect on survival, numbers of litters and pups borne per litter or on the microscopy of the organs examined (Heicklen et al., Environm. Res. 20 (1979) 450-454).

A teratology study on mice indicated that the exposure to the mixture (DEHA 9 ppm, nitroethane 14 ppm, diethylamine HS at vapour pressure) produced no effect on the dams, and there was no evidence of treatment-related terata, embryotoxicity, or inhibition of fetal growth and development (Beliles et al., Environm. Res. 17 (1978) 165-176).

Evaluation

A concentration of 1 % stabiliser in the propellant gas of aerosol cans may result in a daily human exposure to 105 mg, if the can contents contain 35 % propellant, and the consumer uses up to 30 g can contents per day. If it is assumed that the amounts of stabiliser used, is present in 1 m^3 of air surrounding the consumer, the concentration is 34 ppm, which is considerably below the TLV. Because long-term exposure of rats and mice to c.10 ppm did not result in overt toxicity it is unlikely that daily exposure of humans to the can contents for a few minutes would be harmful.

Classification: A.

October, 1988.

5. <u>1-nitropropane</u> CAS Reg.No 108-03-2

Use level 1 % in propellant gas

TLV 25 ppm = 90 mg/m^3 STEL (35 ppm) = (135 mg/m^3)

This substance is more toxic upon acute exposure than NE and NM. The oral LD_{50} varied with the species between 250 and 800 mg/kg. The inhalation LC_{50} for rats was 3100 ppm/8 h. Exposures at 5000 ppm for 3 h. killed rabbits and guinea pigs.

Irritating properties are noticed by humans at concentrations above 100 ppm after brief periods of exposure. Headache and nausea are noted upon prolonged exposure to excessive levels.

One out of two Ames tests conducted was negative, the other was marginally positive. A micronucleus test in mice with up to 300 mg 1-nitropropane/kg bw was negative, as was 2-nitropropane in the same test. (Kliesch and Adler, Mutation Research 192 (1987) 181-184).

In a long-term (21 1/2 months) inhalation study in rats, daily exposure to 100 ppm 7 hr/day, did not induce changes in appearance or behaviour, growth rate, organ weights, serum chemistry or haematology, or in the gross or microscopic appearance of the organs. If rats were exposed to 100 ppm 2-nitropropane (CAS Reg.No 79-46-9) instead of 1-nitropropane, an increased incidence of hepatocarcinomas was observed (Griffin et al. Toxicol. Environm. Safety 6 (1982) 268-282).

Evaluation

A use level of 1 % in the propellant may result in a daily human exposure to 105 mg stabilizer. If this amount is present in 1 m^3 of air around the user, it would be an exposure concentration of 29 ppm, which is close to the TLV. Daily exposure to such a level for only a few minutes is not considered to represent a significant health hazard.
Although the information on mutagenicity is incomplete, the absence of carcinogenic potency in the chronic study may justify acceptance of this compound. However, because no specification was available, and no information on possible contamination of 1-nitropropane by the more toxic coumpound 2-nitropropane, classification was postponed.

Classification: No classification.

October, 1988.

6. Methoxypropane, 2 ol = propyleneglycol monomethylether CAS Reg.No 107-98-2

TLV 100 ppm or 360 mg/m^3 STEL 150 ppm
(1 ppm = 3.68 mg/m^3; 1 mg/l = 272 ppm)

Oral LD_{50} values (in mg/kg bw) for rats vary between 5.2 and 11.9, the value for mice is 10.8, for dogs 9.2, for rabbits 5.3. The LC_{50} for rats and guinea pigs was 10.000 ppm during respectively 5-6 and > 7 h. exposure. The dermal LD_{50} in rabbits was 13-14 g/kg.

The substance causes only very mild skin irritation after prolonged contact. Eye irritation is mild to moderate.

Dermal absorption is rather high (1.17 $mg/cm^2/h$) and dermal application of 7 or 10 ml/kg for 90 days has caused narcosis and mortality in rabbits, and mild narcosis also with 2 and 4 ml/kg. Increased kidney weight was the only organ change observed with 10 ml/kg.

In sub-chronic oral studies in rats and dogs at dose levels of 0.5 up to 4.0 ml/kg b.w./day, depression of the CNS was observed in a dose-related manner. In rats, liver damage occurred with 2.0 and 4.0 ml/kg. Minor kidney damage was seen in both the rats and the dogs at the higher doses (Rowe and Wolf. In: Pathy 1981, pp 3977-3981).

Sub-chronic inhalation studies in rabbits and monkeys with 800 ppm (2.910 mg/m^3) did not induce adverse effects. In rats and guinea pigs no ill-effects were seen with 1500 ppm (5460 mg/m^3). At higher levels there was slight growth depression, slight liver and lung changes and mild CNS-depression.

Teratogenicity studies in mice, rats and rabbits with 0.04 up to 2.00 ml/kg b.w. by gavage or by s.c. injection revealed only minor foetotoxic effects. Inhalation teratogenicity studies in rats and rabbits with 500, 1500 and 3000 ppm induced growth retardation and CNS depression, but no statistically significant increases in malformations. Both 500 and 1500 ppm were no-effect levels.
A reproduction study over 2 generations is being conducted by NTP.
No indications of mutagenic properties were observed in the Ames test or in a chromosomal aberration test in CHO cells. There was no induction of unscheduled DNA synthesis in rat hepatocytes in vitro.

Evaluation

The substance is low in both single- and repeated-dose oral, and inhalation toxicity, but can be absorbed through the skin in toxic amounts if exposure is extensive and prolonged. Because repeated exposures are very disagreeable, as a result of their irritant action on the eyes and mucous membranes, the hazard from inhalation is negligible.

Classification: A.

October, 1988.

7. <u>n-propanol = 1-propanol = propylalcohol</u> CAS Reg.No 71-23-8

TLV 200 ppm = 500 mg/m^3 (skin) STEL 250 ppm = 625 mg/m^3

Use level in propellant 2 %

Oral LD_{50} values for rats vary from 1.87 to 6.5 g/kg. The s.c. and i.v. LD_{50} for mice was 3.2 and 1.1 g/kg respectively. A 4-h. LC_{50} for rats was c. 4000 ppm. Signs of intoxications are ataxia, paralysis, hypothermia, dyspnoea and narcosis.

Irritation to the skin is moderate, but contact with the eye causes moderate to severe irritation.

Sub-acute (4-day) oral administration of 2160 mg/kg/day to 6 rats did not induce mortality or gross liver lesions. Liver damage occurred however, in long-term oral rat studies (Rowe and Mc Collister 1981. In: Patty pp 4556-4561). The substance is well-absorbed through the skin, the intestinal wall and the lungs.

An Ames test was negative. A test in E.coli was positive. Tests for gene mutations, SCE's and micronuclei in mammalian cells in vitro were negative. (Obe and Ristow, Mut. Res. 56 (1977) 211-213; Lasne et al., Mut. Res. 130 (1984) 273-282; von der Hude et al. Env. Mutagenesis 9 (1987) 401-410; WHO Food Add. Series (1980) no. 16).

In long-term studies with limited numbers of rats treated either orally or by subcutaneous injection, an increased incidence of tumours was reported. A long-term dermal study in mice, with applications of 40 mg/mouse (or 800 mg/kg b.w.), 3 x/week, was negative (NTP-study).

Inhalation of 10.000, 7.000 or 3.500 ppm by groups of 15 pregnant rats on days 1-19 of pregnancy, for 7 hr/day, caused lowered food intake and body weights of the dams, and implantation losses, retardation of foetal growth and increased incidences of external - and skeletal malformations in the two high dose groups. Visceral malformations were increased in the highest dose group. The low dose group only showed slightly reduced food intake of the dams. Thus n-propanol was teratogenic at concentrations that also produced maternal toxicity (Nelson et al., Fd Chem. Toxicol. 26 (1988) 247-254).

Evaluation

The substance is low in acute oral and inhalation toxicity, but markedly injurious and irritating to the eye of rabbits. Inhalation of toxic levels was teratogenic in rats, but 3500 ppm was not toxic and not teratogenic.
A use level of 2 % in propellant gas may result in a daily human exposure to 210 mg stabilizer. If this amount is present in 1 m^3, the exposure level in 84 ppm, which is less than half the TLV. Daily exposure to this level for only a few minutes is not considered a health hazard.

Classification: A.

October, 1988.

8. Isobutanol = 2-methyl-1-propanol = isobutyl alcohol CAS Reg.No 78-83-1

TLV 50 ppm = 150 mg/m^3 STEL (75 ppm) = (225 mg/m^3)

The oral LD_{50} for the rat was 2.46 g/kg and for the rabbit 3.4 g/kg; the dermal LD_{50} for the rabbit was 4.24 g/kg. In inhalation tests, rats survived 16.000 ppm for 2 h. suggesting a 4 h. LC_{50} of around 10.000 ppm. Isobutanol acts primarily as a narcotic.

The substance is not considerably irritating to the skin of rabbits, but, may cause severe eye irritation.

Short-term oral exposure of rats to 7.4 % in drinking water for 4 months induced an enlarged stomach filled with food and gas, and signs of constipation and bleedings of the small intestine. Increasing the concentration to 15 % for 2 months induced decreased liver size and decreased contents of fat, glycogen and RNA of the liver (WHO/IPCS Environmental Health Criteria for Isobutanol, 1985).

A long-term oral rat study with a total dose of 29 ml/rat (0.2 ml/rat twice weekly during 495 days) and a long-term subcutaneous rat study with a total dose of 9 ml (0.05 ml/kg, twice weekly for 544 days) were reported to have shown an increased incidence of malignant tumours, viz. in 3 out of 19 test rats in the oral study, and 8 out of 24 test rats in the subcutaneous study. Control rats had no malignant tumours. In the oral study one test rat had a forestomach carcinoma, and a liver cell carcinoma, another had a forestomach carcinoma and myelogenous leukaemia, and the third a myelogenous leukemia. In the subcutaneous study the tumours comprised two forestomach carcinomas, two liver sarcomas, one spleen sarcoma, one mesothelioma and two retroperitoneal sarcomas. A total of three benign tumours was recorded (Patty, p. 4580).

A mutagenicity test in E.coli CA274 was positive.

Humans exposed to low concentrations, show irritation of the respiratory tract and the eyes. Chronic inhalation of 3 mg/m^3 and 0.5 mg/m^3 affects the CNS. Oral ingestion of alcoholic drinks induce behavioural and subjective changes (WHO/IPCS Environmental Health Criteria for Isobutanol, 1985).

Evaluation

Animal studies have shown the substance to be low in acute oral toxicity, appreciably irritating to the eyes, not appreciably irritating to the skin, and low in acute inhalation toxicity. Systemically, isobutylalcohol has a narcotic action. Long-term studies with small numbers of rats suggest a carcinogenic potential. Since, moreoveer, a mutagenicity test was positive, and very low levels in the air have been reported to affect the NS of humans, it was not considered justified to accept this substance as a stabiliser in pressure cans without further studies.

Proposed classification: D.

November, 1988.

9. Tertiary butylalcohol = 2-methyl-2-propanol = trimethyl carbinol

CAS Reg.No 75-65-0

Use level 3 %

TLV 100 ppm = 300 mg/m^3 STEL 150 ppm = 450 mg/m^3
1 mg/l = 330 ppm 1 ppm = 3.03 mg/m^3

The oral LD_{50} in the rat was 3.5 g/kg, in the rabbit 3.6 g/kg. The i.v. LD_{50} for the mouse was 1.5 g/kg. The primary effect of the substance is that of a narcotic. The oral narcotic dose (ND_{50}) for rabbits was 1.4 g/kg.

The substance was not irritating to the skin of rabbits. When applied to human skin, only slight erythema and hyperemia were observed. Excessive exposure of humans to vapour may cause irritation of the airways, and headache, nausea, fatigue and dizziness.

Information on systemic toxicity is not available.

No mutagenic properties were detected in a test with Neurospora crassa.

Oral studies in rats and mice with 0.5, 1.0 and 2.0 % in drinking water are in progress in the framework of NTP. The program also includes inhalation studies.

Evaluation

The absence of information on systemic toxicity and the nearly complete absence of mutagenicity data, makes an evaluation impossible. It is suggested to await the results of ongoing studies by NTP.

Classification: C.

March, 1988.

10. <u>Tertiary pentyl alcohol</u> = tertiary amyl alcohol = 2-methyl-2-butanol = tertiary pentanol CAS Reg.No 75-85-4

Use level 2 %

1 mg/l = 278 ppm   1 ppm = 3.60 mg/m^3

Oral LD_{50} values are 1.0 g/kg and 1-2 g/kg for rats, 2.0 g/kg for rabbits. The dermal LD_{50} for rabbits was 1.7 g/kg. The LC_{50} for rats (exposed 6 h.) was between 3000 and 5700 ppm; with 3000 ppm all rats became unconcious but recovered and survived; with 1100 ppm only slight motor incoordination was observed. The substance exerts a narcotic action. The narcotic dose 50 (ND_{50}) in rabbits was 0.7 g/kg.

Topical, non-occluded application to the rabbit skin did not induce irritation, but severe narcosis occurred when applied under occlusion.

Marked eye irritation occurred in rabbits and was accompanied by signs of pain, swelling of the lids, corneal cloudiness and iritis. Healing was complete after 14 days.

A sensitization test in guinea pigs by a modified Maguire method was negative, whereas all 10 animals of a positive control group were sensitized.

Short-term (4-wk) dermal toxicity in rabbits was examined at dose levels of 0, 344 and 3440 mg/kg/day, 5 days a week. No adverse effects occurred in the low-dose group, except the expected local skin changes. The high-dose caused more extensive skin injury and serious systemic effect. Three of each sex became comatose and died or were sacrificed when moribund. Gross and microscopic examination did not reveal morphologic lesions. Decreases in body size, adipose reserves, and glycogen of hepatocytes were the only changes detected.
Short-term (7-day) inhalation exposure of rats to 0, 150, 500 or 1500 ppm, 6 h./day, for 7 consecutive days revealed motor incoordination and lethargy with 1500 ppm. Liver and kidney weights were increased.

Sub-chronic (90-day) inhalation exposure of rats, mice and dogs, to 0, 50, 225 or 1000 ppm, 6 h./day, 5 days/wk, caused transient motor incoördination in dogs and rats of the high dose group, and an increased alkaline phosphatase activity in top-dose dogs. No mortality occurred. Extensive microscopic examination did not reveal treatment-related changes (Rowe and Mc Collister (1981). In: Patty, pp 4602-4608).

The metabolism of tert. amylalcohol is very slow. An appreciable amount of absorbed material is excreted unchanged in the expired air.

Mutagenicity tests with Salmonella typhimurium (three strains) and with Saccharomyces cerevisiae were negative.

Evaluation

The substance is of moderate acute oral toxicity, is severely irritating to the eye and moderately irritating to the skin. Toxic amounts may pass the skin. Repeated inhalation of 100 ppm, induced relatively minor effects in the dog, lesser effects in the rat, and none in the mouse. No systemic effects were established. The substance is very slowly metabolized and largely excreted in the expired air. It was not found mutagenic in tests with microorganisms. Other mutagenicity tests or a long-term study are not available.
A 2 % concentration of the stabilizer in the propellant gas of aerosol cans and the use of 30 g can contents/day results in a potential daily exposure to 210 mg. If this amount is present in 1 m^3 of air the concentration is 58 ppm. In view of the available data, this level of exposure is not likely to present a health hazard.

Classification: A.

October, 1988.

11. <u>3-Butyn-2 ol, 2 methyl</u> =dimethylacetyl carbinol=2-hydroxy-2-methyl-3-butyn
CAS Reg.No 115-19-5

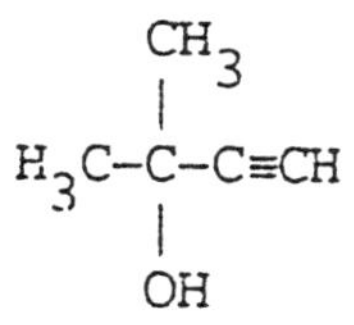

Used at a concentration of 2 % as stabilizer in TCE.

LD_{50} values in mice are 1.8 g/kg oral, 2.34 g/kg s.c., and 3.6 g/kg i.p.

No sensitization occurred when the substance was tested in guinea pigs by the Landsteiner/Draize method.

Because no further information is available an evaluation is not possible.

Classification: C.

March, 1988.

<u>REFERENCES</u>
- H.G. verscuuren, C.G. De Rooij "Health Risk assessment of environmental exposure to 1,1,1-TCE" Regulatory Toxicology & Pharmacology 11, 90-99 1990
- Abstracts on developmental toxicity of 1,1,1-trichloroethane in CD rats. Studies done at Research Triangle Institute NIEHS/NTP Sponsored Chemical Mixture Projects/Studies Annual Plan for the fiscal year 1989
- Federal Register Volume 54 N°163 Aug. 23, 1989

CONTENTS

40th meeting of SCC - January 16, 1989

REPORTS BY THE SCIENTIFIC COMMITTEE ON COSMETOLOGY

40th REUNION

JANUARY 16th, 1989

LIST OF PARTICIPANTS

Present

Mr. DE GROOT
Mrs. DONY
Mr. FIELDER
Mr. GOULDING
Mrs. KNAAP
Mr. LOPRIENO
Mr. MUSCARDIN
Mr. O'MAHONY
Mr. PONS-GIMIER
Mr. RAMOS-MORGADO
Mr. SCHOU
Mr. STUTTGEN
Mr. SOMOGYI

Not Present

Mr. AGACHE
Mrs. ENJOLRAS
Mr. KAPOULAS

COMMISSION

Mrs. MASSE
Mr. GONTIER (Serv. "Politique des Consommateurs")
Mr. COLLIN (expert)

REPORTS BY THE SCIENTIFIC COMMITTEE ON COSMETOLOGY CONCERNING THE USE OF TWO PRESERVATIVE AGENTS

. THE COMMITTEE'S MANDATE

To give its opinion on the question whether the use of :

- HEXAMIDINE at levels of 0.1% and 0.3%
- BENZETHONIUM CHLORIDE at the level of 0.1%

in cosmetic products is admissible from the health point of view.

1. HEXAMIDINE SCC/204/89

. FORMULA

EEC n° 20
Colipa P8

1,6-di(4-amidino phenoxy)-n-hexane and its salts including di-isethionate and di(p-hydroxybenzoate)

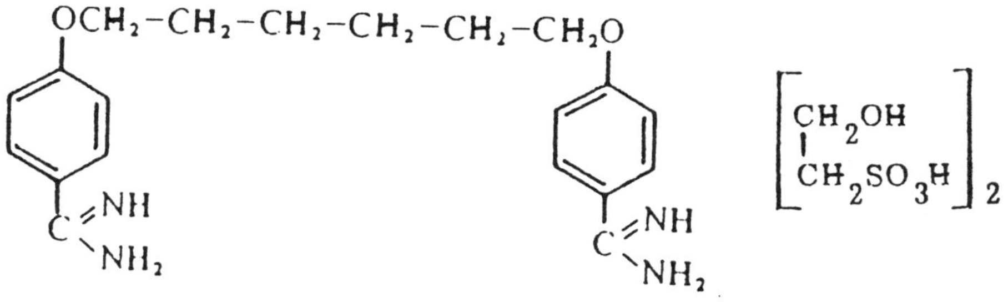

$C_{20}H_{26}N_4O_2$

. CHARACTERISTICS

Soluble in water
Insoluble in organic solvent

. USE

Hexamidine is used in cosmetics as a preservative at a maximum dose level of 0.1%, and for uses at concentrations up to 0.3% in non-rinsed skin products.

. RECAPITULATION OF THE STUDIES OF TOXICITY

The acute toxicity of hexamindine is considerable.

LD50	oral	(rat)	:	750 mg/kg bw
		(mouse)	:	710 - 2500 mg/kg
		(rabbit)	:	500 mg/kg
	i.p.	(rat)	:	57 mg/kg
		(mouse)	:	17 - 51 mg/kg
	i.v.	(rat)	:	8 mg/kg
		(mouse)	:	17 mg/kg
	dermal	(rat)	:	> 4000 mg/kg

ORAL TOXICITY

A recent short-term (4-wk) oral study was conducted by gavage administration of 50, 100 and 200 mg/kg bw/day to groups of 5 rats/sex. All test animals showed post-treatment symptoms (salivation, wet fur, brown oral staining). The top-dose rats also showed abnormal position and locomotion, and increased counts of white blood cells and lymphocytes. In the two higher dose groups there were increases in the values for GPT, GOT and calcium in blood plasma. All treated rats showed caecal enlargement. The lungs, heart, liver, kidneys and caecum did not reveal treatment-related microscopical changes. Other organs (including spleen and adrenals) were not examined.

The clinical signs and the caecum enlargement were not considered to be of toxicological significance. The no-toxic effect level was established at 50 mg/kg, but the study showed several deficiencies (Colipa subm. III, ref. 10).

In a 90-day oral study in male rats, daily doses of 400 and 800 mg/kg by gavage induced mortality, growth depression, signs of anaemia, increased liver weight and decreased liver and kidney function. The lower dose of 200 mg/kg was not a clear no-effect level.

CUTANEOUS AND MUCOUS TOLERANCE

DERMAL TOXICITY

A concentration of 0.1% was slightly irritating to the skin and eye of the rabbit.

Hexamidine did not produce sensitization in guinea pigs and was not photosensitizing.

A subacute (28-day) dermal toxicity study in rabbits showed that solutions of up to 2% were only slightly irritant. Daily application of 4 ml/kg bw of a 0.05, 0.1 and 2.0% solution revealed no systemic toxicity.

A 90-day dermal study in rabbits with the very low dose level of 16 mg/kg bw revealed no systemic toxicity.

MUTAGENICITY

An Ames test with 4 S. typhimurium strains (TA 1535, 1537, 98 and 100) using concentrations up to 500 µg/plate was negative (ref. 11).

No evidence of clastogenic activity was observed in an *in vitro* test for chromosomal aberrations in CHO cells exposed up to 34 µg/ml in the absence of metabolic activation and 420 µg/ml in its presence. The negative result was, however, not convincing, because the effect found at the low dose cannot be disregarded (ref. 12).

CONCLUSION

Information on dermal absorption is not yet available, but a dermal absorption study is ongoing and will be finished by the end of this year (1989).

Classification : B.

REFERENCES

1. to 8. Manufactured data.

9. Toxicité à tèrme chez le Rat, d'une suspension d'Hexamidine diisothio nate, per os (14.05.1974) Communication personnelle de Mesdames SCHMUTS et GHAYE et de Messieurs J. GAILLOT et A. BIDER ,des Laboratoires de Recherche de la Société THERAPLIX réf. LP. LRB N°9415

10. MX 202 Toxicity to Rats by repeated oral administration for 4 weeks. Huntingdon Research Center.14.07.1987.

11. Ames metabolic activation test to assess the potential mutagenic effect of MX 202. Huntingdon Research Centre, January 12, 1987.

12. Analysis of metaphase chromosomes obtained from CHO celles cultured in vitro and treated with MX 202. Huntingdon Research Centre, April 10, 1987.

2. BENZETHONIUM CHLORIDE

COMPLEMENTARY OPINION (April 11, 1989)

PRECEDING OPINION EXPRESSED ON APRIL 12, 1988

CONSULT THE REPORT OF THE 37th REUNION

. OTHER NAMES

EEC n° 15
Colipa P70
Hyamine 1622
Phemerol chloride

4'-(1,1,3,3-tetramethylbutyl)phenoxy-ethoxyethylene-dimethyl-benzyl-ammonium chloride

. USE

As a preservative agent in cosmetics at the level of 0.1%.

. PRECEDING CONCLUSIONS (11-4-88)

Classification C.
Adequate information on dermal absorption was needed.

. FURTHERING INFORMATIONS

A supplementary 28-day study in rats with the same feeding levels was conducted to verify and extend certain findings in the previous study. The results confirmed most of the changes seen at the top-dose, including caecal enlargement.

The latter finding was not accompanied by histopathological changes. Decreased levels of serum-P seen at the two higher levels in the previous study did not occur in the present study. Therefore, 500 ppm was the NEL in the supplementary study (Subm. IV, ref. 15).

CONCLUSION

No modification :

The available data from teratogenicity studies in rats indicate that benzethonium chloride produces both maternal toxicity and adverse effects on the developing embryo at oral doses of 3 mg/kg and above.
The NEL was around 1 mg/kg. The safety factor resulting from cosmetic use and assuming a NEL of 1 mg/kg and 10% dermal absorption is very low (around 20). Although the actual extent of absorption through intact skin is likely to be low, it is essential that vertual absence of dermal absorption is adequately demonstrated.

Classification : C.

Reference 1 : S. DE FLORA
"Study of 106 Organic and Inorganic Compounds in the Salmonella/microsome Test"
Carcinogenesis, Vol. 2, n° 4, (1981), pp. 283-298

Reference 2 : Isao KARUBE et al
"Preliminary Screening of Mutagens with a Microbial Sensor"
Anal. Chem., 53, (1981), pp. 1024-1026

Reference 3 : Rohm and Haas
Report n° 82R-265, 16/12/1982

Reference 4 : Hazleton laboratories America, Inc,
November 16, 1977

Reference 5 : Report from July 16, 1976
J.C. KILLEEN, W.E. RINEHART
Bio/dynamics - New Jersey

Reference 6 : Report from May 5, 1976
J.C. KILLEEN, W.E. RINEHART
Bio/dynamics - New Jersey

Reference 7 : Report from December 2, 1976
G.K. HOGAN, W.E. RINEHART
Bio/dynamics - New Jersey

Reference 8 : Report from July 16, 1976
J.C. KILLEEN, W.E. RINEHART
Bio/dynamics - New Jersey

Reference 9 : Report from April 22, 1976
J.C. KILLEEN, W.E. RINEHART
Bio/dynamics - New Jersey

Reference 10 : CIR Final Report - June 5 1984

Reference 11 : Reference : 32 :
Expert opinion on :
Studies of Benzethonium Chloride in pregnant rats and rabbits
by E. MARSHALL JOHNSON. July 17, 1985

Reference 12 : Dermal absorption study on 14C Hyamine 1622, 1974

Reference 13 : Rat Maternal and Fetal Absorption of 14C-BTC - February 23, 1977

Reference 14 : CIVO INSTITUTES TNO, ZEIST, NETHERLANDS
Report n° V 86.537/260702
Study director : Dr. H.P. TIL
October 1986

Reference 15 : "Addendum to Report n° 86.537/260702" Subacute (28-day) oral toxicity study with Benzethonium Chloride in rats"
Dr. H.P. TIL et al. TNO-CIVO institutes Zeist - the Netherlands - May 1988

REPORTS OF THE SCIENTIFIC COMMITTEE ON COSMETOLOGY CONCERNING THE USE OF CERTAIN COLOURING AGENTS IN COSMETICS

. THE COMMITTEE'S MANDATE

To give its opinion on the question whether the use of the following colouring agents in cosmetic products is admissible from the health point of view.

CI 13065
CI 42045
CI 44045
CI 42535
Solvent Yellow 98

1. CI 13065

COMPLEMENTARY OPINION

For preceding informations and opinions, consult :

- Reports of SCC, seventh series, EUR 11303 P.23-25
- Report of the 35th reunion October 13th, 1987.

. PRECEDING CONCLUSION : Classification : D.

. USE

Up to 0.5% in now rinsed off products.
Up to 0.1% in rinsed off products.

. FURTHERING INFORMATIONS

In a third short-term (31-day) oral study, conducted in 1988, groups of 10 rats/sex were treated by gavage with 0, 10, 30, 100 or 300 mg/kg/day (in a hydrogel of 2% polysorbate 80, sterilized water and 0.01% activated dimeticone as an antifoam). The top-dose induced decreased growth rate and food intake, ocular changes, changes in haematology, blood biochemistry, urinary ocnstituents, and changes in the weight of several organs. Liver enlargement occurred with 300 and 100 mg/kg. Ocular changes (haemorrhages, cataract and/or opacity) were seen with 300, 100 and 30 mg/kg. Histopathological changes were seen in the liver and bone marrow. The NEL was 10 mg/kg bw/day (ref. 18, subm. VI).

Skin absorption test

An in vitro test with human abdominal epidermis, in which 35 mg of 0.05% colourant in a hair dye formulation was applied to 2.0 cm^2 for 30 min., only 0.045% of the applied dose was recovered in the penetration chamber (ref. 19, subm. VI).

CONCLUSION

The substance has shown considerable systemic toxicity and was found to induce several changes, e.g. testicular and ocular damage. Different oral studies in rats showed marked differences in the toxic effects observed. Chromosomal aberrations and sister chromatid exchanges were reported both from in vivo, and from in vitro studies. In a negative chromosomal aberration test reported recently only very low concentrations (1.6 ug/ml) could be used. In view of the high toxicity of this colourant, included the ability to induce catarract at 30 mg/kg bw within 30 days, and testicular toxicity, and the in vivo genotoxic properties, its continued use as a colourant in cosmetics is not justified.

Classification : D.

Reference 1 : Dr. Leuschner, Hamburg
"Akute orale Toxizität von C-ext Gelb 10 an Sprague-Dawley-Ratten"
Lab. für Pharmakologie W. Toxikologie
Report dated Gebruary 15, 1971

Reference 2 : G. Tonfoy 1978
"Sel de sodium de l'Acide (Anilino-4 Pheylazo) -1 Benzene
Sulfonique-3, détermination de la DL 50 par voie orale chez le rat"

Laboratory of "Institut Français de Recherches et Essais Biologiques" (IFREB) St-Germain s/l'Arbresle
Report IFREB - R 801261 of 25 January 1978

Reference 3 : Dr. Leuschner, Hamburg
"Uber die schleimhautverträglichkeit von C-ext Gelb 10 bein Kaninchen"
Lab. für Pharmakologie u. Toxikologie
Report dated February 10 1971

Reference 4 : J.P. Guillot 1978
"COLORANT N° 63
Sel de sodium de l'Acide (Anilino-4 Phenylazo) -1 Benzene Sulfonique
(CI 13065 Acid Yellow 36)
Numéro du lot : AJ 7436 - 15.9.77"

Laboratory of "Institut Français de recherches et essais Biologiques (IFREB) St-Germain s/l'Arbresle
Report IFREB - R 801239 of January 1978

Reference 5 : Dr. Leuschner, Hamburg
"Prüfung der lokalen Verträglichkeit von C-ext Gelb 10 an Kaninchen (Patch-Test)"
Lab. für Pharmakologie u. Toxikologie
Report dated December 20 1970

Reference 6 : J.P. Guillot 1978
"COLORANT N° 63
Sel de sodium de l'Acide (Anilino-4 Phenylazo) -1 Benzene Sulfonique-3
(CI 13065 - Acid Yellow 36)
Numéro du Lot : AJ 7436 - 15.9.77"

Laboratories of "Institut Français de recherches et essais Biologiques (IFREB) St-Germain s/l'Arbresle
Report IFREB - R 802265 of 2 March 1978

Reference 7 : Testicular lesions induced by metanil yellow in mice
GB SINGH and SK KHANNA
Exp. Path. Bd. 9, S. 251-255-1974

Reference 8 : Haematological studies in rats fed with Metanil Yellow
NK. Mehrotra, SK Khanna cend 6 B Singh
Environ. Physiol. Biochem. 1974 4, 232-235

Reference 9 : S. Ivancovic, R. Preussmann
"90 Tage Test"
Deutsches Krebforschungszentrum, Heidelberg
Report dated November 4 1973

Reference 10 : Mutagenicity test of dyes used in cosmetics with the Salmonella / Mammalian-microsome
Test by Jeanne M. Muzzall and Warren L. Cook
Department of Biology, Georgia State University, Atlanta GA 30303 USA
(Accepted 11 December 1978)

Reference 11 : C. Gloxhuber, F. Bartnik, F. Wingen
C-ext Gelb CI 13065 : Resorptionsscreening an Humanhaut in vivo
Henkel KGaA, 12.5.1980

Reference 12 : Prof. Dr. Gloxhuber, Dr. Bartnik
C-ext Gelb 10, CI 13065
Resorptionscreening mittels der TesafilmabriB-Methode nach Applikation einer CI 13065 enthaltenden Basiscreme an Humanhaut in vivo
Henkel KGaA, 11.1.1983

Reference 13 : Khanna, S.K., Singh, G.B.
"Anti-testicular effect of Metanil Yellow in Guinea pigs"
J. Food Sci. Techn. 1973, 10, 75-76

Reference 14 : Jolly, P.N., Michael, W.R.
20 day Percutaneous study
Procter & Gamble Internal. Report V-1439-20, August 2 1965

Note : The report on this study was stored on micro-fiche. The resulting photocopy is of poor quality. It has nevertheless been submitted but in addition the relevant pages have been retyped.

Reference 15 : Jolly, P.N., Michael, W.R.
90 day Percutaneous study
Procter & Gamble Internal. Report V-1439-1
August 8, 1965

Reference 16 : Niewenhuis, R.J., Jolly, P.N.
2 year Mouse Painting study
Procter & Gamble Internal Report V-1606-1
December 15 1964

Reference 17 : M. McConville, D.B. McGregor
Inveresk Research International Report to Procter & Gamble Ltd. March 1979

Reference 18 : C. Bezancon et al 1988
"Evaluation de la toxicité du produit "jaune Acetacid 4R Extra" administré chez le rat pendant 1 mois par voie orale"
Laboratoires d'Etudes et de recherches SYNTHELABO (LERS) - Gargenville - FRANCE
Report 1352 TMR/88.047 - 29 August 1988

Reference 19 : H. Schaefer 1988
"Pénétration in vitro du Colorant CI 13065"
Centre International de recherches dermatologiques (CIRD) - Valbonne - FRANCE
Report 28 August 1988

Reference 20 : A.T. Natarajan & F.H. Sobels 1988
"Evaluation of compound CI 13065 in the chromosome aberration test with chinese hamster ovary cells (in-vitro)"
Department of radiation genetics and chemical Mutagenesis
University of Leiden, THE NETHERLANDS
Report 13 June 1988

REPORT BY THE SCIENTIFIC COMMITTEE ON COSMETOLOGY

CONCERNING THE USE OF

THE COLOURANT CI 42045

COMPLEMENTARY OPINION EXPRESSED ON JUNE, 20th 1989.

For preceding informations and expressed opinion, consult :

- Reports of SCC, seventh series, EUR 11303 P79.

. FORMULA AND SYNONYMES

CAS Reg. n° 129-17-9
Acid Blue 1
C-ext Blau 13
Blue VRS

4,4'-bis (N-diethylamino)-2",4"-disulfotriphenyl methane, sodium salt

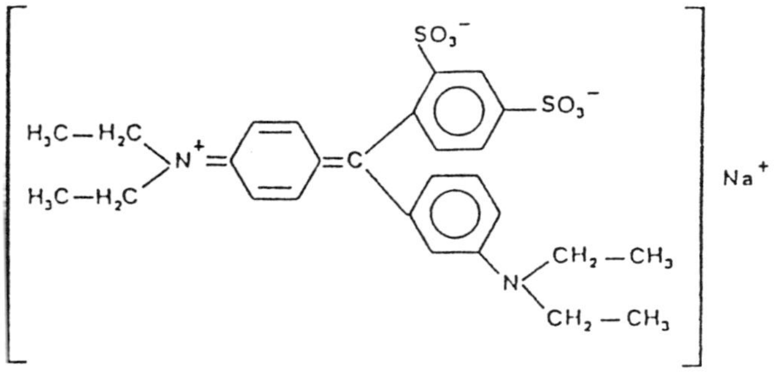

$C_{27}H_{32}N_2O_6S_2$,Na
MW 567

. USE

Use level 0.1% both for rinsed off and for not rinsed off products.

. PRECEDING CONCLUSIONS (1st July 1986)

Further information on genotoxicity is needed. Moreover a sensitization test should be conducted and a short-term dermal toxicity study is needed to examine local skin changes as well as systemic effects.

Classification : A.

. FURTHER INFORMATIONS

Since the 40th reunion (April 11th, 1989), Classification A was in suspense for the time being because the Committee was waiting for information about incidences of leucemia in tested female animals.
Only very short and insufficient informations were found on the increased incidences of leucemia in treated mice. This effect has not been confirmed.

A recently conducted chromosomal aberration test with CHO cells in vitro using concentrations of 625 up to 5000 µg/ml, was also negative (Colipa subm. II, ref. 13).

CONCLUSION

Although, tumours occurred in rats at the site of subcutaneous administration, the induction of injection site tumours has little relevance for hazard assessment in man. Because the oral toxicity is low and mutagenicity tests were negative, the use of only 0.1% of this colourant in cosmetics is considered acceptable.

Confirmation of Classification : A.

REFERENCES

1. Dr. Weygand, Gewerbe und Arzneimittel
   Toxikologie Farbwerke Hoechst.
   Report from 10.7.1961

2. Dr. E. Hall, I.F. Gaunt, M. Fanner, P. Grasso
   British industrial Biological Research Ass., Woodmansterne Road
   Carshalton, Surrey, England
   Fd; Cosmet. Toxicol. Vol 5 pp. 165-170
   Pergamon Press 1967

3. Mannuell, W.A. & Grice, H.C. (1964)
   Chronic toxicity of brilliant blue FCF, blue VRS and green S in rats
   J. pharm. Pharmacol. 16, 56-59

4. Grasso, P. & Golber, L. (1966)
   Early changes at the site of repeated subcutaneous injection of
   food colourings
   Fd. Cosmet. Toxicol. 4, 269-282

5. Grasso, P. Gangolli, S.D., Golberg, L. & Hooson, J. (1971)
   Physicochemical and other factors determining local sarcoma
   production by food additives
   Fd. Cosmet. Toxicol 9, 463-478

6. Hooson, J., Grasso, P. & Gangolli, S.D. (1973)
   Injection site tumours and preceding pathological changes
   in rats treated subcutaneously with surfactants and carcinogens
   Brit.J. Cancer, 27, 230-244

7. Grice, H.C. & Mannell, W.A. (1966)
   Rhabdomyosarcomas induced in rats by intramuscular injections of
   blue VRS
   J. nat. Cancer Inst. 37, 845-857

8. Grice, H.C., Dupuis, I., Dennery, M. & Mannell, W.A. (1966)
Blue VRS induces rhabdomyosarcomas (Abstract N° 25)
Toxicol. appl. Pharmacol. 8, 342-343

9. The technique of bladder implantation further results and an assessment
D.B. Clayson, J.A.S. Pringle, G.M. Bonsor and M. Wood
from the department of experimental Pathology and Cancer Research, The school of Medicine, Leeds 2
Br. J. Cancer 22, 825

10. S.D. Gangolli, P. Grasso, L. Golberg, J. Hooson
"Protein binding by food Colourings in relation to the Production of subcutaneous Sarcoma"
Fd. Cosmet. Toxicol. Vol. 10, pp. 449-462
Pergamon Press (1972)

11. S.D. Gangolli, P. Grasso, L. Golberg
"Physical Factors Determining the eraly local Tissue Reaction produced by food Colourings and other Compounds Injected Subcutaneously"
Fd. Cosmet. Toxicol. Vol. 5, pp. 601-621 (1967)

12. Prof. Kramer - Dr. Weigand
Stellungnahme zu Karzinogenität von Acid Blue 1
Hoechst. 19 January 1978
Statement on the carcinogenicity of Acid Blue 1

13. Blue solid W 6007 CI 42045 - OECD 473 metaphase analysis in CHO cells in vitro
Safepharm Laboratories Limited P.O. Box n° 45 - Derby U.K.
January 1988

2. CI 42535

COMPLEMENTARY OPINION

For preceding informations and expressed opinions consult :

- Reports of SCC, seventh series, EUR 11303 P90-91
- Report of the 35th reunion ; October 13th, 1987.

. PRECEDING USE AND CONCLUSIONS

It was used at levels up to 0.1% in rinsed off and non rinsed off cosmetics, and up to 2.5% in hair setting lotions.

Classification : D: Because the colourant was chemically not well defined. Results of tests depend on the source of the sample (samples contain CI 42555 a colourant with carcinogenic and genotoxic properties).

. NEW PROPOSED USE LEVEL

It is proposed to use only 0.01% and only in hair colouring products.

. FURTHERING INFORMATIONS

A recent Colipa submission (IV, November 1988) included a new Ames test conducted with a commercial sample of CI 42535 (containing 5% CI 42555) and a purified sample (not containing CI 42555). Using test levels up to 0.1 mg per plate the results with the strain TA 1537 were positive both with the purified and the non-purified material, in the absence as well as in the presence of the metabolic activation system (ref. 14).

Negative results were obtained in a recent chromosomal aberration test with CHO cells exposed in vitro up to 1.6 µg/ml. Higher levels were not tested because these did not permit 50% survival of the cells (Colipa subm. IV, ref. 15).

CONCLUSION

However, positive results were obtained also with the purified material in the Ames test, the use of 42535 has to be banned in cosmetics.

Classification : D.

3. CI 44045

COMPLEMENTARY OPINION

For preceding informations and expressed opinons consult :

- Reports of SCC, seventh series, EUR 11303 P97-98
- Report of the 35th reunion, October 13th, 1987.

. PRECEDING CONCLUSION

In the only short-term study available, no treatment-related changes were observed at an exposure level of 10 mg/kg bw on 5 days/wk. Because higher levels were not examined this study did not meet the requirement that several levels should be administered of which at least one should exert an effect.

Classification : C.

. USE

Use level 0.1% in rinsed off and non rinsed off cosmetics and up to 0.5% in hair dye formulations.

. FURTHERING INFORMATIONS

A second oral study in rats with 0, 10, 30, 90 or 270 mg/kg/day for 30 days showed 100% mortality in the top-dose group. With 90 mg/kg 7/20 rats died, and haematological changes and increased relative weights of the liver and adrenals were observed. With 30 mg/kg 2/20 rats died.
A blue discolouration of some organs, and histiocytosis of lymph nodes occurred with 30 mg/kg and more.
No changes were seen with 10 mg/kg (Subm. IV, ref. 10).

Dermal penetration was examined in an in vitro test with human abdominal epidermis, exposed to 40 mg 0.1% formulation on 2.0 cm^2 for 30 minutes. Only 0.02% passed into the penetration chamber (Subm. IV, ref. 11).

CONCLUSION

The colourant showed considerable systemic toxicity upon short-term oral exposure of rats to 30 mg/kg and more but 10 mg/kg was a NEL. An in vitro test for dermal absorption showed negligible penetration. Mutagenicity tests were negative.

Classification : A.

Reference 1 : U. HACKENBERG, D. KUHN
Inbifo Institut für biologische Forschung, Köln
Report from 18.1.1978

Reference 2 : G. RONDOT (1978)
"Détermination de la DL50 par voie orale chez le rat -
Chlorure de (N',N'-Dimethyl)amino-4' (N"-Phenyl) amino-4"
naphto (N,N-dimethyl)fuchsonimonium-4 "
Laboratories of Institut Français de Recherches et Essais
Biologiques (IFREB), Saint Germain s/L'Arbresle
Report IFREB-R 801227 - 10th January 1978

Reference 3 : U. HACKENBERG, D. KUHN (1978)
Inbifo, Institut für biologische Forschung, Köln
Report from 17.1.1978

Reference 4 : U. HACKENBERG, D. KUHN (1978)
Inbifo, Institut für biologische Forschung, Köln
Report from 13.1.1978

Reference 5 : J.P. GUILLOT (1978)
"Détermination de l'Indice d'Irritation Cutanée Primaire,
chez le lapin -
Colorant n° 65 : Chlorure de (N', N'- Dimethyl) amino-4'
(N"-phenyl)amino-4" naphto(N,N-dimethyl) fuschsonimunium-4
(CI 44045 Basic Blue 26)
Numéro de lot : AE 7542 - 15.9.77"
Laboratories of "Institut Français de Recherches et d'Essais
Biologiques" (IFREB) - Saint Germain s/L'Arbresle
Report IFREB-R 802264 - 2nd March 1978

Reference 6 : J.P. GUILLOT et al, (1979)
"Evaluation of the Sensitizing Potential of a Test Substance by Topical Applications in the Guinea-Pig - IFG 69-79 : Colorant (Prelt : 8.1.0889)"
Laboratories of "Institut Français de Recherches et d'Essais Biologiques (IFREB) - Saint Germain s/L'Arbresle
Report IFREB-R 911365 - 27th November 1979

Reference 7 : P.E. FOURNIER (1987)
"Etude de Toxicité Chronique chez le Rat d'un Produit Dénommé Bleu Victoria".
Société de Recherches Biologiques - Bonny sur Loire - FRANCE
Report 13 January 1987

Reference 8 : M.M. SHAHIN (1986)
"Mutagenic Evaluation of the Hair Dye CI Basic Blue 26 in AMES Salmonella typhimurium/Microsome Plate Test"
Division Recherche Fondamentale - Departement de Mutagénèse - L'OREAL - Aulnay-sous-bois - FRANCE
Report 27.11.1986

Reference 10: C. BOUTEMY et al, (1988)
Laboratoires d'Etudes et de Recherches SYNTHELABO (LERS) - GARGENVILLE - FRANCE
Report 1310 TMR/030.88 - 6 June 1988

Reference 11: H. SCHAEFFER (1988)
"Pénétration In Vitro du Colorant CI 44045"
Centre International de Recherches Dermatologiques (CIRD)
Valbonne - FRANCE
Report 28 August 1988

4. SOLVENT YELLOW 98

. FORMULA AND SYNONYMES

HOSTASOL - GELB 3G

1H-thioxantheno [2,1,9-def]-isoquinoline-1,3(2H)dione, 2-octadecyl

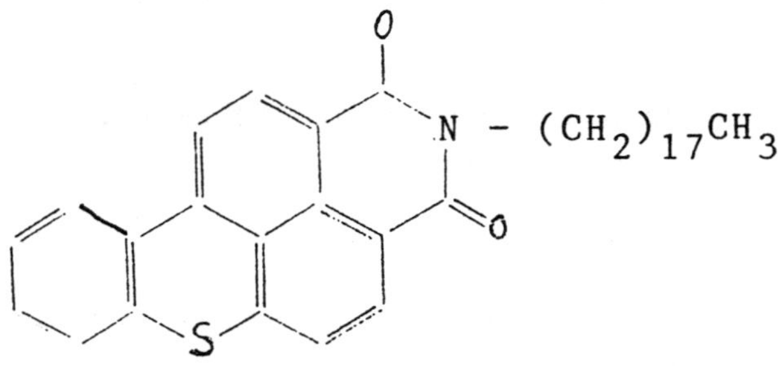

MW: 543

. CHARACTERISTICS

Insoluble in water.
Soluble in organic solvents and mineral oils.

. USE

This new colourant is intended for use only in nail varnish at a concentration of 0.5%.

. RECAPITULATION OF THE STUDIES OF TOXICITY

LD50 oral (rats) : 15 gr/kg

ORAL TOXICITY

A dose of 15 gr/kg did not cause any signs of intoxication. The colourant was excreted with the faeces.
A short-term (30-day) feeding study was conducted with 3 groups of 10 rats/sex fed a basal diet with 0, 1, or 5% of the colourant. There was no mortality or any other sign of intoxication. Growth rate, composition of blood and urine, or the weights of 6 of the major organs were not affected.
A yellow discolouration of the depot fats and liver was observed in test animals at autopsy. Microscopy of the liver, kidneys, adrenals, heart, spleen and lungs did not reveal treatment-related changes (summary report only).

CUTANEOUS AND MUCOUS TOLERANCE

DERMAL TOXICITY

An eye irritation test

Test in rabbits with 0.1 ml of a 5% or a 10% dilution in sesame oil did not produce any response.

Skin irritation

Did not occur in rabbits treated by intracutaneous injection of 0.02 ml of dilutions in sesame oil varying from 0.001% to 10%. Repeated topical application of the 10 and 5% dilution daily for 5 days was also negative.

A phototoxicity

Test in mice by a single intraperitoneal injection of 2.5 g/kg bw of a 25% dilution in sesame oil, followed by UV irradiation for 6 hours, on the same day and on the next day, did not provide indications of phototoxicity as compared with a control group treated with the test substance in the same way but without subsequent irradiation. A similar test was conducted in rabbits by application of 1 ml 10% dilution in sesame oil to the back skin, followed by UV irradiation for 8 hours. Two control rabbits received only irradiation on the neck skin. There were no indications of phototoxicity.

CONCLUSION

The substance has shown little systemic toxicity. An evaluation is not possible because information on genotoxicity or sensitization is not available. An Ames test and a maximization test in the guinea pig should be conducted.

Classification : C.

REFERENCES

1 . Fed. Reg., 43, 166, 38277, 1978.
2 . Data Sheet, Council of Europe, 1978.
3 . Farbenfabrik Bayer AG, Wuppertal, 1971.
4 . Leuschner, F., Laboratorium fur Pharmakologie und Toxikologie, Hamburg, 1979.
5 . Hopf, G., Hamburg, 1971.
6 . Givaudan SA, Vernier, 1973 (Summary).
7 . Givaudan SA, Vernier, 1974 (Summary).
8 . Givaudan SA, Vernier, 1975 (Summary).
9 . Bonin, A.M. et al., Mutation Research, 105, 303, 1982.
10 . Camponovo (1984). Hoffman La Roche, Basle.
11 . Avon and D, September 10, 1982, study ATO 187.
12 . Dr. Buergin (1983) Hoffman La Roche, Basle.
13 . Kistler A. (1983). Hoffman La Roche, Basle.
14 . Klecak, G. (1980). Hoffman La Roche, Basle.
15 . Klecak, G. et al., (1982). Hoffman La Roche, Basle.
16 . Stuttgen, G. (1982). Universitatsklinikum Charlottenburg.
17 . Darragh, A. (1980). Biological and Medical Research Institute.
18 . Schuepbach, M. (1980). Hoffman La Roche, Basle.
19 . Strobel (1983). Hoffman La Roche, Basle.
20 . Chetelat, A. (1983). Hoffman La Roche, Basle.
21 . Hugentobler (1984). Hoffman La Roche, Basle.
22 . Strobel, R. (1985). Hoffman La Roche, Basle.
23 . Strobel, R. (1985). Hoffman La Roche , Basle.
24 . Schuepbach, M. (1983). Hoffman La Roche, Basle.
25 . Hugentobler, A.L. (1983). Hoffman La Roche, Basle.
26 . Schuepbach, M. (1984). Hoffman La Roche, Basle.
27 . Forbes et al., (1986). Center for Photobiology, Philadelphia.

First version July 1984.

Revision August 1987.

Revision March 1988.

REPORTS BY THE SCIENTIFIC COMMITTEE ON COSMETOLOGY CONCERNING THE USE OF CERTAIN U.V. FILTERS

. THE COMMITTEE'S MANDATE

To give its opinion on the question whether the use of the following U.V. filters :

- S_3 Ethoxylated ethyl-4-aminobenzoate up to 10%
- S_8 2-ethylhexyl-p-dimethylaminobenzoate up to 8%
- S_{13} 2-ethylhexylsalicylate up to 5%
- S_{40} 2-hydroxy-4-methoxybenzophenone-5-sulphonic acid up to 5%

in cosmetic products is admissible from the health point of view.

1. ETHOXYLATED ETHYL-4-AMINOBENZOATE

. FORMULA AND SYNONYMES

EEC n° 2.2

Colipa S3

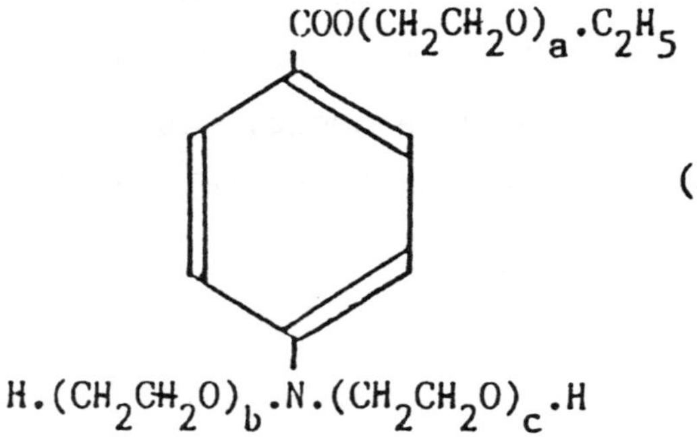

$(a + b + c) = 25$

. CHARACTERISTICS

A clear slightly viscous yellow liquid at room temperature. Absorption maximum 307 nm. Soluble in water, poorly so in ethanol or anhydrous isopropanol.
The compound is manufactured by reacting the ethyl ester of para-aminobenzoic acid with ethylene oxide.
Free ethylene oxide is then blown away by a stream of nitrogen. Purity : not less than 99%.

. USE

Proposed use level : up to 10%.

. RECAPITULATION OF THE STUDIES OF TOXICITY

LD50 oral (rat) : >1.9 gr/kg
i.p. (mouse) : >1.9 gr/kg

Exposure of rats to air saturated with a.i. for up to 8 hrs gave no abnormality.

CUTANEOUS AND MUCOUS TOLERANCE

Irritation

In the rabbit, 3 sets of experiments were carried out, using undiluted and diluted a.i., applied both to scarified and intact skin, for up to 24 hrs. Some animals showed slight erythema ; the tests were otherwise negative.
Two animals had 5 applications within a week of a 50% aqueous solution to an area of 36 cm^2 of skin. Each application was allowed to act for 8 hrs. There was no evidence of irritation.

In 20 human subjects, undiluted and diluted a.i. was applied to areas of 1 cm^2 for 20 hrs. No irritation was produced. A test for mucous membrane irritation has been carried out. This was found to be satisfactory.

Sensitization

Ten guinea pigs were used. Induction was by application of a 50% solution in acetone to the skin. An 80% solution was then applied 5 days a week for 9 days. After a 12-day rest, a challenge application of a 50% solution was made.
No reaction was found.

Phototoxicity

A 10% aqueous solutiona of a.i. was applied to the skin of 10 human subjects. Control applications were also made. The treated areas were exposed to UV radiation, increasing in a step-wise manner to determine the m.e.d. There was no evidence of phototoxicity. The compound had a protective effect.

Photosensitisation

Ten male and ten female guinea pigs were used. A daily application of 0.5 ml of a 20% aqueous solution of a.i. was applied to the shaved skin of one flank for 5 days. Each application was followed by 15 mins of UV irradiation from a lamp with a maximum at 260 nm. The procedure was repeated on the opposite flank after a 10-day rest.
The test was negative.

MUTAGENICITY

The substance tested was designed "LUSANTAN 25" (stated to be ethoxylated ethyl-4-aminobenzoate).
An Ames test was normal.

CONCLUSION

The Committee was satisfied with the data presented on acute toxicity, inhalational toxicity, and tests for capacity to irritate mucous membranes. It is noted that the tests for sensitisation were not carried out using a maximisation technique. The test described as a photosensitisation test should more properly be regarded as a test for photo-allergenicity or phototoxicity. A test for chromosomal aberration in vitro should be carried out.

It will then be possible to consider whether tests for carcinogenesis are necessary. It is understood that a 90 day oral test and a test for percutaneous absorption are in progress. When the result of the latter is available, the need for tests for teratogenicity and pharmacokinetics can be decided. Ethylene oxide should be below the limits of detection.

Classification : C.

2. 2-ETHYLHEXYL-p-DIMETHYLAMINOBENZOATE

. FORMULA AND SYNONYMES

EEC n° 2.5
Colipa S8
"Padimate O"
"Escalol 507"

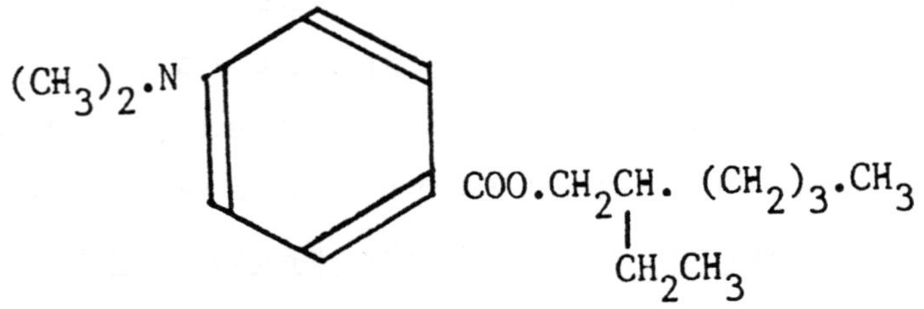

. CHARACTERISTICS

Yellow fluid, stated by manufacturer to contain not less than 98.5% of a.i.
Maximum absorption 310 nm.
Not known to polymerise.
Soluble in isopropyl alcohol, mineral oil, and ethanol.
Insoluble in water.

. USE

Use level up to 8%.

. RECAPITULATION OF THE STUDIES OF TOXICITY

LD50 oral (rat) : 3-15 gr/kg bw.

CUTANEOUS AND MUCOUS TOLERANCE

DERMAL TOXICITY

Irritation of mucous membranes

A Draize test in the rabbit at concentrations of 2% and 5% in mineral oil showed slight transient irritation.

Irritation of skin

Rabbit : Solutions of 5% a.i. were applied to both intact and abraded skin for 24 hrs under occlusion. The test was negative.

Man : Occlusive patch tests with 5% a.i. in yellow soft paraffin were applied for 48 hrs. The test was negative.

Sensitisation

Guinea pig : Ten male animals had an initial intracutaneous injection of 0.05 ml of a 0.1% solution of a.i. in saline, followed by 9 injections of 0.1 ml 3 days a week. After a 12 week rest period, a challenge dose of 0.05 ml was given. There were no adverse effects. The report is somewhat ambiguous as to the concentration used : it is possible that a 0.005% solution was used.

Man : (a) Fifteen applications of a 4% solution of a.i. in soft paraffin were made under occlusion over 3 weeks. A challenge application was made after a 2 week rest. There was no adverse reaction.

(b) A mixture of 7% a.i. with 3% oxybenzone was used in 150 subjects in a repeated insult patch procedure. No abnormality was found.

(c) Ninety subjects were similarly tested using 8% a.i. and 3% benzophenone. The test was negative, although there were occasional slight irritant responses during the induction.

(d) A panel of 156 subjects was similarly tested with 7% a.i. in soft paraffin. The test was negative.

Phototoxicity

Guinea pig : The ears of 10 animals were stripped and a formulation containing 7% a.i. and 3% oxybenzone was applied several times to one ears with vigorous rubbing. The untreated ear served as a control ; 2 of the animals had 8-methoxypsoralen applied as a positive control. Thereafter the animals were exposed to UV radiation (wavelenght not stated) for 2 hr. The test was negative ; the positive controls showed marked effects.

In another test, a similar preparation was applied to the nuchal area with occlusion for 2 hrs. This was followed by irradiation with 3 J/cm^2 at 320-400 nm. Suitable positive and negative controls were used. The test was negative.

Man : In a poorly reported tes, a mixture of 7% a.i. and 3% oxybenzone was tested in 26 human subjects. No adverse effects were seen.

In another similar test, a 5% ehtanolic solution was used. At 30 J, the control area showed more damage than the test area. Ten fair-skinned subjects were treated with a mixture of 7% a.i. and 2% oxybenzone under occlusion for 24 hrs. A control was similarly applied. After removal of the patches a further application was made to the skin and irradiation was carried out using 1 m.e.d. of UVB followed by 12 minutes of UVA. The test was negative

Percutaneous absorption

Man : An 8% ethanolic solution of ^{14}C a.i. was applied over 100 cm^2 of forearm skin in 4 female subjects. After the ethanol had dried, the areas were covered with a gauze pad for 24 hrs. No radioactivity was found in the blood ; the urine containded between 1.2% and 2.5% of the applied radioactivity.

Subchronic toxicity

A 13-week dermal toxicity study was carried out in groups of 20 rabbits at dose levels of 140 and 280 mg/kg bw. No significant abnormality was detected.

MUTAGENICITY AND GENOTOXICITY

A standard Ames test was negative. A second similar test is also reported negative, but figures are given only for plates with activation.
A micronucleus test was carried out in the mouse, using a dose which caused disorders of gait and hypotonicity. The a.i. was given intraperitoneally in a dose of 5000 mg/kg bw to 3 groups of 10 animals. Positive and negative control groups were included. Sacrifice was at 30, 48 and 72 hrs. The test was negative.

TERATOGENICITY

Rat : Dermal applications of 2 ml/kg bw of a preparation (concentration of a.i. not specified) were made daily from days 6 to 16 of pregnancy. In the test group 7/56 foetuses had bilateral wavy ribs and 2/56 had unilateral wavy ribs. There were no such findings in the control group. This effect is not regarded as indicating teratogenic activity, and is a common finding in rats of this strain, but the reason for its appearance in foetuses of the test group only is unexplained.

CONCLUSION

The Committee notes that sensitisation tests were carried out at levels less than the proposed use level. Testing up to irritant level would have been desirable to reveal any potential for sensitisation. Tests for phototoxicity and photosensitisation are poorly reported, but seem to be negative ; since the compound is widely used, they may be accepted. Tests for absorption suggest that the amount absorbed may be about 1 mg/kg bw/day.
A chromosomal aberration test and a 28-day or 90-day oral toxicity study should be carried out.

Classification : C.

REFERENCES
++++++++++

1. Avon Products Inc. R & d Toxicology Dept. (April 8, 1983)

2. William P. Jordan, Medical College of Virginia, Richmond, Virginia. Report P86-15 28/02/86

3. Study n°4080 Lab. Notebook 10012 31/01/79

4. Harold Shwartz and Marvin Kirchner, FDA Research Laboratories 60 Evergreen Place, East Orange, New Jersey, USA, 27/01/82

5. Food and Drug Research Laboratories Inc., September 6,1977 Gibraltar Biological Laboratories.

6. Pharmakon Research International Inc., Ames Salmonella/ Microsome plate test (EPA/OECD) PH 301-002-86. 25 June 1986

7. PH-309-VD-001-86 , Pharmakon Research International Inc., Waverly, Pennsylvania 18471 11/03/87

8. Avon Products Inc., R & D Toxicology Dept. September 17, 1975

9. Study n°RB 434-1559, F. J. AIKIN, Director Pharm & Tox. 27/07/77

10. Study n° RB A49-170, BUNNEL & ROSETTI, Septembre 6, 1985

11. A. M. KLIGMAN, Hospital of the University of Pennsylvania, Nov. 23, 1977

12. TKL Research Inc., New York. Study n°846003, July24, 1984

3. 2-ETHYLHEXYLSALICYLATE

. FORMULA AND SYNONYMES

EEC n° 2.6
Colipa S13
OCTYLSALICYLATE
"SUNAROME"

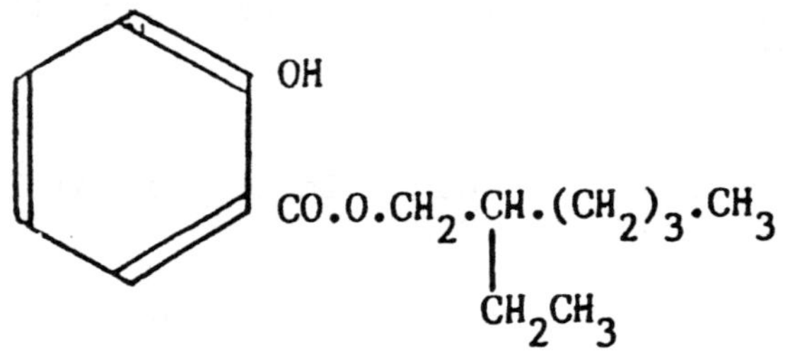

$C_{15}H_{22}O_3$ MW 250·33

. CHARACTERISTICS

Clear odourless liquid.
Insoluble in water.
Soluble in ethanol, mineral oil, other organic solvents.
Specific gravity 1.013 - 1.022.
Maximum absorbance at about 300 nm.
There is evidence that this compound dissociates only very slowly and to an extent of less than 0.5% at pH values of 8.9 ; there is no dissociation below pH 8.0, approximately (8).

. USE

Proposed use level : up to 5%.

. RECAPITULATION OF THE STUDIES OF TOXICITY

LD50 oral (rat) : 4.8 ± 0.3 gr/kg bw.

CUTANEOUS AND MUCOUS TOLERANCE

DERMAL TOXICITY

Irritation of mucous membranes

A Draize test in the rabbit was negative. The concentration of a.i. used is not clear from the report ; it may have been used undiluted.

Irritation of the skin

In the rat, an 18.5% solution caused irritation in the sub-chronic dermal test noted above.
In man, a 5% solution of a.i. in mineral oil was applied for 24 hrs with occlusion in 10 subjects. After a 7-day rest, the procedure was repeated. No adverse effect was found.
A repeated insult patch test was carried out in 25 subjects. The initial application was allowed to remain on the skin for 48 hrs ; thereafter 24 hr applications were made every second day until 10 applications had been made. The concentration of a.i. used is not clear from the report : it may have been undiluted. After a 10-14 day rest, a 48 hr test patch was applied. There was no adverse reaction.

Phototoxicity

In 10 human subjects, a 5% solution of a.i. in ethanol was applied for 1 hr on stripped skin, and for 24 hrs on intact skin. In each case, the area was then exposed to UV radiation at 322-410 nm. There were no adverse effects. Declomycin as a positive control under the same conditions caused phototoxicity.

Photocontact allergy

Twenty-five subjects were tested with an ointment containing 15% of a.i. Applications of 10 $\mu l/cm^2$ were applied for 24 hrs with occlusion. After removal, the areas were exposed to 3 m.e.d. of simulated solar radiation from a xenon arc. This sequence was repeated after 48 hrs and thereafter twice weekly for 3 weeks, always at the same site.
After a 10-day rest, applications of 5 $\mu l/cm^2$ were made to a fresh site with occlusion for 24 hrs, followed by irradiation. No adverse reactions were seen.

Subchronic toxicity

A 13-week dermal study was carried out in the rat. The a.i. was applied as an 18.5% ethanolic solution at doses of 0, 55.5, 277 and 555 mg/kg bw. There were changes in biochemical values in the animals of the intermediate and top dose groups, suggesting adverse effects on the liver, but there was no histological evidence of liver damage. At the low dose, in females, the absolute and relative weights of the kidney and spleen were increased, and the absolute weight of the heart ; in males, the relative weith of the lungs was increased. At higher doses, both sexes showed changes, with a fall in absolute weight of the liver in males and of the kidney and heart in females. There was a dose-related effect on the skin : histological changes of hyperkeratosis were found at the low dose, progressing to marked hyperkeratosis and inflammatory foci at the top dose.

MUTAGENICITY

An Ames test was negative.

CONCLUSION

Are required :
- a subchronic oral toxicity test
- a test for percutaneous absorption
- investigation of pharmacokinetics and teratogenic activity
- a chromosomal aberration test in vitro
- photomutagenicity test
- inhalation toxicity (if used in aerosols and confirmed spaces).

Classification : C.

REFERENCES
++++++++++

1. 3. United States Testing Company, Inc., Biological Services Division, 1415 Park Avenue, Hoboken New Jersey 07030 April 9, 1976

2. 4. BIO-TOXICOLOGY LOBORATORIES INC., Toxicity and applied Research Studies - Animal and Humans (Sunarome WMO-lot 63203) MOORESTOWN, New Jersey, Feb. 17, 1976

5. Avon Products Inc., Division Streer Suffern - New York File ATO 189 - Final Report 08/13/1982

6. LITTON BIONETICS INC., Kensington Maryland, LBI project n°2833, Sept. 1977

7. IVY Research Laboratories, Inc., Philadelphia, USA, July 1982

4. 2-HYDROXY-4-METHOXYBENZOPHENONE-5-SULPHONIC ACID

. FORMULA AND SYNONYMES

EEC n° 2.19

Colipa S40

Sulisobenzone (INN)

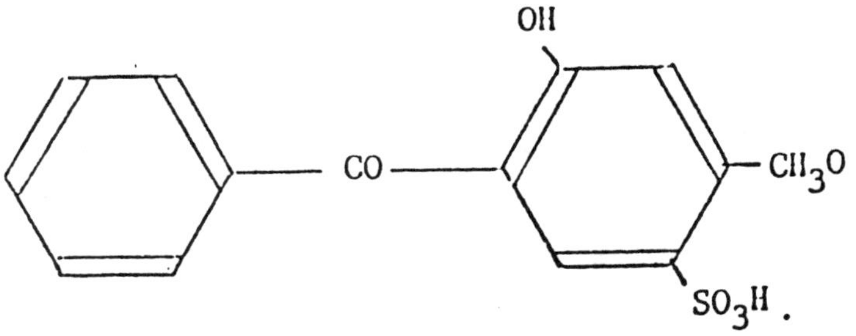

$C_{14}H_{12}O_6S$

MW = 308·3

. CHARACTERISTICS

A free-flowing powder, soluble in water and alcohol. Maximum absorbance at 285 nm.

. USE

Use level : up to 5%.

. RECAPITULATION OF THE STUDIES OF TOXICITY

LD50 oral (rat) : > 6.4 gr/kg bw.

CUTANEOUS AND MUCOUS TOLERANCE

Irritation of mucous membranes

The compound was non-irritant in a 4% solution in dimethylphthalate and slightly irritant at 8%. A 5% aqueous solution was non-irritant. A 16% solution in dimethylphthalate was markedly irritant.

Irritation of skin

Solutions of 4%, 8% and 16% of a.i. in dimethylphthalate were tested on the skin of the rabbit. There was slight irritation at the highest concentration.
In 14 human subjects, an occluded patch test was carried out. There was no reaction at 4% at 8%, 1/14 showed irritation, and at 16% 4/14 showed irritation.
A Shelanski repeated insult patch test was carried out in 50 subjects, using a 5% aqueous solution. Occluded patches were applied for 24 hrs, repeated every 48 hrs, to a total of 15 applications. After a 2 week rest, a challenge application was made.Ther was no evidence of irritation or sensitisation. Daily application of a 1% solution in alcohol for 21 days was non-irritant.

Sensitisation

The Shelanski test, noted above, was negative.
A 10% alcoholic solution applied as a cumulative test over 21 days showed the substance to be a marked contact allergen. (No further details given).

MUTAGENICITY

An Ames test was carried out using 7 batches of a.i.
Each batch was tested for toxicity, using filter paper discs, in two strains, TA 1537 and TA 100. Concentrations ranged up to 10,000 µg/plate. One sample (400) showed inhibition at both 1000 and 10,000 µg.
The highest concentration used in mutagenicity testing was 1000 µg/plate. Sample 400 showed toxicity, with and without activation, as low as 100 µg, but not consistently.
Sample D-49 also showed some toxicity as low as 100 µg, but again, not consistently. Marked increases in revertants were found in some strains with samples numbered N-35 (without activation) and D-49 (with activation).
The toxicity of samples D-49 and 400 confined testing to concentrations of 10 to 100 µg/plate and less.
A standard chromosomal aberration test was carried out with Chinese hamster ovary cells. There was no evidence of clastogenic activity with or without metabolic activation.

CONCLUSION

A test for percutaneous absorption should be carried out.
The problem of batch effects in the Ames test should be resolved, and testing carried out up to toxic concentrations in each strain.
A 28-day or 90-day oral study at suitable dose levels, and tests for phototoxicity and photosensitisation, should be carried out. The full report of the positive allergenicity test in man should be provided, and further tests carried out if necessary.

Classification : C.

REFERENCES

++++++++++

1. Hill Top Research Inc., Miamiville, Ohio, July14, 1980

2. Dr. H. D. F. M. TAALMAN "Clastogenic evaluation of 2- hydroxy-4-methoxybenzophenone-5-sulphonic acid in an in vitro cytogenetic assay measuring chromosome aberration frequencies in Chinese Hamster Ovary (CHO) cells".
Hazelton - Biotechnologies - The Netherlands - April 29, 1988

CONTENTS

42nd meeting of SCC - October 30-31, 1989

43rd meeting of SCC - January 23, 1990

REPORTS BY THE SCIENTIFIC COMMITTEE ON COSMETOLOGY

42nd REUNION

OCTOBER 30th - 31st, 1989

LIST OF PARTICIPANTS

Present

Mr. AGACHE
Mrs. ENJOLRAS
Mrs. DONY
Mr. FIELDER
Mr. GOULDING
Mr. KAPOULAS
Mr. LOPRIENO
Mr. O'MAHONY
Mr. PONS-GIMIER
Mr. SCHOU
Mr. STUTTGEN
Mr. SOMOGYI
Mr. RAMOS-MORGADO

Not Present

Mrs. KNAAP

Mr. MUSCARDIN

COLIPA : Dr. WILSON, Dr. TAYLOR, Mrs. RICHOLD,
Dr. STERZEL, Mrs. VANDEVORST

COMMISSION

Mrs. MASSE
Ms. DE CRAWHEZ
Mr. GONTIER

1. CHLORPHENESIN

. FORMULA AND SYNONYMES

Chlorphenesin

EEC n°3
Colipa P4
p-chlorophenyl-glycerol ether
3-(p-chlorophenoxy) propane-1,2-diol

COMPLEMENTARY OPINION

For preceding informations and expressed opinions consult :
- Report of the 37th reunion, April 12th, 1988.

. PRECEDING CONCLUSION (12-4-88)

Although short-term studies have been conducted, insufficient information on the studies were available as required :

- a repeated-dose oral study (with special attention for the immune system)
- a chromosomal aberration test in vitro
- information on dermal absorption
- a teratogenicity study (depending on the degree of dermal absorption).

Classification : C.

. FURTHERING INFORMATIONS

Proposed use : up to 0.3%.
A chromosomal aberration test with human lymphocytes exposed in vitro up to 0.325 mg/ml was negative (subm. IV, ref. 22).

CONCLUSION

Although short-term studies have been conducted, insufficient information on these studies is available. A repeated dose oral study should be conducted with special attention for the immune system. Information on dermal absorption is needed. Depending on the degree of dermal absorption a teratogenicity study may be needed.

Classification : C.

2. DIAZOLIDINYLUREA

. FORMULA AND SYNONYMES

EEC n°17

Colipa P79

Germall II

COMPLEMENTARY OPINION

For preceding informations and expressed opinions consult :

- Report of the SCC, fifth series, EUR 11080 (25-6-85)
- Report of the 34th reunion June 30th 1987.

. PRECEDING CONCLUSION (30-6-87)

Classification : B.

In the previous discussion on this substance (June 1987) the Committee requested an oral teratogenicity study and a gene mutation test in mammalian cells.

. FURTHERING INFORMATIONS

In a teratogenicity study using the dermal route, rats were exposed to 30, 95 or 300 mg/kg body wt/day on days 6-15 of gestation. Compound was applied directly to the shaved skin as a 30% solution in deionised water (no occlusion). The only toxic effects noted were due to a direct effect on the skin (erytherma and scabbing). Foetuses of 26 gravid rats in each group were examined for abnormalities in bone and soft tissue. Only one soft tissue abnormality was found by the Wilson technique, and this was in a control foetus.

There were no significant differences in skeletal variations among the groups. Germal II was not teratogenic or embryotoxic under the conditions of the test (Subm. X, ref. 29).

In a chromosomal aberration test with CHO cells the highest concentration tested was only 1.5 µg/ml because even 1 µg/ml caused 70% mitotic inhibition (Submission X, ref. 27). The positive control substances triethylene amine and cyclophosphamide were tested at 50 and 0.5 µg/ml respectively. Under the conditions of the test Germal II was negative (Submission X, ref. 28).

CONCLUSION

Industry has now provided a dermal teratogenicity study and a chromosomal aberration test. The dermal study is of little significance, because there is no information on dermal absorption, which is probably very low. The chromosomal aberration test is of little value for the evaluation because the levels tested were very low, while negative results had already been obtained at very high dosages in an *in vivo* micronucleus test. Moreover the SCC requested a gene mutation test instead of a chromosomal aberration test.

Classification : B. confirmed.

3. DECOMINOL

. FORMULA AND SYNONYMES

Colipa P87
3-decyloxy-2-hydroxy-1-aminopropane-hydrochloride
Ster 4
PVA 44

COMPLEMENTARY OPINION

For preceding informations and expressed opinions consult :
- Report of the 33rd reunion, February 20th 1987.

. PRECEDING CONCLUSION (20-2-87)

The Committee required informations on :
- percutaneous absorption
- results of tests of chromosomal aberration and mutagenicity in mammalian cells.

Classification : C.

. FURTHERING INFORMATIONS

Dermal absorption of ^{14}C-Decominol was examined in young rats by single application of 5 mg a.i./kg bw in water or in a water/oil emulsion to groups of 3 males and 3 females for 24 h. One similar group of 3 rats/sex received 5 mg/kg intravenously in aqueous solution. Urine faeces and expired air were collected at intervals over 96 h. Dermal absorption from the emulsion was 7.2 and 10.7% for males and females respectively, while for the aqueous solution these figures were 2.4% and 4.1%.

The major route of elimination was the urine ; faeces and expired air were minor routes. The mean recoveries of the dose were 91.3% for the emulsion, 95.3% for the aqueous solution and 83.8% for the intravenous dose.
Radioactivity occurred in all tissues after percutaneous administration in either of the formulations mainly in the liver and residual carcass (ref. 16).

A chromosomal aberration test in human lymphocytes in vitro was also negative. Toxicity of the compound precluded testing of concentrations higher than 0,1 mg/ml (ref. 17).

CONCLUSION

The tests presented show a compound which has some irritant action on the skin and mucous membranes at 1%, and which may also cause liver and kidney damage when given systemically. In an oral 13-week study in the rat, the NEL was less than 10 mg/kg bw/day with local irritant effects on the gastric mucosa being seen at this level ; uraemia was seen at 40 mg/kg. The material does not seem to be well absorbed from the gastrointestinal tract.
Percutaneous absorption in rats was relatively low (<11%) from an oil emulsion and lower from aqueous solution. Tests for genotoxic properties in vitro and in vivo were negative, but only low concentrations could be examined in vitro due to the toxicity of the compound. The use of 0.5% Decominol in all types of cosmetics may result in a human dermal exposure of up to 2.5 mg/kg bw/day. Assuming 10% dermal absorption, the systemic exposure may be up to 0.25 mg/kg bw/day. The NEL is below 10 mg/kg effects being limited to local irritation of the gastric mucosa at this level. There was evidence of nephrotoxicity (uraemia) at 40 mg/kg, and also hepatoxicity at 140 mg/kg.

Classification : A.

REFERENCES

Colipa Subm. I

1-2-3-4. LD50
LABO. TOXIPHARMA - MONTROUGE
1974-1979 Monique HEROLD

5. Eye irrit. LABO TOXIPHARMA - MONTROUGE

6. Skin irrit. HUNTINGDON RESEARCH CENTER
SHEENA R - KYNOCH - MP LIGGET
5508/40D176 15-1-1976

7. id.
CENTRE D'ETUDES POUR LA RECHERCHE ET LE CONTROLE - PARIS
Jacques GUILLAUME

8. Sensitization
TOXICOL. LABORATORIES LTD
Bromyard. Road Ledbury Herefordshire
J.D. MIDDLETON 10-1978

9. Oral Short Term TOXIPHARMA - MONTROUGE

10. i.v. TOXIPHARMA - MONTROUGE

11. 90-day TOXIPHARMA - MONTROUGE

12. 90-day TOXIPHARMA - MONTROUGE

13. MUTA. AMES
H.T. Bui INCRA 91170 Vert le Petit
Etude b7828 11-1983

14. Micronucleus
FOURNIER-MASSON Labor. EVIC-CEBA
33290 Blanquefort 30-9-1983

15. Phototoxicity
DUFOUR Labor. EVIC-CEBA 3-1984

REFERENCES

Colipa Subm. II 3/89.

16. HAZLETON. Otley Road - Harrogate - North Yorkshire England HG3. IIPY.
"Absorption, distribution and excretion following percutaneous and I.V. administration to the rat".
Report n° 5778.324/3 1/89.

17. HAZLETON. Institut Français de Toxicologie Test pour l'évaluation de l'induction d'aberrations chromosomiques dans le lymphocytes humains.
Report n° 806490 24/6/88.

18. HAZLETON. I.F.T.
"Evaluation de l'induction de mutations géniques sur cell CHO (Locus HGPRT).
Report n° 804462 29/4/88.

REPORTS OF THE SCIENTIFIC COMMITTE ON COSMETOLOGY

CONCERNING THE USE OF CERTAIN COLOURANTS

. THE COMMITTEE'S MANDATE

To give its opinion on the question whether the use of certain colourants :

- CI 73900 (46500)
- CI 74180
- Solvent Yellow 98

is admissible from the health point of view.

1. CI 73900 (46500)

. COMPLEMENTARY OPINION

For preceding informations and expressed opinions consult :

- Reports of SCC, seventh series, EUR 11303 P103
- Report of SCC 35th reunion October 13th 1987.

. PRECEDING CONCLUSION (13-10-87)

Information on possible sensitizing properties, and a chromosomal aberration test in vitro were required as well as a detailled report on the short-term oral study.

Classification : B.

. FURTHERING INFORMATIONS

In a second short-term (33-day) oral study, groups of 8 rats/sex were fed diets with 0, 1, 5 or 10% of a purified form of the colourant to examine systemic toxicity (Five rats of each group were also used for examining chromosomal aberrations and 4 males were added to each diet group for an UDS test in hepatocytes. See mutagenicity data). There was no growth depression even at the high dose level, but food intake was increased in top-dose males.
A large number of haematological and clinical chemical characteristics were examined in week 2 and 4, but there were no treatment-related changes.
Organ weights and gross and microscopic examinations failed to reveal abnormalities attributed to the colourant (Subm. III, ref. 8).

An in vivo chromosomal aberration test conducted with bone marrow cells from rats fed 0, 1, 5 or 10% of the purified colourant for 33 days was negative (Subm. III, ref. 8).

An in vivo/in vitro test for unscheduled DNA synthesis with hepatocytes from rats fed 0, 1, 5 or 10% of the purified colourant for 14 days did not show an increased number of nuclear grain counts, while rats dosed in a similar way with the positive control substance 2-acetylaminofluorene showed significant increases over that in the negative controls (Subm. III, ref. 9).

CONCLUSION

The purified substance did not show systemic toxicity even at 10% in the diet. Information on possible sensitizing properties is requested. A report on the bio-availability study, which was in progress already in 1987, has not yet been provided.

Classification : B.

REFERENCES

Reference 1 : Avon Acute Oral Toxcity Test
Avon Products Inc, New York, June 8th 1979

Reference 2 : Hoechst Aktiengesellschaft, Pharma Forschung Toxikologie
15th May 1979

Reference 3 : Avon Draize Eye Irritation Test
Avon Products Inc, New York, June 8th 1979

Reference 4 : Hoechst Aktiengesellschaft, Pharma Forschung Toxikologie
15th May 1979

Reference 5 : Avon Primary Rabbit Skin Irritation Test
Avon Products Inc, New York, June 8th 1979

Reference 6 : Hoechst Gewerbetoxikol. Labor. Report of 12.9.60
Full report not available

Reference 7 : Mutagenicity Evaluation of Quinacridone Red Y
Avon Products Inc, Suffern, New York
Assay n° 3885 - February 1979

Reference 8 : B.S. LEVINE
"Subchronic oral toxicity study in rats"
Microbiological associates, Bethesda, Maryland 23.12.86

Reference 9 : R.D. Curren "In vivo - in vitro rat hepatocyte unscheduled DNA synthesis assay"
Microbiological Associates, Bethesda Maryland 29.01.1988

2. CI 74180

. COMPLEMENTARY OPINION

For preceding informations and expressed opinions consult :
- Reports of SCC, seventh series, EUR 11303, P129.

. PRECEDING CONCLUSIONS (1st July 1986)

The Committee saw no objection to maintaining the use of this colouring agent for the time being in rinsed off products only. For now rinsed off products, no opinion could be expressed.
The Committee required :

- a dermal penetration test *in vitro* or *in vivo*
- a short-term oral study to establish the NEL (if penetration is considerable).

. FURTHERING INFORMATIONS

A dermal absorption test was conducted *in vitro* using skin from the domestic pig or the hairless rat, by applying 0.05% solutions of the labelled compound in water or in 50% aqueous ethanol in an amount of 6 $\mu l/cm^2$ (which is 0.5 μg a.i./cm^2) to an area of 5 cm^2.
The total penetration after 16 h. exposure was about 10% in rat skin and 7% in pig skin without a clear difference between the different solvents (Colipa subm. II, ref. 4).

CONCLUSION

No data are available on mutagenicity.
An Ames test should be conducted and a chromosomal aberration test with mammalian cells *in vitro*.

Classification : C.

REFERENCES

Reference 1 : K.H. LEIST

"Subacute Toxicity Studies of Selected Organic Colorants"
Ecotoxicology and Environmental Safety 6, 457 - 463
(1982)

Performed by : Industriels Français du Médicamment (IFM)
Centre d'Etudes Biologiques,
Miserey-Evreux, FRANCE

PCUK : Etude Toxicologique chez le Rat pendant 30 jours du Produits Direct Blue 86

Report from 29 August 1979

Reference 2 : Prof. Chr. GLOXHUBER

Henkel Hauptabteilung Toxikologie
Report from 19.3.1979

Reference 3 : Prof. Chr. GLOXHUBER

Henkel Hauptabteilung Toxikologie
Report from 17.7.1980

Reference 4 : Dr. G. KLECAK

"Penetration studies with RO 42-8247/001 ^{14}C sulfonated Copper Phtalocyanine - CI 74 180 ^{14}C - on intact skin of naked rat and pig "in vitro"
Hoffmann La Roche, Basle, December 29 1988

3. SOLVENT YELLOW 98

. COMPLEMENTARY OPINION

For preceding informations and expressed opinions consult :
- Report of SCC, 40th reunion April 12th 1989.

. PRECEDING CONCLUSIONS (12-4-1989)

The substance has shown little systemic toxicity. An evaluation is not possible because informations on genotoxicity or sensitization is not available.
An Ames test and a maximization test in the guinea pig should be conducted.

Classification : C.

. FURTHERING INFORMATIONS

Mutagenic potential was examined in 5 strains of S. typhimurium and in E. coli WP2 uvr A at dose levels up to 10 mg/plate, with and without metabolic activation.
There were no indications of a positive response (Colipa subm. II).

CONCLUSION

A chromosomal aberration test in vitro, and a maximization test in the guinea pig should be conducted.

Classification : B.

REFERENCES

1. Fed. Reg., 43, 166, 38277, 1978.
2. Data Sheet, Council of Europe, 1978.
3. Farbenfabrik Bayer AG, Wuppertal, 1971.
4. Leuschner, F., Laboratorium fur Pharmakologie und Toxikilogie, Hamburg, 1979.
5. Hopf, G., Hamburg, 1971.
6. Givaudan SA, Vernier, 1973 (Summary).
7. Givaudan SA, Vernier, 1974 (Summary).
8. Givaudan SA, Vernier, 1975 (Summary).
9. Bonin, A.M. et al., Mutation Research, 105, 303, 1982.
10. Camponovo (1984). Hoffman La Roche, Basle.
11. Avon & D, September 10, 1982, study ATO 187.
12. Dr. Buergin (1983) Hoffman La Roche, Basle.
13. Kistler A. (1983). Hoffman La Roche, Basle.
14. Klecak, G. (1980). Hoffman La Roche, Basle.
15. Klecak, G. et al. (1982). Hoffman La Roche, Basle.
16. Stuttgen, G. (1982). Universitatsklinikum Charlottenburg.
17. Darragh, A. (1980). Biological and Medical Research Institute.
18. Schuepbach, M. (1980). Hoffman La Roche, Basle.
19. Strobel (1983). Hoffman La Roche, Basle.
20. Chetelat, A. (1983). Hoffman La Roche, Basle.
21. Hugentobler (1984). Hoffman La Roche, Basle.
22. Strobel, R. (1985). Hoffman La Roche, Basle.
23. Strobel, R. (1985). Hoffman La Roche, Basle.

24. Schuepbach, M. (1983). Hoffman La Roche, Basle.

25. Hugentobler, A.L. (1983). Hoffman La Roche, Basle.

26. Schuepbach, M. (1984). Hoffman La Roche, Basle.

27. Forbes et al., (1986). Center for Photobiology, Philadelphia.

First version July 1984.

Revision August 1987.

Revision March 1988.

SCIENTIFIC COMMITTEE ON COSMETOLOGY

43rd MEETING

JUNE 20th, 1989

LIST OF PARTICIPANTS

Present

Mr AGACHE
Mrs DONY
Mrs ENJOLRAS
Mr FIELDER
Mr GOULDING
Mr KAPOULAS
Mrs KNAAP
Mr LOPRIENO
Mr O'MAHONY
Mr PONS-GIMIER
Mr RAMOS MORGADO
Mr SCHOU
Mr SOMOGYI

Apologies for absence
Mr MUSCARDIN
Mr STÜTTGEN

Commission
Mr GONTIER
Ms DE CRAWHEZ

OPINION OF THE SCC CONCERNING

BERGASOL SUNTAN PRODUCTS CONTAINING 5-MOP

(Opinion expressed on 23.1.90)

. THE COMMITTEE'S MANDATE

5-MOP occurs naturally in oil of bergamot (up to 3000 ppm).
The use of this oil is at present allowed in suntan products by the Cosmetic Directive, but the SCC have recommended that the maximum level of 5-MOP arising from the presence of natural assences in suntan products should be 1 ppm.

The SCC have now been asked to consider data on a specific range of suntan products marketed by Bergasol and containing 15-60 ppm 5-MOP (as the natural product Bergapten). These also contain sunscreens and it is claimed that such formulations behave differently to 5-MOP alone, particularly with regard to the photomutagenicity and photocarcinogenicity which gave rise to the concern with 5-MOP itself.
Bergasol have provided a substantial amount of data to support this submission.

The question that the SCC have to consider is difficult and raises points that the Committee does not, in general, have to consider. There is a finely balanced argument regarding health risks and benefits, whereas in general the health risks of cosmetic ingredients are minimal.

In the case in question, the active melanogenic ingredient, 5-MOP, clearly gives rise to concern regarding its photomutagenic and photocarcinogenic potential.

However, sunlight itself induces skin cancer and the suntan products in question are designed to enhance tanning. More rapid tanning leads to a reduced cumulative dose of UVA and UVB radiation whilst attaining a tan that is protective against subsequent exposure.

There is also the additional problem as to how 'clearance' could be given to specific products under the Cosmetics Directive which is designed to consider either allowed or prohibited ingredients rather than products.

. <u>SUMMARY OF THE AVAILABLE DATA ON THE TOXICITY OF 5-MOP ITSELF</u>

5-MOP itself is well absorbed orally and may be absorbed through the skin. It is widely distributed in tissue, extensively metabolised, and rapidly excreted in the urine. It has very low oral toxicity, but is strongly phototoxic, producing marked skin reactions after exposure to UV radiation. Only limit data are available on sub-acute toxicity.
The combination of 5-MOP plus UV radiation is clearly mutagenic, much information being available on the interaction of psoralens with DNA in the presence of UVA. Linear psoralens such as 5-MOP are bifunctional, forming DNA adducts that react further producing DNA cross-linking.
5-MOP plus UVA has been shown to produce positive results in mutagenicity studies in bacteria, yeasts and mammalian cells, and this combination is also clastogenic.

Several studies have been carried out to investigate the skin carcinogenicity of 5-MOP plus UVA in animal bioassays.
Such treatment has clearly been shown to produce skin cancer in mice. There are no adequate data available to assess the carcinogenicity of 5-MOP in the absence of UV radiation, nor are any adequate data available in man.
However the combination of 8-MOP plus UVA (PUVA treatment) is clearly carcinogenic in man, and in the IARC Group I Carcinogens (i.e. known human carcinogens).

. BERGASOL PRODUCTS : DATA SUBMITTED BY BERGASOL

Studies in animals and human volunteers showed no significant skin irritation or sensitization, nor any significant photo-irritancy or photo-toxicity. Furthermore the data provided by Professor Agache indicates that the absorption of 5-MOP from such formulations is minimal (less than 1%), although greater amounts of 5-MOP were present in the stratum corneum itself.

The crux of the problem is however the potential photomutagenicity and photocarcinogenicity with respect to the induction of skin cancer and this is considered in some detail below :

Several studies have been carried out to investigate the mutagenicity of Bergasol preparations using Salmonella typhimurium TA 102, but interpretation of these are difficult since somewhat conflicting results were obtained, possible due to differences in the amount of material diffusing from the oil emulsion formulations through the agar medium.
Some studies indicated a reduction in the number of revertants but positive results were obtained with commercial preparations and mutagenic potential cannot be ruled out from these studies.

In addition it was pertinent that data from limited studies to investigate the mutagenicity of 'blister fluid' obtained from skin preparations of human volunteers treated dermally with a Bergasol preparation suggested that this had some photomutagenic potential when tested using the yeast strain Saccharomyces cerevisiae D7 (studies by Professor Dubertret).
However, in view of the very limited nature of these studies (no dose response, no repeat of the results in an independent experiment) no firm conclusions can be drawn.

Long term bioassays for skin cancer in mice have given conflicting results.
Limited data from an early study showed no protection using a UV filter (ethyl hexyl paramethoxy cinnamate). More extensive studies using a Bergasol formulation containing 5-MOP plus 2 UV filters (2 ethylhexyl, 4 methoxy cinnamate and 1.7, 7-trimethyl, 3-benzylidene-biocyclo (2,2,1) - 2 heptanone) revealed substantial protection against skin tumour induction in albino mice, but only during the treatment period (45 weeks). Animals exposed to SSR after treatment with 5-MOP plus sunscreens then went on to develop tumours sooner than those treated with sunscreen alone.
In another study in pigmented mice of the same strain, much less protection was seen, but it was argued that these mice were not an appropriate model for human pigmentation due to the absence of epidermal melanin. It is thus difficult to assess the significance of these studies or to identify the mechanisms involved. It is not possible to assess the increased risk of skin cancer in those using formulations containing 5-MOP plus UV filter. The data do not allow concerns regarding this aspect to be discounted.

More recent studies have been reported using an 'accelerated' *in vivo* model for skin tumourigenicity based on hairless mice treated with UV radiation (with or without bergapten containing 50 ppm 5-MOP) as 'initiator' for 4 months together with croton oil, as promoter, for the first 2 months. At the end of the 4 month period a similar incidence of tumours was seen following treatment with UV radiation alone, *ca* 60% incidence of skin papillomas. A marked reduction of the incidence of skin tumours was seen when sunscreens were incorporated in the treatment formulation (20% incidence) but the most impressive results were obtained by the additional incorporation of antioxidants such as BHT or B carotene or by the addition of an inhibitor of polyamine (putrescene) synthesis.

. BENEFITS OF INCORPORATION OF BERGAPTEN (5-MOP)

Considering the protective effect of 5-MOP in sunscreens, data has clearly established that the inclusion of 5-MOP enhances pigmentation. In addition the series of studies by Dr. Young showed that the 5-MOP induced tan protected against UV radiation induced DNA damage in humans as measured by unscheduled DNA synthesis (UDS) in individuals of skin type I to V.
It was particularly noteworthy that the most sensitive population for UV induced skin cancer (ie. types I and II) became like types III and IV in this regard.

However the mechanisms underlining the effects seen are not known and it is important to note that UDS is an indirect measure of damage, and a decrease may reflect a reduction in repair.

The situation is complex and it is clear that there is no direct proportionality between UDS and SSR induced DNA damage. The extent to which 5-MOP induced cross-linking is detected is unknown. It is premature to draw any conclusions from this work. No meaningful assessment can be made of the relative risk of induction of a tan by 5-MOP plus SSR compared to SSR alone.

CONCLUSIONS

In summary consideration of the health risks of products containing 5-MOP gives rises to concern. 5-MOP itself plus solar simulated radiation (SSR) is clearly mutagenic, and produces an increased incidence of skin cancer compared by SSR alone. Data provided on the Bergasol products containing 5-MOP plus UV filters gives a less clear-cut picture, but does not allow concerns about this aspect to be completely discarded.

Concerning the benefits, 5-MOP is clearly efficacious at inducing a tan. The data provided on the protective effect of this tan on subsequent DNA damage induced by SSR, as measured by unscheduled DNA synthesis, are difficult to interpret, in view of the lack of knowledge regarding the mechanisms involved. It is not possible to draw any meaningful conclusions regarding the relative risk of inducing a tan by SST plus 5-MOP plus sunscreens, as compared to SSR suscreens alone.

Much of the information provided is of research interest, but was not believed to be directly relevant to the Bergasol products on the market. For example the studies of Khettab et al relating to the incorporation of antioxidants with sunscreens. Such combinations were shown to offer complete protection against the development of skin tumours in the 4 month initiation/promotion model in hairless mice.

The Committee were concerned about procedures whereby specific cosmetic products could be considered acceptable. In this instance it was argued that certain products were less harmful tan would be expected from the 5-MOP content alone, because of the protective effect of other substances present eg. sunscreens, but this would entail a need to consider not only the concentration of the component of concern (5-MOP but also the presence of other specific compounds at specified levels. This would, in effecct, result in the need for clearance of specific products). The question of balancing health risks versus health benefits had been raised. The Committee felt that it was questionable whether it was appropriate to consider such arguments for cosmetics ; it is a requirement that cosmetic products are not harmful.

On the available data the Committee felt that it was not possible to conclude that sunscreen products containing 15-60 ppm or more 5-MOP were without risk of producing harm due to their potential for photomutagenicity and photocarcinogenic effects.
It was the opinion of the Committee that the maximum amount of 5-MOP in all sun tan products should be 1 ppm.

REFERENCES

I ADDITIONAL DATA ON THE RISKS OF 5-MOP IN SUNSCREENS

Photomutagenic effect induced by psoralen deviatives in the yeast Sacchoromyces cerevisiae.
Cassier - Chauvat C. and Averbeck D. pp 329-335.

Study of the protective activity against photomutagenicity of 5-MOP on Salmonella typhimurium TA 102 strain by UV filters in suntan preparations.
Marzin D. and Olivier PH. pp. 337-43.

Study with 5-MOP, Bergamot oil and Bergasol in mouse skin carcinogenicity assays.
Chouroulinkov I. Lasne C. Nguyen-By G. pp. 345-55.

Experimental photocarcinogenesis of psoralens.
Young A.R. and Walker S.L. pp. 357-66.

II. PROTECTIVE VALUE OF PSORALEN INDUCED PIGMENTATION

Enhanced skin tanning by psoralens and a sunscreen psoralens product in minature pigs.
Urbach F. and Sambuco C.P. pp. 389-97.

Comparitive photoprotection in humans by tans induced either by SSR or after a psoralen - containing sunscreen
Kligman A.M and Farlot P.

III. EVIDENCE FOR A PROTECTIVE EFFECT OF 5-MOP TANNING ON UV INDUCED DNA DAMAGE IN HUMANS

STUDIES BY Dr. A.R. YOUNG et al.

Pigment Cell Research I 350-4 1988

Paris Synposium 1989
A 5-MOP induced tan protects against DNA damage from a subsequent exposure to SSR in human skin pp. 431-432.

CONTENTS

45th meeting of SCC - October 10-11, 1990

REPORTS BY THE SCIENTIFIC COMMITTEE ON COSMETOLOGY

45th REUNION

OCTOBER 2nd, 1990

LIST OF PARTICIPANTS

Present

Mr. LOPRIENO
Mr. CARSTENSEN
Mr. CELLENO
Mr. COTTE
Mrs. DONY
Mr. FIELDER
Mr. KAPOULAS
Mrs. KNAAP
Mr. LINA
Mr. O'MAHONY
Mr. PARRA-JUEZ
Mr. PONS-GIMIER
Mr. SHOU
Mr. SOMOGYI
Mr. WHITE

Excused

Mrs. GUEDES-BAHIA
Mrs. MASSE

Not Present

Mr. KEMPER

Commission

Mr. ANGELIS
Mr. DILLEN
Miss. de CRAWHEZ

Expert

Mr. AGACHE

COMPLEMENTARY OPINIONS CONCERNING

THE USE OF CERTAIN PRESERVATIVE

AGENTS IN COSMETIC PRODUCTS

1. OXABAN A

. FORMULA AND SYNONYMES

EEC n° 60
Colipa P75
4,4-dimethyl-1,3-oxazolidine
dimethyl oxazolidine
oxadine A

. For preceding informations and expressed opinions consult :

- Reports of the SCC second series EUR 8634, 26/6/1982.
- Reports of the SCC, 38th reunion 10-11/10/1988.

. PRECEDING CONCLUSION

Classification : B.

. FURTHERING INFORMATIONS

MUTAGENICITY

Regarding mutagenicity data negative results were consistently obtained when the ability of the compound to produce gene mutation was investigated in Salmonella.
Results in one case showed a reproducible dose-related increase in revertants in TA98 and TA100, but this was less than twice background and was not significant (ref. 10 ; ref. 18). Negative results were obtained in a second assay using TA98 and TA100, that has been briefly reported (Monte et al, Fd Chem Tox 21 (5) 695-696 (1983)).
A fluctuation test in the S. typhimurium strains was also negative (ref. 17). A chromosomal aberration test in human lymphocytes in vitro with and without S-9 mix was however positive at all doses (ref. 19).

An in vitro test in CHO-cells showed dose-related increases in activity, and a mouse lymphoma assay was also positive (file dated 11-5-83). The compound thus has mutagenic potential, as demonstrated from in vitro studies ; the notifiers argue that in view of the rapid reactions with protein, this mutagenic potential would not be expressed in vivo.

Data from 3 in vivo studies are available. Negative results were obtained when the compound was investigated for its ability to produce chromosome damage in the bone marrow of rats treated with 20, 40 or 80 mg/kg using the intraperitoneal route.

Similarly negative results were obtained in a bone marrow micronucleus test the compound being given as a single oral dose of 500 mg/kg (estimated to be the MTD) and harvesting at 24, 48 and 72 hours. Thus the clastogenic potential seen in vitro does not appear to be expressed in vivo, at least in the bone marrow.

Similarly no effects were seen in a study to investigate single strand DNA breaks in testicular DNA following treatment with 5 and 50 mg/kg compound, as a single dose, or when given on 5 successive days ; (50 mg/kg was estimated to be the MTD).

Cellular preparations were obtained from the testes at 2, 6, and 24 hours post dose. No increase in the amount of low molecular weight DNA could be seen cf controls.

Negative results were thus obtained in this assay for DNA damage in the testes. These 3 studies indicate that the compound does not express its mutagenic potential in vivo in either the bone marrow or germ cells.

CONCLUSION

In conclusion the material was found to be only moderately toxic in short-term oral and dermal studies in rats. The no-effect levels were 50 and 25 mg/kg respectively (expressed as commercial product). Studies on dermal absorption are not available but on acute oral and dermal toxicity suggest dermal absorption to be considerable.
No evidence of embryotoxicity or teratogenicity was found in rabbits treated dermally with up to 300 mg/kg. The product showed clastogenic properties in mammalian cells *in vitro*, indicating mutagenic potential, but this was not expressed *in vivo*. Negative results were obtained in studies to investigate clastogenicity in bone marrow of rats, using both metaphase analysis and the micronucleus test, and using maximum tolerated doses of compound.
In addition negative results were obtained in an assay for DNA damage in the testes (alkaline elution method to detect single strand breaks).
It should however be noted that the material is nitrosatable.

Classification : A.

REFERENCES
++++++++++

1. Altech Laboratories Report n° AL/1/54 20/06/1973

1a. Altech Laboratories Report n° AL/1/59 31/06/1973

2. Altech Laboratories Report n° AL/1/53 19/06/1973

3. Altech Laboratories Report n° AL/1/118 06/08/1974

4. CSC Bioassay Report 30775 01/07/1975

5. Altech Laboratories Report n° AL/1/56 10/07/1973

6. IMC Pharmacology Laboratory PL Report n°60 04/08/1978

7. Products Investigations, Inc.,Report n°PI/09/46/571C 24/07/1979

8. IMC Pharmacology Laboratory Report n°PLR/68 23/02/1979

9. Avon Products, Inc., 24/04/1980

10. EG & G Mason Research Institute 23/05/1980

11. 12. & 14 MANUFACTURERS' DATA

13. International Research and Development Corporation November 30, 1973

15. International Minerals and Chemicals Corporation Report n°PLR-79 22/08/1979

16. NOTOX 0542/697 June, 1987

17. Toxicol Laboratories Ltd. Report n°36/8303 (April 1983)

18. Toxicol Laboratories Ltd. Report n°35/8303 (April 1983)

19. Toxicol Laboratories Ltd. Report n° SR 634. Reported 9/832 (March 1983)

20. Toxicol Laboratories Ltd. Report n°216/8308 dated 22.3.84

21. International Research and Development Corp., 13/02/1986

22. NOTOX b.v., Project n°01522, July 1989, 's Hertogenbosch, The Netherlands

23. Microbiological Associates, Inc., Study n° T5562.390 June 1989, Bethesda, MD, USA

2. DIAZOLIDINYLUREA

. FORMULA AND SYNONYMES

EEC n° 17
Colipa P79
N-(hydroxymethyl)-N-(dihydroxymethyl-1,3-dioxo-2,5-imidazolidinyl-4)-N'-(hydroxymethyl) urea
GERMALL II

. For preceding informations and expressed opinions consult :

- Reports of the SCC - fifth series EUR 11080, 25/3/'85
  - 34th reunion 30/6/'87
  - 42nd reunion 31/10/'89

. PRECEDING CONCLUSIONS

Classification : B.

The confirmed NEL of 100 mg/kg justifies the continued approval of the substance Classification A. for this relatively new preservative that finds increasing use in cosmetics may be considered if results :
- of an oral teratogenicity study
- of a gene mutation test in mammalian cells become available.

. FURTHERING INFORMATIONS

A number of studies have been carried out to investigate the mutagenic potential of this compound. Its ability to produce gene mutation in bacteria has been investigated using Salmonella typhimurium TA98, 100, 1535, 1537 and 1538. Negative results were obtained in all instances except in one study with 1537, when there was some evidence of weak activity in the presence of rat S-9.

However this was not reproducible in further studies using both rat and hamster S-9 (various concentrations) and negative results were also obtained against TA97. It can be concluded that Germall II is essentially ne-negative in these assays. Negative results were also obtained in a test for chromosomal damage (metaphase analysis) in CHO-cells, but concentrations up to only 1.5 µg/ml could be used, because of toxicity (1 mg/ml caused 70% mitotic inhibition (submission X, ref. 27), limiting the value of this study. However negative results were also obtained in an in vivo bone marrow micronucleus test, using oral doses of 1200, 2000 and 2800 mg/kg and multiple harvest times (30, 48 and 72 hours), indicating that Germal II is not clastogenic in bone marrow in vivo.

CONCLUSION

The Committee had requested a study using the oral route, since this would have provided much more information on the teratogenic potential of Germal II than the dermal study actually carried out. It is likely that (eg from physiochemical considrations) the compound is poorly absorbed through the skin, although there is little data on this aspect. The results do however indicate that compound is unlikely to give rise to any concern re teratogenic effects following dermal exposure.

CLassification : A.

LABORATORY REPORT REFERENCES

. Acute Oral Toxicity in Rats ; Leberco Laboratories
Assay n° 11857, January 19, 1981

. Acute Dermal Toxicity in Rabbits ; Leberco Laboratories
Assay n° 11858, January 28, 1981

. Primary Skin Irritation in Rabbits ; Leberco Laboratories
Assay n° 10942, November 21, 1980
Assay n° 11850, December 29, 1980

. Primary Skin Irritation in Guinea Pigs ; Bio/dynamics, Inc.
Project n° 4230-82 ; January 28, 1983

. Eye Irritation in Rabbits ; Leberco Laboratories
Assay n° 11859, December 30, 1980

. Contact Sensitization in Guinea Pigs ; Leberco Laboratories
Assay n° 11861, February 4, 1981

. Contact Sensitization in Guinea Pigs ; Bio/dynamics, Inc.
Project n° 4230-82, January 28, 1983

. 90 Day Subacute Oral Toxicity in Rats ; Leberco Laboratories
Assay n° 16545, December 30, 1981

. Ames Test/Standard Assay ; Microbiological Associates
Report Study n° T2039.501, July 29, 1983

. Ames Tests/Modified Assay ; Microbiological Associates
Report Study n° T2039.501008, September 15, 1983

. Repeat-Insult Patch Test with Lotion ; Charles of the Ritz
Report Protocol n° CRG-38, June 4, 1982

. Elicitation Patch Test ; Dr. Alexander Fisher
Personal Communication, April 30, 1984

. Phototoxicity : Unpublished Company date 81-82-84

. Additional date

. Dr. G. Foster ; Independant consultive
Safety Assessment Summary, May 14, 1985

. 90 Day Oral ; Sutton Laboratories New Jersey

. Release of formaldehyde ; J.F. Walker ; American Society

. Monograph ; Series 3rd ed ; P 471-472

LITERATURE REFERENCES

1. Cosmetic Ingredient Review Expert Panel, Final Report of the Safety Assessment for Imidazolidinyl Urea, J. Environmental Pathology and Toxicology, 4(4) : 133, 1980
2. Berke, P.A. and Rosen, W.E., J. Soc. Cosmetic Chemists, 31:37, 1980
3. Rosen, W.E., Berke, P.A., Matzin, T. and Peterson, A.F., J. Soc. Cosmetic Chemists, 28:83, 1977
4. Berke, P.A. and Rosen, W.E. J. Soc. Cosmetic Chemists, 29:757, 1978
5. Richardson, E.L., Cosmet. Toilet., 92(3):85, 1977 and 96(3):91, 1981
6. Litchfield, V.T. and Wilcoxon, F., J. Pharmacol. Exp. Ther., 96:99, 1949
7. Draize, J.H., Woodward, G., and Calvery, H.O., J. Pharmacol. Exp. Ther., 82:377, 1944
8. Landsteiner, K. and Jacobs, J., J. Exp. Med. 61:643, 1935
9. Magnusson, B. and Kligman, A.M., J. Investigative Dermatol. 52:268, 1969
10. Foster, G.V. in "Safety and Efficacy of Topical Drugs and Cosmetics", eds. A.M. Kligman and J.J. Leyden, Chapter 6, Pg 107, 1982
11. Ames, B.N., et. al., Mutation Research, 31:347, 1975

3. OXABAN E

. FORMULA AND SYNONYMES

Colipa P90
7-Ethylbicyclooxazolidine
Bioban/Amine CS-1246
Zoldine ZE
Oxazolidine E
P1601

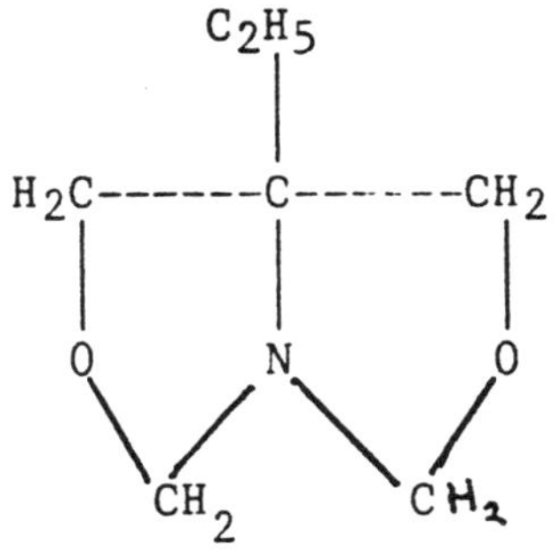

. CHARACTERISTICS

The material is specified by COLIPA as being>97.5% pure, however the substances tested did not reach this specification, with purity being as low as 96%. The major impurities were 4,4-dimethyl oxazolidine and 4-ethyloxazolidine (2%) and water (2%).

The compound is soluble in water and most organic solvents, exceptions are cyclohexane and 1,4-dioxane.

. USE

The proposed use in cosmetics is in non-rinsed-off products, excluding oral hygiene and mucous membrance products, at concentrations up to 0.3%.

. RECAPITULATION OF THE STUDIES OF TOXICITY

The compound has low acute toxicity.
Oral LD50 values in rats are 3.7 g/kg bw (females) and 5.3 g/kg bw (males) (ref. 1).
The dermal LD50 value in rabbits was 1.95 g/kg bw (combined value for abraded and unabraded skin) (ref. 1).
An inhalation (aerosol) LC50 of 3.1 mg/litre air was obtained from a 4 h. whole body exposure of male and female rats to an aerosol mist of mass median aerodynamic particle size in the range 3.9-4.7 µm (ref. 2).
Based on the acute toxicity data, dermal and oral absorption appears to be of a similar order, though no studies of absorption/permeation were performed.

ORAL TOXICITY

In a 28 day oral study, groups of 5 rats/sex received 100, 300 or 1000 mg/kg bw/d given in deionised water. Both 1000 and 300 mg/kg bw/d produced local effects on the stomach, indicative of an irritant effect. Significant changes in many haematological and clinical chemistry parameters were seen in both sexes receiving 1000 mg/kg bw/d : evidence of anaemia, increased WCB count with neutrophil and lymphocyte numbers increased, thrombocyte numbers were increased, serum potassium and phosphate levels were increased, with glucose levels reduced.
Increases in relative oragan weights were seen for liver, adrenals, testes and kidneys ; absolute values for adrenals were increased in both sexes, despite the reduced body weight seen at 1000 mg/kg bw/d.

Microscopic examination was limited with no changes reported in spleen, liver, kidney, adrenals, heart and testes. Total protein levels were reduced in males.
In the 300 mg/kg bw/d groups there was evidence of anaemia in males and of increased thrombocyte counts in both sexes. Serum phosphate levels were increased in both sexes whilst glucose levels were reduced in both sexes (not statistically significant in females). Adrenal weight was significantly increased in females. At 100 mg/kg bw/d non significant decreases in serum glucose and WBC numbers were seen in both sexes together with evidence of increased adrenal weights. Though there was some evidence of compound related effects at the lowest dose these were not statistically significant and 100 mg/kg bw/d may be taken as the no effect level (ref. 5).

CUTANEOUS AND MUCOUS TOLERANCE

DERMAL TOXICITY

A preparation of unknown specification (probably >96%) was severe irritant to the rabbit eye, with washed eyes showing more damage than unwashed.
The effects persisted for 7 days. No dilutions were tested (ref. 1).

A test of primary skin irritancy in rabbits using 0.5 g of undiluted material of unknown purity produced severe irritation following 24 h. occluded exposure (ref. 1).
A 4 h. occluded exposure produced signs of mild irritation, but no tissue destruction. Repeated application (21 days) of a 0.7% solution (6.7 mg/ml) produced erythema in rats (ref. 4). Repeated application of a 0.3% solution (12 applications over 3 weeks) produced mild irritation in one of 100 human volunteers (ref. 3).

A skin sensitisation test performed by the Buehler method was defined as inconclusive (no details given) (ref. 1). A second test using the Landsteiner and Jacobs procedure (a non-adjuvant technique) showed a 0.5% solution to be non-sensitising (ref. 1). A study in 100 human volunteers found no sensitising potential following repeated applications of a 0.3% solution (ref. 3).

A 21 dermal toxicity study in rats (n=6/sex/group) at dose levels of 30, 100 or 300 mg/kg bw/d applied daily for 5d/week in a deionised water solution (0.75, 1.25 or 3.75% w/v) produced dose-related irritation. Eschar formation was observed in the majority of top dose animals and in 3 males and 2 females from the mid-dose group. Dose-related increases were seen in relative adrenal weights in females and GPT in both sexes, though these did not achieve statiscal significance. A dose-related increase in relative kidney weights was seen in females, reaching statistical significance at the top dose (ref. 4).

MUTAGENICITY

Regarding mutagenicity studies, results from a well performed Ames test showed no evidence of mutagenicity at concentrations between 6 and 600 µg/plate ± rat liver S-9 ; cytotoxicity was evident at 300 µg/plate and above (ref.6).

An assay for chromosomal aberrations in CHO cells showed no evidence of clastogenicity with rat liver S-9 mix a slight increase in the numbers of cells with aberrations and in the numbers of aberrations per cell were reported.

No details of individual cultures or types of aberrations are given, making it difficult to assess the significance of the results. The concentrations used were 0.5 to 4 µl/ml, with only slight cytotoxicity (about a 15% decrease in mitotic index) seen at the top dose. Furthermore only one harvest time was investigated, and the results were not confirmed in an independent experiment.
This study was inadequate to draw any definite conclusions (ref. 7).

An assay for unscheduled DNA synthesis in rat primary hepatocytes showed increased activity at all concentrations (0.25 - 4 µl/ml) but this was not dose-related or significant and results were negative (ref. 8).

TERATOGENICITY

In a teratogenicity study groups of 25 mated female rats received 50, 250 or 650 mg/kg bw/d on days 6-15 of gestation, by gavage in deionised water.
There were clear effects at 650 mg/kg bw/d : decreased maternal body weight and increased incidences of malformation (eg cleft palate and umbilical hernia of intestines). The fetal effects were concentrated in 4 litters including the dam with lowest body weight gain. No increases in malformations or variations were recorded at the mid and low doses. The NEL in this study was 250 mg/kg bw/day (ref. 9).

CONCLUSION

A further study to assess whether Oxeban E can produce chromosome damage in vitro in mammalian cells, by metaphase analysis, using a more rigorous protocol, to current standards is necessary before any definite conclusions can be drawn regarding the clastogenic potential of the compound.
In addition skin sensitization potential should be investigated using a more sensitive method, the magnussen - kligmann maximisation Test being the recommended method.

In addition information on the amount and rate of formaldehyde generation from this preservative should be obtained.

Classification : B.

REFERENCES

1. "Acute toxicity profile of P-1601", PLR-124, June 25, 1980. IMC Toxicology Laboratory, Terre Haute, Indiana, USA.

2. "Acute mist inhalation toxicity of ZOLDINE ZE in rats", PLR-355/AMR-218, Jan 4, 1983. Bio/dynamics, Inc., Project No. 82-7567, Terre Haute, Indiana, USA.

3. "Skin irritation and sensitization potential in human volunteers", PLR-91, Jan. 4, 1980. Product Investigations, Report No. 632B, Terre Haute, Indiana, USA, October 1979.

4. "Assessment of the subacute dermal toxicity of BIOBAN CS-1246 in the rat: 21-day study", NOTOX 0170/241, 's-Hertogenbosch, The Netherlands, October 1985.

5. "Assessment of the toxic potential of BIOBAN CS-1246 following oral administration in rats over 28 days", RCC NOTOX 0879/1097, 's-Hertogenbosch, The Netherlands, January 1989

6. "Determination of mutagenicity potential of ZOLDINE ZE - Ames Test", PLR-341/AMR-144, October 13, 1982. Microbiological Associates, Study No. T 1758.501, Bethesda, Maryland, USA, September 1, 1982.

7. "Chromosome aberration in Chinese hamster ovary (CHO) cells", Toxicon Project No. 87G-0140. Toxicon Corporation, Woburn, Massachussetts, USA, January 6, 1988.

8. "Unscheduled DNA synthesis in rat liver primary cell cultures", Toxicon Project No. 87G-0139. Toxicon Corporation, Wobum, Massachussetts, USA, December 28, 1987.

9. "A Teratology study in rats with Amine CS-1246", May 30, 1989. WIL Research Laboratories, Inc., Ashland, Ohio, USA.

4. OMADINE MDS

. FORMULA AND SYNONYMES

EEC n° 41

Colipa P50

Dithio-2,2'-bispyridine-dioxide 1,1' with magnesium sulphate trihydrate

Pyrithion disulphide + magnesium sumphate

. FOR PRECEDING INFORMATIONS AND EXPRESSED OPINIONS CONSULT :

Reports of SCC : 33rd reunion 2/2/87.

. PRECEDING CONCLUSION

Classification : D.

The toxicity of the substance, the dermal exposure and the potential dermal absorption occur with a too tight margin of safety to permit the use of this substance in cosmetics (even limited to anti-dandruff shampoos).

. FURTHERING INFORMATIONS

In a second 90-day study, rats received 15, 0.8 or 0.5 mg/kg bw/day for 13 weeks in a water vehicle. Due to signs of toxicity the top-dose was reduced to 7.5 mg/kg bw/day after 2 weeks In the top-dose group 9/15 females and 1/15 males died. At autopsy, animals from the top-dose group showed muscle wasting, reduced body weight gain, hind limb muscular atrophy, hepatocyte vacuolation, thickening of the non-glandular stomach mucosa and pancytopaenia. In the mid-dose animals, hind limb muscular wasting was seen in 2/15 females, alterations in various clinical chemistry and haematological parameters were reported. At the lowest dose reductions were seen in total serum bilirubin and leucocyte counts and the incidence of hepatocyte vacuolation was slightly increased. Histology was limited at mid and low-doses but did include liver, muscle and sciatic nerve. The lowest dose used in this study is not a clear NEL.

- In man 1.26% of a dermal dose was recovered in the urine over 120h when 160 µg/cm2 were applied in a shampoo formulation to the forearms of 3 volunteers. The treated area was rinsed after 4 minutes. Total absorption was less than 3%.

- Dermal retention in shaved and unshaved rats receiving 0.15 ml of shampoos containing 1% omadine was 0.85% and 0.54% of the dose respectively. 17% of the dose was found bound to the hair. Treated areas were rinsed after 4 minutes with samples taken at 1h. No measurements of absorption were made.

The Company have calculated using several assumptions that intakes of omadine by a 60 kg person would be between, 1.9 and 16 µg/kg bw/shampooing. A realistic figure based on 12 ml shampoo, 1% omadine and 3% absorption would be 60 µg/kg bw/shampooing, only a small margin below a minimal effect level of 500 µg/kg bw/day in a 90-day oral study.

CONCLUSION

Omadine possesses considerable toxic potential and is possibily genotoxic. Clear adverse effects were seen at oral doses of 0.8 mg/kg bw/day and 10 mg/kg bw/day dermally with a report of foetotoxicity following dermal exposure to 5 mg/kg bw/day. The margins between proposed uses and toxic levels are inadequate to ensure safety in use, therefore Omadine MDS should not be used in cosmetic products.

Classification : D.

REFERENCES

Colipa subm VIII Dec 89.

49. OLIN CORPORATION
13-w oral tox. study (rat)
ref n° OLA/12/89
April 89 - Toxicol. Laboratories England

50. The dermal absorption of 14 C (man)
HRC report n° HRC FOLN 5/89890 Sept 25, 1989

51. Deposition of 14C (on rat skin in presence and absence of hair)
HRC report n° HRC FOLN 6/89888 Sept 18, 1989
Final report by Huntingonn Research Center Ltd

5. C.T.A.B.

. FORMULA AND SYNONYMES

EEC n° 4

Colipa P72

Alkyl (C_{12}-C_{22}) trimethylammoniumbromide chloride

. FOR PRECEDING INFORMATIONS AND EXPRESSED OPINIONS CONSULT :

Reports of the SCC : 34th reunion 30/6/87

. PRECEDING CONCLUSION

Classification : B.

A chromosomal aberration test on mammalian cells was needed.

. FURTHERING INFORMATIONS

Negative results were obtained in an assay to investigate the ability of the compound to induce chromosome aberrations in Chinese hamster V79 cells both in the presence and absence of exogenous metabolic activation. Only very low concentrations could be used, up to 10 µg/ml, due to cytotoxicity. The highest concentration was associated with a 50% reduction in mitotic index.

Negative results were also obtained in a morphological transformation assay with cells of 13 day old Syrian hamster embryos, and up to 300 µg test substance/ml.

CONCLUSION

This substance has shown considerable systemic toxicity including indications of teratogenicity. However, absorption through the intestinal wall and through the skin is low. There was no evidence of mutagenicity in *in vitro* assays for gene mutation and chromosome damage.
Not to be used on a damaged skin.

Classification : A.

<u>REFERENCES</u>

Data sheet National Institure of Public Health,
the Netherlands, Sept 1978

Colipa Subm I April 1984

1. Skin absorption
BARTING and WINGEN (79)
Fd. Cosmet. Toxicol. Vol <u>17</u> p. 633-637

2. Terato
B. ISOMAA and K. EKMAN
Food Cosmet. Toxicol. Vol <u>13</u> (1975) p. 331-334

Colipa Subm II February 1987

3. Ames test
D.J.N. ROASACK and al (1977)
HUNTINGDON RESEARCH CENTER England

4. GLOXHUBER & WALLAT (1979)
Hauptabteilung Toxikologie
HENKEL - Düsseldorf Germany

5. K. INQUE and al (1980)
Food Cosmet. Toxicol. Vol <u>18</u> (1980) p 289-296

6. Reverse mutation test
U. MULLER (1976)
Battelle Institut Frankfurt Germany

7. Terato
C.A. LEWIS (1977)
Consumer Product Testing Fairfield, New Jersey USA

Colipa Subm III 12-1989

8. Chromosome Aberration Assay Chinese Hamster V79 cells in vitro
A. HEIDEMANN Cytotest cell Research GmbH & Co. KG
In den Leppteinswiesen 19, D. 6101 ROßDORF 15-9-1989

6. CHLORPHENESIN

. FORMULA AND SYNONYMES

EEC n° 2
Colipa P4
3-(p-chlorophenoxy) propane-1,2-diol
p-chlorophenyl-glycerol ether
CAS n° 104-29-0

. FOR PRECEDING INFORMATIONS AND EXPRESSED OPINIONS CONSULT :

Report of SCC : 37th reunion 12/4/88
42nd reunion 30/10/89

. PRECEDING CONCLUSION

Classification : C.

A long-term oral study and available information on dermal absorption were required.
Depending on the results, a teratogenicity study may be required.

. FURTHERING INFORMATIONS

Full details are however available of a 28-day oral toxicity study in rats given doses of 10, 100 and 1000 mg/kg compound by gavage as an aqueous suspension. Detailed autopsies were performed at the end of the exposure period and in addition serum immunoglobulin levels and B ; T lymphocyte ratios in blood and spleen were determined.
Compound related mortality was seen at the top dose, 1/5 male animals dying. Other effects noted at this level were reduced weight gain, abnormal posture and gait, reduced spleen and thymus weight and evidence of nephrotoxicity.

The only significant effects seen at 100 mg/kg were a slight reduction in haemoglobin levels. No pathology was seen in the spleen, lymph nodes, thysmus or bone marrow at any dose level. The no effect level was 10 mg/kg with only marginal effects at 100 mg/kg.

CONCLUSION

Chlorphenesin has low acute toxiicty, no significant irritant properties and has relatively low toxicity on repeated exposure by the oral route.
No data are however available on dermal absorption.
This is needed, and depending on the results, a teratogenicity study may be required.

Classification : B.

REFERENCES

Colipa Subm I, September 1983

1. MARTINDALE Ed 1978 (other uses)

2. HARRY, RALPH G.
Preservative for Cosmetic emulsions
Amer. Perfumer and Arom. 1956, Vol 67, pt 6, P. 33, 34, 48

ACUTE TOXICITY

3. Quarterly Journal of Pharmacy and Pharmacology
F. JARTLEY 1947 20(3) 388-395

4. BERGER F.M. (1948) p 470-481

5. BERGER F.M. G.M. FUKUI
Wallace labor. New Jersey p 552-563 J. Arteriosclerosis

MUCOUS and SKIN IRRITATION

3. 6-7-8 : HUMAN PATCH TEST 1982

40-DAY TOX.

3.

90-DAY ORAL

5.

IMMUNOSUPPRESSANT and ANTIANOPHYLACTIC ACTION

2.

9. H.Y. WHANG and E. NETER
Infection and Immunity ; July 1970 p 60-64
American Society for Microbiology

10. BERGER F.M., FUKUI G.M. and al
Selective suppression of Penicillin Hypersensitivity by certain Phenoxypropanediols
PROC. SOC. EXP. BIOL. MED. 124:303, 1967

11. BERGER F.M., FUKUI G.M. and al
The effect of Chlorphenesin on the immune response
J. IMMUNOL. 102:1024, 1969

12. HOOK W.A., SIRAGANIAN R.P.
Complement-induced Histamine Release from Human basophils
J. IMMUNOL. 118:679, 1977

13. LICHTENSTEIN L.M., ADKINSON N.F.
Chlorphenesin : a new inhibitor of IgE - Mediated Histamine Release
J. IMMUNOL. 103:866, 1969

14. MALLEY A., BAECHER L.
Inhibition of Histamine and SRS - A from Monkey Lung Tissue by Chlorphenesin
J. IMMUNOL. 107:586, 1971

15. J. EDEKSON, J.F. DOUGLAS and B.J. LUDWIG
Biochemical Pharmacology, Vol 18, p 2331-2338
Pergamon Press 1969
Wallace Laboratories Cranbury USA

REFERENCES

Colipa Subm IV 3/89

22. P.C. BROOKER, J.D. KING, V.M. GRAY
Chlorphenesin - Metaphase chromosome Analysis of Human Lymphocytes cultured in vitro
Huntingdon Research Center : Rep. SR098/881424 Feb 9, 1989

Colipa Subm III 12/87

Mutagenicity

20. Ames Test
Laboratories sérobiologiques Nancy
Dr. M. Pauly 15-10-1987

21. CHO - TEST
Laboratories sérobiologiques Nancy

REPORTS BY THE SCIENTIFIC COMMITTEE ON COSMETOLOGY CONCERNING THE USE OF CERTAIN COLOURANTS IN COSMETIC PRODUCTS

. THE COMMITTEE'S MANDATE

To give its opinion on the question whether the use of :

CI 12075
CI 15585
CI 45170
CI 26100

in cosmeticsproducts is admissible from the health point of view.

1. CI 12075

. FORMULA AND SYNONYMES

1-(2,4-dinitrophenylazo)-2-naphthol

(D and C Orange 17, Pigment Orange 5)

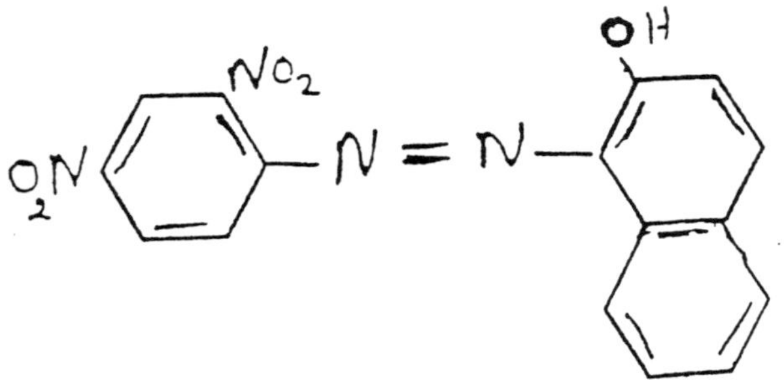

. USE

The colourant has been in use in cosmetics and drugs for external application since 1936. It is listed as an allowed colourant for all uses in the EEC Cosmetics Directive.

. RECAPITULATION OF THE STUDIES OF TOXICITY

Preamble : The Colipa submission I (May 1989) consists of a review of data and a risk assessment prepared by The Cosmetic, Toiletry and Fragrance Association, Inc of the US, dated April 15, 1983. This concentrated on long-term studies, particularly relating to carcinogenicity, and no data on acute toxicity, skin and eye irritation, sensitisation, and short-term studies for systemic toxicity were provided.
More importantly no data on the mutagenic potential of CI 12075 were given ; this is essential when assessing the relevance of the carcinogenicity data.

LONG-TERM ORAL TOXICITY AND CARCINOGENICITY

The CTFA - review contains a few details of 4 long-term studies conducted by HAZLETON Laboratories and summarised by the US-FDA in 1972.

In a rat-feeding study, dose levels of 0, 0.025, 0,1 and 1.0% were fed in the diet for 2 years. There was growth depression in the high-dose group, and dose-related haematological changes consisting of decreased haematocrit, haemoglobin content and red blood cell count. The urine showed a dose-related increase in bilirubin levels.

In a second rat-feeding study, 0, and 80 ppm was fed in the diet for 26 months, without inducing any adverse effect. This study was inadequate, in view of the low single dose used, to assess the carcinogenicity of CI 12075.

A dog-feeding study comprised diet groups fed 0, 0.025, 0.125 or 1.0% for 2 years. The top dose induced anaemia, enlargment of the thyroid, liver and spleen and changes in the skin. Increased weights of the thyroid and liver also occurred with 0.125% bilirubin excretion in urine was increased at all dose levels. There were no tumours related to treatment, but no conclusions can be drawn from this study in dogs, in view of its relatively short duration. A skin painting study in mice with a 1% aqueous suspension for 18 months produced acanthosis and hyperkeratosis, but no evidence of carcinogenicity. From these studies FDA estimated the ADI to be 0.0625 mg/kg (August 4, 1972).

More recent long-term studies were conducted by Bio/dynamics for CTFA. In one long-term study the rats were exposed from *in utero* by feeding diets with 0, 0.02, 0.05 and 0.1% from 60 days before mating through the gestation and lactation period and then to the descendents for 27 or 28 months, for males and females respectively. Slight growth retardation occurred with 0.1 and 0.05%. Decreased levels of creatinine and elevated levels of fasting glucose occurred in all treated groups. The incidence of lymphoreticular tumours was increased in high-dose females. (14% *of* 4% in controls).

A second long-term rat study was conducted in a similar was (using in utero exposure) but with only one high exposure level of 1.0%. This study lasted 26 months for the males and 30 months for the females. Growth depression, increased food intake and anaemia were observed in treated rats. A clear increase in benign hepatocellular tumours (adenomas) was seen in the female rats, with a slight increase in hepatocellular carcinomas. No increase in lymphoreticular tumours was seen in this study. There was also an increased incidence of proliferative non-neoplastic alterations in the liver (clear cell foci) as well as enlargement of liver cells.

A feeding study in mice used dietary levels of 0, 0.025, 0.25 and 1.0%. Increased mortality was observed with 0.25 and 1.0%. Anaemia occurred in the high-dose group. Pigment deposits were seen in the spleen, caecum and liver of the top-dose mice. The incidence of chronic myocarditis was increased in high-dose females. A slight increase (but not statistically significant) was seen in liver tumours (both hepatocellular adenomas and carcinomas) in the male animals only at the high dose level (1% in the diet).

The increased incidences of tumours in the foregoing studies have been evaluated in the CTFA review, which resulted in the conclusion that none of these studies indicated that D and C orange 17 is a potent or primary carcinogen. Therefore CTFA requested permanent listing of the colourant only for external cosmetics and drugs that are not subject to incidental ingestion.

DERMAL TOXICITY

Studies are briefly reported in the CTFA review on the absorption of CI 12075 through half thickness human abdominal skin using different vehicles. Rates from 0.009 ug/cm2/24 hour (estimated to represent ca 0.006% of applied dose) from a cream base to 0.0012 ug/cm2/24 hour (equivalent to ca 0.00009% applied dose) from a talc formulation.

MUTAGENICITY

However no consideration was given in the CTFA review to the mutagenicity data on this colourant. The published data are summarised below.

Several groups have reported positive results in bacterial assays various Salmonella typhimurium strains. Muzzall and Cooke (1979) obtained positive results using Salmonella typhimurium TA 98 both in the presence and absence of S9, and vs TA 100 in the absence of S9. Brown et al (1979) obtained positive results vs TA 98, 100, 1537 and 1538 both in the presence and absence of S9. Green and Pasteuka (1980) reported positive results were consistently obtained in these bacterial assays. Since activity was seen in the absence of S9, this was likely to be due to activation by endogenous nitroreductase known to be present in Salmonella. Mutagenicity data from other systems would be helpful, but no further data are available. The colourant clearly has mutagenic potential.

CONCLUSION

In summary CI 12705 clearly has mutagenic potential, as demonstrated by the induction of reverse mutations in Salmonella typhimurium, although no further data are available and it is not possible to fully assess the mutagenicity of the compound. However it has been demonstrated to have carcinogenic potential in long-term animal bioassays. In one study in rats an increase in lymphoreticular tumours was seen in females, and in a second study an increase in hepatocellular adenomas. In a study in mice an increase in liver tumours (both adenomas and carcinomas) was seen, but this was not statistically significant. It is accepted that the carcinogenicity data is relatively weak, but in view of the mutagenic potential of the compound, it cannot be discounted. The colourant should not be allowed in cosmetics.

Classification : D.

REFERENCES

"Final Review and Analysis of Scientific Studies and Risk Assessments supporting the Safety of D.C.Orange N°17 for use in External Cosmetic and Drug Products not subject to Incidental Ingestion."
CTFA , April 15, 1983.

2. CI 15585

. FORMULA AND SYNONYMES

Monosodium salt of 1-(4-chloro-O-sulpho-5-tolylazo-2-naphtol
(Pigment Red 53, C-Rot 55)

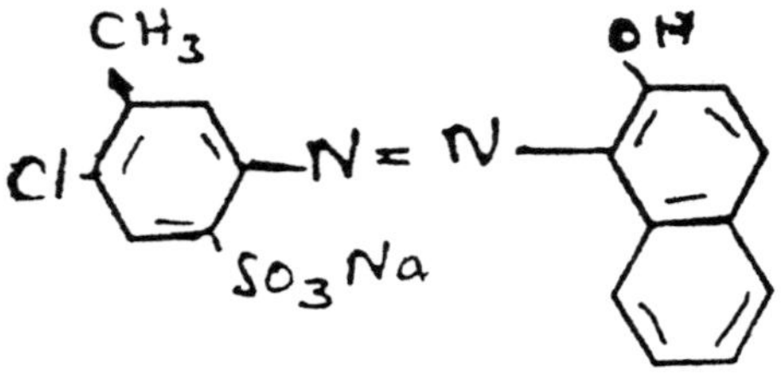

The barium salt of 1-(4-chloro-O-sulpho-5-nolylazo-2-naphtol)
D & C red n°9
Cas. Reg. n° 5160-02-01
MW: 376,7
$C_{17}H_{13}O_4N_2ClS^1$

. USE

This colourant has been used in external cosmetics and drugs, including those subjects to incidental ingestion, for very many years (since the late 1930s).
It is used widely in lipsticks.
Use levels are up to 5%.
It is allowed as a colourant under the EEC Cosmetics Directive for all uses except around the eye.

. RECAPITULATION OF THE STUDIES OF TOXICITY

The Colipa submission I (May 1989) consists of the final review and analysis of studies on D & C Red n° 9 prepared by the Cosmetic, Toiletry and Fragrance Association (CTFA) of the US, dated, August15, 1983.

The review does not contain any information on acute toxicity, skin, or eye irritation, or sensitisation, but concentrates on long term toxicity, especially carcinogenicity, and includes mutagenicity data.

ORAL TOXICITY AND CARCINOGENICITY

A subacute (90-day) feeding study in Fischer 344 rats with 0, 0.25, 0.50, 1.0 and 2.0% in the diet revealed enlargement of the spleen in all dose groups, and abnormal red blood cells. In a 20 week study in rats using the same dietary levels splenomegaly, and low haemoglobin levels and haematocrit values were observed (in the review these changes are evaluated as 'no significant adverse effects').

The review tabulates a few details of a 2 year rat feeding study and a 2 year dog feeding study conducted before 1976.

The 2 year rat study (Osbourne Mendel strain) was conducted with dietary levels of 0, 100, 500, 2500 and 10.000 ppm. Splenomegaly and abnormal RCBs were seen at the two high dose levels. The NEL was 500 ppm, or about 25 mg/kg bw/day. No neoplasia were observed.

The 2 year dog study comprised groups fed 0, 150, 1000 or 5000 ppm.

Splenomegaly and destruction of RBCs occurred with 1000 and 5000 ppm, increased liver weights with 5000 ppm. The NEL was 150 ppm or about 4 mg/kg. Again no neoplasia were observed, but the study was of too short a duration to allow any conclusions to be drawn regarding carcinogenicity in this species.

Carcinogenicity bioassays were carried out by Battelle (1977-79) as part of the NTP programme. Fischer 344 rats were given dietary levels of 1000 and 3000 ppm, and B6C3F1 mice were given levels of 1000 and 2000 ppm. Animals were exposed to treated diet for 103 weeks and the study terminated 1 week later. D & C Red 9 was carcinogenic in male rats causing an increased incidence of sarcoma of the spleen at the top dose, and a dose related increase in neoplastic nodules of the liver ; the significance of the latter is however questionable. There was no evidence of carcinogenicity in female rats. Nor was there any evidence of carcinogenicity in the B6C3F1 mice of either sex. The significance of the spleenic tumours seen in the F344 rats was questioned, as it was argued that the mechanism by which these arose was secondary to marked toxic effects on the spleen at the high dose (fibrosis) (see later for detailed discussion).

The CTFA sponsored long-term studies conducted by Litton are considered below.

A 2 year mouse study at dose levels of 0.50, 250 and 1000 ppm produced changes in cell count, haemoglobin and haematocrit in the high dose group. Other changes were not considered related to treatment.

A second mouse study using feeding levels of 0 and 2000 ppm. The only change attributed to treatment was chronic inflammation of the stomach.

A chronic study in Sprague Dawley rats, and involving in utero exposure, used the relatively low feeding levels of 0, 100, 200 and 500 ppm for 30 months. The only change observed was an increased spleen weight accompanied by decreases in red cell parameters, at 500 ppm and observed only at 12 months. The no effect level was 200 ppm.

A second chronic (30-months) rat study again in Sprague Dawley rats and with in utero exposure used only one high feeding level of 10.000 ppm. This level caused changes of the spleen, kidney, liver, pancreas and pituitary. The spleen changes were accompanied by 4 mesenchymal tumours of the spleen in the treated animals two of which were very uncommon in the rat-strain used ; two spleenic tumours were seen in the control. The incidence was, however, not statistically significant. In the adrenals of the treated rats, the incidence of hyperplasia and of phaeochromocytoma were increased in both sexes but the increase in the latter was not statistically significant.

Mechanism of induction of spleenic tumours in rats by D & C Red 9

Chronic studies in rats have shown that exposure to high dietary levels of D & C Red 9 results in the induction of spleenic tumours, particularly in Fischer F344 animals, with a high incidence of both fibrosarcomas and angiosarcomas being observed. There was no clear evidence for an increase in any other tumour type in rats, nor of any carcinogenic effect in mice. The spleenic tumours in rats were associated with levels that also resulted in marked toxicity to the spleen (capsular and parenchymal fibrosis), effects being observed in most animals.
A number of suggestions regarding potential mechanisms have been made ; these have recently been reviewed (Bus and Popp 1987). It was suggested by Goodman et al (1984) that spleenic toxicity, arising from an initial toxicity to erythrocytes (probably by an amine metabolite), followed by sequestration of the damaged erythrocytes in the spleen, leading to haemosiderin deposition coupled with enhanced delivery of toxic metabolites in the spleen, resulted in fibrosis and subsequently the formation of fibrosarcoma. A sligthly different mechanism was proposed by Weinburger et al (1985) nameley that acute vascular congestion, resulting from spleenic scavenging of chemically damaged erythrocytes may be an important initial toxic lesion to the spleen.

The vascular congestion would lead to spleenic haemorrhage, formation of fibrous tissue mass and, again in conjunction with accumulation of toxic metabolites within the spleen (derived from scavening erythrocytes) transformation of mesenchymal cells of the spleen, resulting in the expression of spleenic fibrosarcomas and a variety of other lesions. Haemosiderin deposition was not critical for the latter hypothesis, which is pertinent as there was no evidence of increased intra-splenic accumulation of iron containing pigment in the animals treated with D & C Red 9.

These data support the hypothesis that the sarcomas seen at high dose levels were secondary to toxcity, and that a threshold thus exists.

However a crucial question in this regard is whether the compound is acting by a "non-genotoxic" mechanism.

MUTAGENICITY

There are a number of reports of this colourant being investigated for its ability to produce gene mutation in bacteria using Salmonella typhimurium. Negative results were obtained by Brown et al (1979) using TA 98, 100, 1535, 1537 and 1538, both in the presence and absence of rat S9. Duplicate plating was used, and the results were not confirmed in an independent experiment. Muzzall and Cooke (1979) reported a negative result using the spot test method and also with a plate incorporation assay with strains TA 98, TA 100, TA 1535, TA 1537 but using a lipstick containing D & C Red 9 rather than the compound itself.
This study was too limited to allow any conclusions to be drawn regarding the mutagenic potential of D & C Red 9. In a recent report on compounds tested as part of the NTP programme (Zeiger et al 1988) the colourant is reported to be weakly positive against TA 97 in the absence of S9.

However examination of the data indicates that this result was at most equivocal. Negative results were obtained with the other strains. Thus D & C Red 9 has essentially given negative results in Salmonella assays.

No data are available on the clastogenicity of D & C Red 9 either in vitro or in vivo. However the ability of the compound to produce UDS in rat hepatocytes has been investigated both in vitro and also in an in vivo/in vitro liver UDS assay using oral doses of 500 mg/kg, with perfusion of the liver and harvesting after 2 and 5 hours (Kornbrust and Barfknecht 1985).
Negative results were obtained in both cases.

There is one brief report of a positive result in a cell transformation assay using Balb/c 3T3 system, but insufficient details were given to assess this study (Trennent et al 1986).

To summarise the mutagenicity data, D & C Red 9 has given negative results in assays to investigate its ability to produce gene mutation in Salmonella. Negative results were also obtained when the compound was investigated for its ability to produce UDS in cultured primary hepatocytes, and also in an in vivo liver UDS assay using an oral dose level of 500 mg/kg. However no data are available on its clastogenic potential. Although the colourant does not appear to have significant mutagenic potential, the data are too limited to allow difinite conclusions to be drawn regarding this aspect.

MISCELLANEOUS

Dermal absorption was examined by an in vitro test with human skin using Franz diffusion cells. The maximum absorption found was 0.06%, and the calculated total absorption of cosmetic use was ca ug/day, or ca 0.07 ug/kg/day.

CTFA calculation of ingested colourant from lipstick (assuming present at 2%) use (assuming ingestion of 50% CTFIA of the amount applied) resulted in a maximum daily intake of 0.4 mg, or 0.008 mg/kg/day. However much higher values would be obtained using other assumptions (0.01 g per application, up to 6 applications per day).

CONCLUSION

CI 15585 has been shown to be carcinogenic in rats at high dose levels producing fibrosarcomas and angiosarcomas of the spleen. No clear evidence of tumour induction at other sites was seen in the rat, nor of any carcinogenic effects in mice. Dietary levels that produced tumours of the spleen in rats was also associated with marked toxic effects (fibrosis), and the tumours may be secondary to toxicity. The no effect level in chronic studies in rats and dog was about 25 and 4 mg/kg/day respectively. The available mutagenicity data are negative in Salmonella assays for reverse mutations, and in both in vitro and in vivo assays for UDS in hepatocytes. However no information is available on its ability to produce chromosome damage either in vitro or in vivo.

In view of the importance of adequate reassurance that a genotoxic mechanism was not involved in the induction of the spleenic tumours, data form a well conducted *in vitro* assay to investigate chromosome damage in mammalian cells (metaphase analysis) are needed. In addition, the relatively low NEZ for toxicity, to the spleen (4 mg/kg) gives rise to concern. Detailed information on capsure levels are needed, and it may prove necessary to set - a permitted concentration in mucous membrane products below 5%.

Classification : C.

REFERENCES

. Bus JS and Popp JA. - Fd Chem Toxic 25 619-26 (1987)
. Goodman DG et al. J Nat Can Inst - 73 265-73 (1984)
. Weinberger MA et al. J Nat Can Inst - 75 681-7 (1985)
. Brown JP et al. Mut Res - 66 181-5 (1979)
. Muzzall JM and Cook WL Mut Res - 67 1-8 (1979)
. Zeiger E et al. Environ Mol Mutagen - 11 Supp 121-158 (1988)
. Kornbrust D and Barfknecht T Environ Mutagen - 7 101-20 (1988)
. Tennant RW et al. Environ Mut - 8 205-27 (1986)

3. CI 45170

. FORMULA AND SYNONYMES

3,6-bis diethylamino - 9(2-carboxyphenyl) xanthylimonim chloride

D & C Red n° 19

C-Rot 59

Rhodamine B

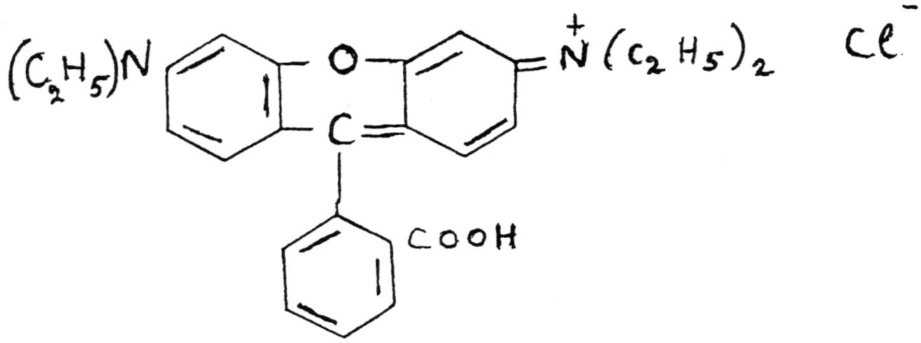

. CHARACTERISTICS

The compound is a red colourant
Soluble in water and ethanol

. USE

It is allowed as a colourant in all cosmetic products under the Cosmetic Directive.

. RECAPITULATION OF THE STUDIES OF TOXICITY

The Colipa submission on this substance consists of the Final Review and Analysis of studies on D & C Red n° 19 and the risk assessment prepared for the Cosmetic Toiletry and Fragrance Association (CTFA) dated February 15th 1983. This review mainly deals with carcinogenicity studies but also briefly reviews mutagenicity data.

The CTFA review tabulates chronic studies in mice, rats and dogs conducted prior to 1976. None of these studies was adequate for an assessment of the carcinogenic potential of CI 45,170.

LONG-TERM ORAL TOXICITY

A lifetime study in mice involved the use of a single concentration given in drinking water (0.05%) for 52 weeks, with the animals then observed for their lifespan. Polyposis of the stomach was seen in 3/30 treated mice, intestinal tumours in 2 and lymphomas in 7. There was no concurrent control group.

A lifetime rat study with 0.13% in the diet given for 7 months and the 0.2% in the diet, did not produce any significant effect. However only 18 animals survived 300 days and only one survived 661 days.

A 24 month study in rats using dietary levels of 0.008, 0.025 and 0.15% showed a dose related increase in thyroid activity and weight and epithelial nodules in the thyroid at the high dose level.

A 26 month rat study using 0.004% in the diet produced pink colouration of the eye lens but no other adverse effect. The single dose level used was very low.

A three generation rat study using doses of 0,0004 and 4 mg/kg/bw/day given in the diet, did not reveal any treatment-related changes.

A 24 month study in dogs using 0,0.0008, 0.3125 and 0.1% levels in the diet showed no adverse effects.

DERMAL ABSORPTION

The CTFA report contained details of skin absorption studies carried out by Dr. Franz using human abdominal skin and the diffusion cell technique and radiolabelled material. Various formulations were investigated containing 0.001 - 0.5% compound. Absorption values for a 24 hour period were calculated to be in the range 0.002 - 1.22% (The worst-case maximum amount absorbed for all combinations of product was calculated to be 3.3 - 3.8 ug/day or 0.06 - 0.07 ug/ kg bw/day).

MUTAGENICITY

Somewhat conflicting results have been obtained in assays for gene mutation using Salmonella typhimurium, due in part to differences in purity of the samples tested. Negative results were reported by Parodi et al (1981) and Muzzall and Cooke (1979). Brown et al (1974) obtained positive results with TA 1537, 1538 and 100 in the presence of rat S-9, but the compound tested was only about 20% 45,170 and it was noted that a purer sample had only 'marginal' acitivity. Nestman et al (1979) found a commercial sample (90% pure) to be positive vs TA 1538 and 98 ; purification by thin layer chromatography revealed that most of the activity was associated with an impurity, but the chromatographically pure material had some activity vs TA 1538. Negative results were obtained using laser-grade rhodamine B. (Douglas et al 1980, Weubbles & Fellan 1985). However this grade material did produce both chromosome aberrations and SCEs in mammalian cells in vitro.

Only very limited data are available from in vivo studies. Negative results were obtained in an assay for DNA fragmentation in the liver following i.p. administration of compound. No data are available from bone marrow assays for clastogenicity.

CI 45,170 clearly has mutagenic potential, particularly the technical material although the compound itself does appear to be mutagenic. Inadequate data are available to fully access this potential.

CARCINOGENICITY

More recent carcinogenicity bioassays done to acceptable protocols are described in detail in the CTFA review. These are considered below.

A mouse study used dietary levels of 0,50,200 and 200 ppm for 2 years and involved groups of 60 animals of each sex at each dose level. No significant effect were seen on mortality or growth rate. A dose-related increase in hepatocellular carcinomas was seen in the females, the incidence being 2/60 (3%) 5/60 (8%) and 14/60 (23%) as compared to 0/115 (0%) in the controls.
Effects were less clear in the males, the incidence being 13%, 9% and 15% in the low, medium and high dose group compared to 8% in the controls. No significant increase in hepatocellular adenomas was seen in either sex.

A study in rats involved dietary levels of 0,20, 50 and 200 ppm, using groups of 70 animals per sex per dose level, and involved in utero exposure (maternal animals being given these diets for 2 weeks prior to mating), followed by feeding for 30 months. No effects were seen on mortality.

A slight reduction in body weight gain was seen in the females given the high dose but no other effects were noted during the study. The only increased tumour incidence seen in any group was a slight increase incidence (but not significant) of tumours of the brain and spinal cord in the high dose males.

A second study was carried out in rats using a single higher dose level (750 ppm in the diet), following concern by the FDA that the 200 ppm level in the above study was not a maximum tolerated dose. At 750 ppm no significant effects were seen on mortality but there was a clear reduction in body weight gain in 30 males and female rats and increased incidence of both adenomas (12/56, 21%) and carcinomas (5/56, 9%) of the thyroid compared to the controls 4% (2/59) adenomas, 2% (1/57) carcinomas. It was suggested by CTFA that the thyroid tumours may be due to chronic hyperplastic stimulation as a result of thyroid hormone imbalance. No increase in tumours of the CNS or any other site was observed.

CONCLUSION

CI 45,170 clearly has mutagenic potential, but inadequate data are available on whether this can be expressed *in vivo*. The compound is however carcinogenic both in mice producing a clear increase in thyroid adenomas and carcinomas. It should not be allowed as a colourant for cosmetics.

Classification : D.

REFERENCE

"Final Review and Analysis of Scientific Studies and Risk Assessment supporting the Safety of D.C.Red N°19 for use in External Cosmetic and Drug products not subject to Incidental Ingestion."
CTFA, February 15, 1983.

4. CI 26100 (complementary opinion)

For preceding informations and expressed opinions consult :
- Reports of SCC, seventh series, EUR 11303 pp 68-69

. FORMULA AND SYNONYMES

CAS Reg n° 85-86-9

1- (4-(phenylazo) phenyl azo) -2naphthalenol

Solvent Red 23

Sudan III, C-Ext

Rot 56

DC Red 17

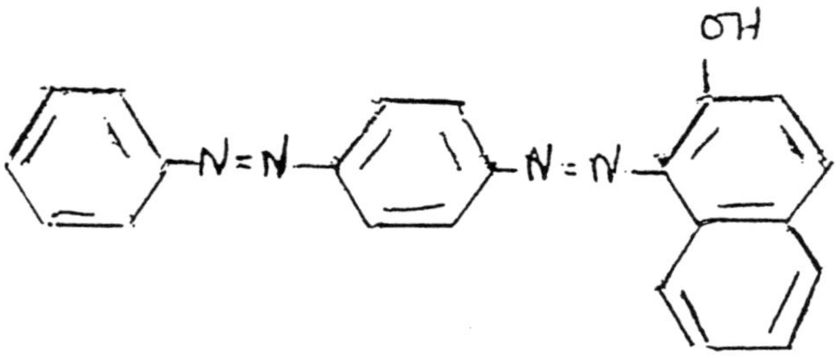

. PRECEDING CONCLUSION

A short-term oral toxicity study to examine systemic effects and further information on mutagenicity were required.

. NEW PROPOSITION FOR USE

Used up to 0.5% for non-rinse applications not in contact with mucous membranes (Colipa subm II).

. FURTHER INFORMATION

A chromosomal aberration test in human lymphocytes exposed in vitro to up to 100 ug/ml was also negative (submission II, ref. 10).

CONCLUSION

CI 26100 has low oral toxicity following single and repeated exposure and marginal irritant properties. The pure material does not appear to be a skin sensitizer. There was no evidence of any mutagenic potential in studies to investigate gene mutation in Salmonella or chromosome aberrations in human lymphocytes.
No evidence of carcinogenicity, or other toxic effects were seen in chronic toxicity studies in rats and mice, using the oral and dermal route, but inadequacies in the design of these studies prevent any definite conclusion being drawn. The compound is very poorly absorbed orally, the Committee do not consider it justifiable to request additional data from a short-term oral toxicity study. It would be expected to be poorly absorbed through the skin. In view of the sensitization seen with impure compound , purity criteria are needed.

Classification : A.

REFERENCES

1. Food and Drug Research Laboratories for the Toilet Goods Association Inc. n°81998 December 6, 1961

2. Carrol R. "Lesions of the Liver Produced by Sudan Dyes" J. of Path. and Bact. 1964, 87,317

3. Pharma Forschung Toxikologie, Bericht n° 213/78 Dr. Weigand und Dr. Leist, HOECHST AG, 5.05.1987

4. Food and drug Researc Laboratries for the Toilet Goods Association Inc. n°81998, January 4, 1963

5. Idem, December 1963

6. T. KOZUKA - M. TASHIRO "Pigmented Contact dermatitis from azo-dyes; cross-sensitivity in humans" Contact Dermatitis 1980 6 p. 330-336

7. MUZZAL and COOK "Mutagenicity en test of dyes used in cosmetics with the Salmonella / mammalian microsome test" Mut. Research 67 (1979) p. 1-8

8. R. WILLHEIM M.D.- AC IVY Gasrto-enterology 23 p.1-18 1953

9. N. WATERMAN - GOE LIGNAC "The influence of feeding of a number of food colours on the occurence of tumors en mice". Acta Physiol. Phamacol. Neerlandica 7 1958 p. 35-55

10. Report n° TX 187132 (11 December 1987) R.P. Everest and V.C. Williams - The Boots Cy PLC - Research Department.

REPORTS BY THE SCIENTIFIC COMMITTEE ON COSMETOLOGY

CONCERNING THE USE OF

CERTAIN U.V. FILTERS

1. MEXENONE	EEC 2.18	S39
2. MEXORYL SL	EEC 2.24	S59
3. MEXORYL SD	EEC 2.26	S61
4. ETHOYYLATED ETHYL-4-AMINOBENZOATE	EEC 2.2	S3
5. MEXORYL SX		S71

. THE COMMITTEE'S MANDATE

To give its opinion on the question whether the use of these UV filters at their usual concentration is admissible from the health point of view.

1. MEXENONE

. FORMULA AND SYNONYMES

EEC n° 2.18
Colipa S39
2-hydroxy-4-methoxy-4'-methylbenzophenone
Methanone (2-hydroxy-4-methoxyphenyl)

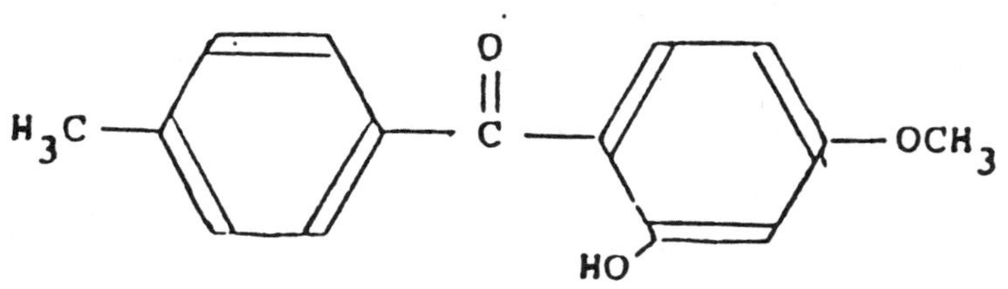

MW: 242

. CHARACTERISTICS

White crystalline powder

Insoluble in water

Soluble in alcohol and acetone

. USE

Proposed for use as a sunscreen agent at levels up to 4%.

. RECAPITULATION OF THE STUDIES OF TOXICITY

LD50 oral (rat) : $>$5 mg/kg

ORAL TOXICITY

Rat. A 28 day study was carried out in groups of 10 animals with doses of 0, 50, 250 and 1000 mg/kg bw/day, given by gavage. No dose related adverse effects were seen. The NEL is put at 1000 mg/kg bw/day.

CUTANEOUS AND MUCOUS TOLERANCE

DERMAL TOXICITY

Test for capacity to produce irritation of mucous membranes.
Rabbit. A test was carried out using a product, "Uvistat Sun Screen", containing 4% of a.i. Six animals were used. A dose of 0.1 ml of the product was instilled into one eye of each animal. The untreated eye served as control. There were very slight transient changes in one animal ; the test was otherwise negative. It was concluded that a 4% formulation of a.i. in a cream base was non-irritant.

Test for capacity to produce irritation of the skin.

Rabbit. The material tested was again a product containing 4% of the active ingredient : "Uvistat Sun Screen". Areas of abraded and non-abraded skin were prepared in each of 6 rabbits, and 0.5 ml of the preparation was applied to each of the areas with occlusion for 24 hours. Four animals showed well defined erythema at both sites; 2 had oedema as well. The severity of the reaction was graded 1.9 according to the US Federal Hazardous Substances Act. The formulation at 4% was judged to be a mild irritant under the conditions of the test.

Guinea Pig. A brief report states that daily application of a 1% ethanolic solution to the shaved skin of the guinea pig for 14 days did not produce irritation. No other details are available.

Test for capacity to induce sensitisation.

Man. A modified Draize-Shelanski test was carried out in 12 healthy volunteers, using the product containing 4% of the active ingredient. The induction consisted of the application of 0.5 ml of the test product to the upper arm, followed by 24 hours occlusion. This was carried out 3 times a week for 3 weeks. The challenge application (which was the same as the induction application) was made (probably) 17 days after the conclusion of the induction phase. Applications were made to new sites and to the challenge sites. There was slight primary irritation in some subjects, but no evidence of sensitization.

Test for capacity to produce phototoxicity.

Guinea pig. Twenty female Dunkin-Hartley animals were used: 10 test and 10 control. Three areas of 2 cm^2 were prepared on the back of each animal. In the animals of the test group, 0.025 ml of solutions of 2% and 4% of a.i. in polyethylene glycol were applied to two of the areas, and 0.1%

8-MOP in ehtanol/acetone to the third. The control animals had no application. Thirty minutes later, the test animals were exposed to UVA 320-400 nm, until 2.5 Jcm^2 had been administered. Control animals were not irradiated. There was no evidence of phototoxicity. The positive control areas showed slight to moderate signs of irritation.

Test for percutaneous absorption.

Man. Three male and three female volunteers were used. Areas of 20 x 20 cm were delineated on the back of each subject, and 800 mg of a cream containing 4% of a.i. was rubbed into the area. Occlusion was not used. Blood was sampled at 0, 1, 2, 4 and 6 hours. There was stated to be no evidence of the presence of a.i. in the blood. Interpretation is difficult because the description of the method is obscure and recovery figures are not given.

MUTAGENICITY

A standard Ames test was carried out. The a.i. was dissolved in DMSO. There was no evidence of mutagenicity.
A test for the production of chromosomal aberrations *in vitro* was carried out using Chinese hamster ovary cells. There was no evidence of mutagenic activity.

CONCLUSION

The Committee was satisfied that the tests for skin and mucous membrane irritation, and for sensitisation potential, were negative, but noted that the concentrations used were the proposed use concentration (4%). Tests for photosensitisation and photo-allergenic potential would be necessary.

The test for percutaneous absorption was not satisfactory. Oral acute and 28 day toxicity studies were judged to be satisfactory. There was no 90 day study. Mutagenic studies were judged to be satisfactory, but photomutagenic studies have not been undertaken.
The Committee felt that a study of inhalational toxicity would not be required unless it was proposed to use the substance in an aerosol formulation. Acute and short term dermal studies are not suggested. When adequate date on percutaneous absorption are available, further testing may be required (e.g. for teratogenic potential).

Classification : C.

REFERENCES
++++++++++

1. Toxicol Laboratories Ltd. 179/8409 November 1984

2. Toxicol Laboratories Ltd. 184/8409 October 1984

3. Toxicol Laboratories Ltd. 183/8409 October 1984

4. SafePharm Laboratories Derby, Novembre 10, 1987

5. Report B151966 Biomedical Services, Basingstoke, Septembre 1987

6. Toxicol Laboratories Ltd. 180/8409, Septembre 1984

7. Toxicol Laboratories Ltd. 181/8409, November 1984

8. Mexenone : Determination of phototoirritation potential in guinea pigs. Dr. CUTHBERT and Dr. D'ARCY BURT. Inveresk Research. International Report n°3964 - Musselburgh, Scotland, September 5, 1987

9. Toxicol Laboratories Ltd. 182/8409 Novembre 1984

2. MEXORYL SL

. FORMULA AND SYNONYMES

EEC 2.24 (previously 2.28)

Colipa S59

Alpha'-(2-oxoborn-3-ylidene)-toluene-4-sulphonic acid
and salts

3-(4'-sulpho) benzylidene bornan-2-one

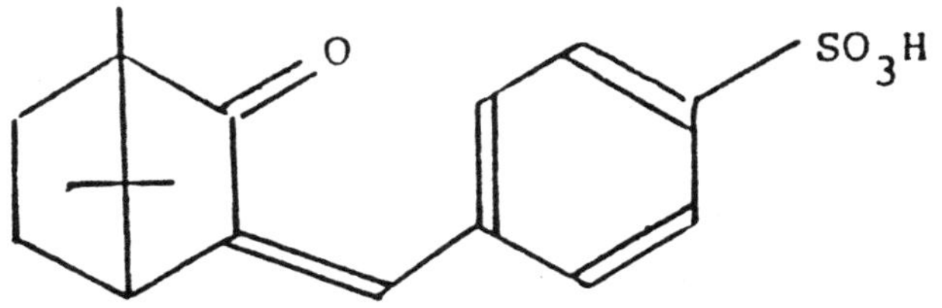

$C_{17}H_{20}O_4S \cdot 3H_2O$

MW 374.5

. CHARACTERISTICS

Crystalline substance

Highly soluble in water and ethanol

. USE

Proposed for use as a sunscreen agent at levels up to 6% (expressed as acid)

. RECAPITULATION OF THE STUDIES OF TOXICITY

LD50	oral (rat) :	2,2 gr/kg (females)	
		3,2 gr/kg (males)	
	(mouse)	1,83 gr/kg (no details given)	
	(rat)	1,29 gr/kg (no details given)	

ORAL TOXICITY

Rat. Oral 28 day study. Following a preliminary 5 day ranging study, doses of 0, 150, 300 and 600 mg/kg bw/day were chosen, to be given by gavage 7 days a week for 28 days. Groups of 5 male and 5 female animals were used. The study seems to have been properly carried out.

The chief findings were as follows. There were no drug related deaths. There was a significant reduction in weight gain in week 2 in high dose males. Food consumption was reduced in both sexes at the intermediate and high doses, particularly in the males.

There was a fall in serum sodium and an increase in chloride in high dose males. The blood glucose was lower than control in all dosed females and in high and intermediate dose males. Organ weights : in males, there was some reduction in relative liver weights at all dose levels, but the absolute changes were small. Histological examination revealed granular cytoplasm in some hepatocytes in high dose males and females. The no effect level was put at 150 kg/bw /day.

CUTANEOUS AND MUCOUS TOLERANCE

DERMAL TOXICITY

Test for capacity to injure mucous membranes.

Rabbit. A convential Draize test was carried out in 6 NZW rabbits. A 6% solution of a.i., neutralised with triethanolamine, was used in a dose of 0.1 ml. The opposite untreated eye served as a control. There was slight conjunctivitis which cleared in 2-3 days. The compound at this concentration was judged to be very slightly irritant.

Test for capacity to cause primary irritation of the skin.

Rabbit. Six NZW animals were used. Two areas of skin, 2.5 x 2.5 cm in area, were shaved in each animal : one was abraded and one not. A volume of 0.5 ml of a 6% solution of a.i., neutralised with ethanolamine, was applied to each area and occluded for 24 hours. No reaction was seen.
A similar test using a 4% solution is briefly reported, without details. The index of irritation was 0.12/8.

Test for capacity to produce irritation of the skin on repeated administration.

Rabbit. Guinea pig, rat. A 4% solution of the compound neutralised with treithanolamine was used. Volumes of 2.5 ml, 1 ml, and 0.5 ml, respectively, were applied daily to the shaved skin of each species. The treatement was carried out for a month. The index of aggressivity is given as : rabbit, 0.56/8 ; guinea pig, 0.2/8 ; rat, 0.5/8.
No details are given.

Test for capacity to produce sensitisation.

Guinea pig. Thirty animals of the Dunkin-Hartley strain were used, 20 test and 10 control. A Magnusson-Kligman procedure was carried out using a 6% aqueous solution of a.i. neutralised with triethanolamine.
There was no evidence of sensitisation.

Test for capacity to produce phototoxicity.

Rabbit. A brief report states that application of a 5% aqueous solution followed by ultraviolet irradiation was carried out each day for 2 weeks. The index of phototoxicity is given as 1.27/8, which is termed "slightly phototoxic".
No further details are available.
Guinea pig. Thirty female Dunkin-Hartley animals were used: 20 test and 10 positive controls. Three areas of skin were prepared in each animal, and each was covered by a patch

of filter paper. In the test animals, two of the patches were saturated with an aqueous solution of a.i. (probably 1%). The third patch was dry. In the positive control animals, two of the patches were impregnated with 20% chlorpromazine in petrolatum.
In both groups, the applications were occluded for 1.5 hours. At the end of that time, the patches were removed from one of the treated areas of skin and from the untreated area, and each animal was exposed to ultraviolett irradiation with a maximum output at 350 nm. There was no evidence of phototoxicity, but the positive control animals gave a very feeble response.

MUTAGENICITY

A standard Ames test was carried out, using an aqueous solution of a.i., neutralised with triethanolamine up to 5000 ug/plate.
There was no evidence of mutagenic effect.

CONCLUSION

The Committee was satisfied with the data on primary irritation of the skin, and irritation of mucous membranes, but noted that the concentrations used for testing were those proposed for use. The test for capacity to produce sensitisation was satisfactory. A test for capacity to produce photosensitisation should be carried out. Acute and 28 day oral toxicity tests were satisfactory. It was not felt that acute and short term dermal toxicity tests would be necessary ; inhalational toxicity tests would not be required unless aerosol preparations were proposed.

A test for percutaneous absorption is required ; depending on the results further testing may be required (e.g. for teratogenic potential). A test for chromosomal aberration in vitro should be carried out. Tests for photomutagenic acitivity had not been carried out.

Classification : C.

REFERENCES
++++++++++

1. HRC Report 861469 D/LCL 2/AC 17/02/1987

2. HRC Report 861340 D/LCL 4/SE 02/02/1987

3. HRC Report 861339 D/LCL 3/SE 02/02/1987

4. HRC Report 87197 D/LCL 5/SS 25/03/1987

5. HRC Report LRL/79/87440 09/06/1987

6. HRC Report LRL/91/A/8617000 09/01/1987

7. HRC Report 861113 D/LRL 84/SS 24/10/1986

3. MEXORYL SD

. FORMULA AND SYNONYMES

EEC n° 2.26

Colipa S61

3-benzylidene-1-camphor

3-benzylideneborman-2-one

1,7,7-trimethyl-3-benzylidene-2,2,1-bicyclo-2-heptanone

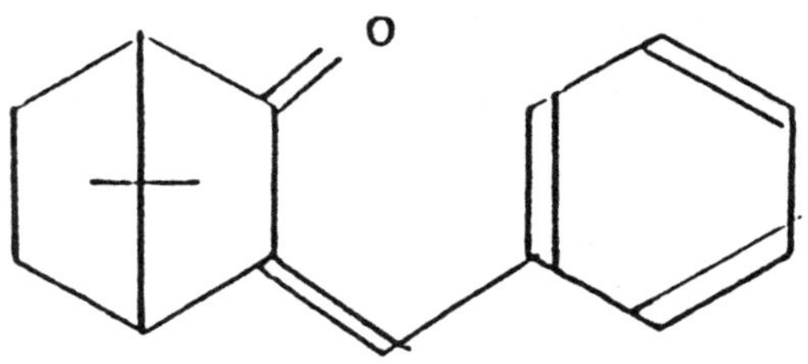

$C_{17}H_{20}O$

MW 240.4

. CHARACTERISTICS

Crystalline powder

Insoluble in water

Soluble in absolute alcohol and isopropanol

. USE

Proposed for use as a sunscreen agent at levels up to 6%

. RECAPITULATION OF THE STUDIES OF TOXICITY

LD50 oral (rat) : $>$5 gr/kg

ORAL TOXICITY

Acute toxicity.

Rat. Five male and five female animals were used. The a.i. was given by gavage suspended in 1% propylene glycol in a dose of 5 g/kg bw. Observation was for 14 days. No abnormality of any kind was seen.

Subchronic toxicity.

Rat. A 13 week study was carried out in which the compound was administered by gavage in doses of 0, 100, 250 and 500 mg/kg bw/day, 5 days a week, to groups of 10 male and 10 female animals. Treatment was stopped after 6 weeks in one half of the animals in each group, and these served as recovery groups. The chief abnormality found was that there was an increase in plasma lipids at 13 weeks in female animals at all dose levels. In addition the plasma cholesterol levels were increased in males at the top dose, and in females at the intermediate and top doses. No abnormality of lipids or of cholesterol was found in any animal after 6 weeks.

Rat. A further study using groups of 20 animals was carried out, using doses of 20 and 40 mg/kg bw day in a similar manner. An abnormally high value for plasma lipids was found in females given 40 mg/kg bw/day ; the increase at the lower dose was not significant.

The NEL is probably 20 mg/kg bw/day.

CUTANEOUS AND MUCOUS TOLERANCE

DERMAL TOXICITY

Test for capacity to cause irritation of mucous membranes. Rabbit. A 6% solution of a.i. in isopropyl palmitate was used. The untreated opposite eye was used as a control. The index of ocular irritation was : After administration, 5.33/10 ; day 1, 2.67/10 ; day 2, 0.67/10 ; thereafter negative. The authors judged the compound under the conditions of the test to be "mildly irritant".

Test for capacity to cause primary irritation of the skin. Rabbit. A 6% solution of the compound in isopropyl palmitate was applied to abraded and non-abraded skin in 6 animals, with occlusion for 24 hrs. The compound was adjusted to be a mild irritant under these conditions.

Test for capacity to cause irritation with repeated application. Rabbit. A 6% solution in isopropyl palmitate was used. The skin was carefully clipped on both flanks and each animal had 2 ml of the test material rubbed into the flank on one side ; a similar procedure, without the solution, was carried out on the opposite side. Applications were made 5 days a week for 6 weeks. The experiment was then continued for a further week without treatment, to study recovery. The weekly mean index of irritation (maximum 8) was : 1.83 ; 1.21 ; 2.23 ; 1.13 ; 2.07 and (recovery) 1.0. The substance under these conditions was judged to be a mild irritant.

Test for capacity to induce sensitisation. Guinea pig. Twenty albino animals of the Hartley strain were used. For induction, 0.5 gm of the compound, as a powder, was applied under occlusion for 48 hrs, 3 days a week, for 10 applications.

On days 1 and 10, an intradermal injection of 0.1 ml of Freund's compete adjuvant, 50% in saline, was given. After a 12 day rest, a challenge application, the same as the induction application, was made to a new site. No abnormality was seen.

Test for capacity to induce phototoxicity.

Rabbit. Six male and six female animals were used for the test, and 6 animals were used a vehicle controls. One site about 5 x 5 cm in area was prepared on the left flank of each animal, and 2 on the right. The compound was made up as a 5% solution in 95% ethanol. One ml of this was applied to the left flank, and to one of the sites on the right flank. The sites on the right flank were irradiated with 1 med of ultraviolet light daily for 10 days. The irradiation was carried out by exposure to an "Osram Vitalux" lamp, but its spectral characteristics are not given. There was no evidence of phototoxicity.

Test for capacity to produce photosensitisation.

Guinea pig. Twelve male and twelve female albino Hartley animals were used in 3 groups.

Induction : Animals of group 1 (test) had injections of 0.2 ml of Freund's complete adjuvant in the foot on day 1 ; on days 1, 3, 5, 8 and 11, 0.5 ml of a 4% solution of the compound in olive oil was applied to a prepared site on the back of the neck, followed by exposure for 30 minutes to 2 lamps covering the range of 285 to 450 nm. Animals of group 2 (vehicle control) were similary treated except that the applications were of olive oil only. Animals of group 3 had the Freund's adjuvant but no other treatment. Challenge. On the 22nd day of the experiment, sites on either side of the lumbar vertebrae were prepared in all animals. Applications of 10 ul of the following solutions were made to each side : olive oil ; 2%

of the compound in olive oil ; 2% of the compound in ehtanol. The sites were then exposed to irradiation by one of the lamps (320-450 nm) for 30 minutes. Readings were made at 24 and 48 hrs. There was no evidence of photosensitisation. There was no positive control, but the test was nevertheless adjudged to be satisfactory.

Test for percutaneous absorption.

Man. Four volunteers were treated. Areas of 100 cm were delineated on the upper arm. The compound was labelled with 14C and made up in a concentration of 5.02% in an o/w emulsion. About 0.5 g of ointment was applied to the delineated areas. Contact was for 6 hrs, without occlusion. At the end of the experiment, the skin was swabbed clean and also stripped. Urine and faeces were collected for five days. The data given permit the calculation that under the circumstances of the experiment, the flux (J) = 139 $\pm$ 30 ug cm^{-2} H^{-1} and the permeability constant (K_p) = 2.8 $\pm$ 0.6 x 10^{-3} g $cm^{-2}h^{-1}$.

Assuming a maximum exposure for 10 hours a day over 1.8 m^2 of body surface, and a concentration of 6%, the total amount absorbed would amount to approximately 0.6 g/kg bw/day.

MUTAGENICITY

An Ames test was carried out using strains TA 1525, TA 1537, TA 99 and TA 100. Toxicity experiments were not carried out ; in strain TA 100 there was a decrease in revertants with increasing dosage, suggesting some toxicity. Testing was carried out up to 1000 ug/plate in each strain. There was no evidence of mutagenic activity.

A culture of Chinese hamster ovary cells was used to test for the production of chromosomal aberrations in vitro. Without activation, doses up to 80 ug/ml were used, and with activation, doses up to 25 ug/ml. There was a highly significant increase in aberrations in the preparation with activation in a dose of 25 ug/ml.
A standard micronucleus test was carried out in the mouse. There was no evidence of genotoxicity.

CONCLUSION

Acute oral toxicity is low. Tests for primary irritation of the skin and for irritation of the skin on repeated application showed only slight effects at 6%, but this is the use concentration. Tests for phototxicity and photosensitivity were negative, although a positive control was not used for the latter.
A 90-day oral test in the rat showed elevated plasma lipids in female rats at doses of 40 and 20 mg/kg bw/day, although this was not significant at the latter dose. The NEL is therefore considered to be 20 mg/kg bw/day.
Percutaneous absorption is marked and was calculated to amount to as much as 0.6 g/kg bw/day.
A test for chromosomal aberration in vitro was positive, the Ames test and a micronucleus test were negative.
There is no sufficient margin of safety for percutaneous absorption.

Classification : D.

REFERENCES
++++++++++

1. Unité de Pharmacologie Clinique. Pr. Fournier
   Toxicité aigüe chez le rat par v.o. du Benzylidène camphre.
   05/02/1981

2, 3, 4. HIFT Report 51033 09/10/1985

5. IFREB Report 006344 19/06/1980

6. L'Oréal Report Contrôles biologiques, 30/01/1984
   Phototoxicité sur lapins - Benzylidène Camphre

7,8. Sociétés de recherches biologiques, Hôpital Fernand Widal,
   Pr. Fournier, 23/03/1982- Mexoryl SD - 13 week oral toxicity
   study in rats.

10. Life Science Research - Pr. LOprieno, Report 088123-M-07785
    January 86 - Gene Mutation Assay on Salmonella typhimurium
    by Benzylidene Camphor.

11. L'Oréal, Contrôles biologiques, Peport T 86/10/210LL 18/01/86

12. Chromosome Aberration in CHO cells in vitro, LSR-RTC,
    Report n°088124-M-07885 January 86 A. NUNZIATA & N. LOPRIENO

13. Report on the Percutaneous Resorption of MEXORYL SD in man,
    Institut für Pharmakodynamik und Toxikologie der Universität
    GRAZ, Pr. KUKOVETZ - July 28, 1987.

4. ETHOXYLATED ETHYL-4-AMINOBENZOATE

EEC 2.2
Colipa S3

CONSULT

Reports of SCC : 40th reunion 11-4-89

CONCLUSION

A reply from Colipa has been received.

a) A test for mucous membrane irritation has in fact been carried out. This was evaluated by the subgroup and found to be satisfactory.
b) The substance "Lusantan 25" is stated to be ethoxylated ethyl-4-aminobenzoate. In the light of this, the Ames test is now regarded as negative.

Confirmation of Classification : C.

REFERENCES
++++++++++

1. Report of the study of LUSANTAN 25 (ZNT Substance n°85/234) in the Ames test. Dr. GELBKE, Dr G ENGELHARDT - BASF AG-Abteilung Toxicologie Ludwigshafen, Germany, Septembre 1985

2. Pr Dr TRONNIER - 1979. "Photopatch test"

3. Dr H. F. BENTHE "Gutachten über die experimentelle Prüfung des akuten Hautreizwirkung und der photosensibilisierenden Eigenschaffen" Universität Hamburg - September 7, 1981

5. MEXORYL SX

. FORMULA AND SYNONYMES

Colipa S71

3,3'-(1,4-phenylenedimethylidyne) bis (7,7-dimethyl-2-oxo- bicyclo(2,2,1)heptane-1-methanesulfonic acid) and its salts

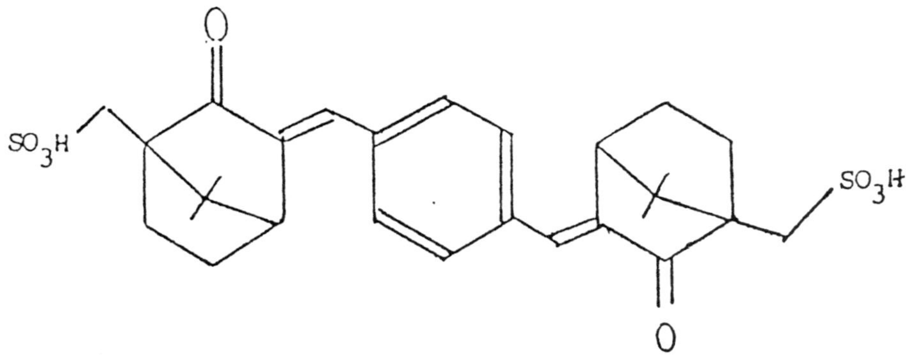

$C_{28}H_{34}O_8S_2$ MW/ 562.7

The material consists of two molecules of camphorsulphonate united by 1,4-divirylbenzene

. CHARACTERISTICS

Freely soluble in water, giving an acidive solution.
In use, it is expected to be neutralised and used as a salt.
Absorption maximum 345 nm.

Test for photostability.

A thin layer of a 4% solution of a.i. was exposed to simulated solar radiation of an intensity substantially greater than which would be found in the south of France, or Corsica. It was found that there was a rapid, though relatively small, fall in the first few minutes : after his, the loss was no more than about 1.5% in 1 hour.

. USE

Proposed for use as a sunscreen agent at levels up to 10% expressed as acid.

. RECAPITULATION OF THE STUDIES OF TOXICITY

LD50 oral (rat) : > 1835 mg/kg

The experiment was repeated 3 times, using the a.i. neutralised with triethanolamine
potassium hydroxide
sodium hydroxide

In each case we found

LD50 : > 2092 mg/kg (expressed as acid)

LD50 dermal (rat) : > 1367 mg/kg

ORAL TOXICITY

Subchronic toxicity.

Rat. Oral. A 13 week study was carried out in groups of 10 male and 10 female SD animals. The a.i. was administered by gavage 7 days a week at doses of 0, 100, 300 and 1000 mg/kg bw/day. There were few signs of abnormality during the study. The rate of weight gain was slightly less in the mid and high dose groups. At autopsy, the absolute and relative weights of the thyroid were reduced at the middle and top dose in males only ; the effect was not dose related, and no histological change was found in the gland. The study seems to have been a well conducted one; the NEL is judged to be 1000 mg/kg bw/day.

Test for capacity to affect the production of thyroid hormones. Rat. Twenty male Wistar animals were used, 10 test and 10 control. The a.i. was neutralised with TEA and given by gavage in a dose of 305 mg/kg bw/day (expressed as acid) daily for 21 consecutive days.
Control animals received distilled water. No abnormalities were observed during the experiment. At the end of the experiment the serum was assayed for its content of thyroxine. There were no significant difference between the groups.

CUTANEOUS AND MUCOUS TOLERANCE

DERMAL TOXICITY

Test for capacity to cause irritation of mucous membranes. Rabbit. Groups of 6 male NZW animals were used. Three experiments were carried out, using, respectively, a.i. neutralised with triethanolamine, potassium hydroxide and sodium hydroxide, all at a concentration of 10.5%. Results were similar in all experiments, and the material at this concentration was judged to be very slightly irritant.

Test for capacity to produce primary irritation of the skin. Rabbit. Three groups of 6 male NZW rabbits were used. Areas of skin were prepared on either flank, one of which was scarified, the other not. The applications were of 0.5 ml of a 10.4% solution of a.i., and occlusion was used for 24 hrs, with readings at 24 and 48 hr. The 3 experiments were identical except that in the first group the a.i. was neutralised with triethanolamine, in the second with potassium hydroxide, and in the third with sodium hydroxide. The results were similar in all experiments : the cumulative irritancy scores were 6, and 9 respectively. The substance under these conditions was deemed to be non-irritant.

Rabbit. Three NZW animals were used, and 0.5 ml of a 36.7% solution of a.i. was placed upon an area of clipped skin and maintained under semi-occlusive conditions for 4 hrs. Reading was at 1, 24, 48 and 72 hrs. There was no evidence of primary irritation.

Test for capacity to produce sensitisation.

Guinea pig. The a.i. was neutralised with triethanolamine. A modified Magnusson-Kligman technique was used. Preliminary testing was carried out on 2 male and 2 female animals the definitive test was carried out on 10 male and 10 female animals. There was no evidence of sensitisation.

Test for capacity to produce phototoxicity and photoallergenicity.

Guinea pig. Following preliminary testing, a 10.4% solution of a.i., neutralised with triethanolamine, was used. Three male and 2 female animals were treated with 0.5 ml of the solution followed by occlusion for 90 minutes. These animals were not irradiated, and served as controls. In the test, 10 males and 10 females were similarly treated and thereafter given a minimum erythema dose from 2 lamps which covered the spectrum from 285 to 400 nm. Reading was at 6 and 24 hrs after irradiation. No differences were found between test and control animals.

The test for photoallergenicity was carried out as follows. Induction was carried out by the use of 4 intradermal injections of 50% FCA followed by occlusive patches, as above, on days 4, 7 and 9. In each case exposure to 15 minutes of irradiation as above was given, and then 40 minutes of irradiation with 310-400 nm. Following a rest period of 14 days, a single occluded application, as above, was made to a previously untreated area.

Following removal of the patch, the animal was irradiated for 90 minutes using the longer wavelength lamp only. Reading was at 6, 24 and 48 hrs. There was no evidence of abnormality in any animal.

Test for capacity to produce photocontact allergy.

Man. A cream formulation was used, containing 10% a.i. neutralised with triethanolamine. The cream also contained butyl-p-hydroxybenzoate 0.2% and imidazolidinyl urea 0.2%. Fifteen white females and 10 white males were used, aged from 18 to 50. Following the determination of a minimal erythema dose for each subject, using a xenon arc solar simulator, 10 microlitre/ cm^2 of the cream was applied to the skin of the back and occluded for 24 hrs. After removal of the patch, the ara was exposed to 3 meds. After a 24 hr rest, the procedure was repeated, and thereafter repeated twice a week for 3 weeks. All these induction applications were made at the same site. After a 10 day rest period, the challenge was carried out by making the above application in duplicate on 2 new sites on eihter side of the midline. One patch was removed after 24 hrs, and the other left in place : the area was then exposed to 4 J/cm^2 of UVA. Reading was at 24 and 72 hrs, and was blinded. There were no adverse reactions of any kind.

Test for percutaneous penetration.

Rat. Five male and five female SD rats had a single application of a 36.7% solution of a.i. applied and maintained under semi-occlusive conditions for 24 hrs. The animals were observed for 14 days, and then subjected to gross necropsy. No abnormality was found.

Rat. Eight female hairless SD rats were used. A cup was glued to the skin of the back which covered an area of 1 cm^2. A 5% formulation of the a.i. doubly labelled with 14C, was made up in "Ambre Solaire" cream. An amount of 2 mg of the cream, containing 178 nmoles of a.i., was applied and allowed to remain in contact for 4 hrs. From the data, it is possible to calculate that the permeability coefficient is 3.8 +/- 0.6 x 10^{-6} $gcm^{-2}h^{-1}$. Assuming the permeability constant is the same in rat and man, and an application to the whole body area (1.8 m^2) over a 10 hour day at the maximum concentration, i.e., 10% this would amount to an absorption of about 68.4 mg, i.e. an absorption of 1.14 mg/kg bw/day.

Man. Human skin was obtained in the form of surgical specimens from two negro women. The skin was sectioned by dermatome to obtain samples 300 to 400 um thick. There was sufficient skin to enable three experiments to be carried out with each specimen. A Franz cell was used, and 30 mg of a 5% cream formulation of the a.i. was applied to the skin. Contact time was 24 hrs. The amount of a.i. penetrating was less than the sensitivity of the HPLC method for its estimation (50 ng/ml). The penetration was judged to be less than 0.12 ug/cm^2. The permeability coefficient calculated from these data would be less than 0.1 x 10^{-6} $gmcm^{-2}$ h^{-1}.

A similar test was carried out with skin from two other negro women. The results were the same.

Rat. Skin from 2 female hairless rats was used in a Franz chamber in the same way as for human skin. The result was the same as with the test using human skin. These <u>in vitro</u> results, when analysed, give a permeability coefficient of not more than 0.8 x 10^{-6}gm $cm^{-2}hr^{-1}$, and the amount absorbe in man, over a 10 hr day (as above) would be less than 12.8 mg, or 0.2 mg/kg bw.

MUTAGENICITY

A standard Ames test was carried out. There was no evidence of a mutagenic effect, with or without activation.

A standard test for chromosomal aberration in vitro. Was carried out using Chinese hamster ovary cells. There was no evidence of induction of aberration.

Test for photomutagenic activity.
The a.i. was neutralised with TEA and tested for its ability to induce tryptophan-independent revertants in *E. coli* WP2. Following a range finding study, concentrations from 8 to 5000 ug/plate were tested, with and without exposure to UVA and UVB irradiation. Suitable positive and negative controls were used. There was no evidence of photomutagenicity.
Chinese hamster ovary. A similar experiment using cultured cells was carried out, except that following a pilot experiment the concentrations tested were 1250, 2500 and 5000 ug/ml. Exposure to UVA and UVB was carried out. Suitable positive and negative controls were used. There was no evidence of clastogenic activity.

CONCLUSION

Evaluation. The a.i. at 10.5% is only slightly irritant to the eye, and at 10.4% it is non-irritant to the rabbit skin. The proposed use level is 10%. The data on sensitisation, phototoxicity, photosensitisation and photocontact allergy are regarded as satisfactory.

Inhalation toxicity testing would be required only if it were proposed to market an aerosol formulation. Tests for mutagenicity and photomutagenicity were negative.
Teratogenic testing showed no evidence of abnormality ; however it was carried out in one species only, and the dose used did not cause toxicity.

Classification : A.

REFERENCES

1. Safepharm Laboratories Report 1016/ 109 / 150 - 21.09.1988
Acute oral toxicity,Rat.
2. HAZELTON - IFT Report 712389 23.12.1987.
3. " " " 712388 23.12.1987.
4. " " " 712387 23.12.1987.
5. " " " 712332 10.12.1987.
6. " " " 712331 10.12.1987.
7. " " " 712330 10.12.1987.
8. " " " 712320 07.12.1987.
9. " " " 712319 07.12.1987.
10. " " " 712318 07.12.1987.
11. " " " 802410 15.02.1988.
12. LERS Report 1335 TTR/ 054.88. 06.02.1989.
13. A.ROUGIER
"Absorption percutanée in vivo chez le rat hairless du MEXORYL SX à 5% dans la crème Ambre Solaire n° 202838 après 4h d'administration topique."
L'OREAL 92117 CLICHY CEDEX, 20.05.1988.
14. HAZELTON - IFT Report 801205 18.01.1988.
15. "Evaluation of compound MEXORYL SX in the chromosome aberration test with CHO (in vitro)."
University of Leiden, June 13,1988.
16. LERS Report 1412 RMR/064.89 20.01.1989.
17. HAZELTON - IFT Report 802409 15.02.1988.
18. IVY Lab. Report, K.KAIDBEY 15.02.1988.

SCC OPINION

BERGASOL SUNTAN PRODUCTS CONTAINING 5-MOP

Background

5-MOP occurs naturally in oil of bergamot (up to 3000 ppm). The use of this oil is at present allowed in suntan products by the Cosmetics Directive, but the SCC have recommended that the maximum level of 5-MOP arising from the presence of natural essences in suntan products should be 1 ppm.

The SCC have now been asked to consider data on a specific range of suntan products marketed by Bergasol and containing 15-60 ppm 5-MOP (as the natural product bergapten). These also contain sunscreens and it is claimed that such formulations behave differently to 5-MOP alone, particularly with regard to the photomutagenicity and photocarcinogenicity which gave rise to the concern with 5-MOP itself. Bergasol have provided a substantial amount of data to support this submission and this is summarised in Appendix I to this report.

The question that the SCC have to consider is difficult and raises points that the Committee does not, in general, have to consider. There is a finely balanced argument regarding health risks and benefits, whereas in general the health risks of cosmetic ingredients are minimal. In the case in question, the active melanogenic ingredient, 5-MOP, clearly gives rise to concern regarding its photomutagenic and photocarcinogenic potential. However, sunlight itself induces skin cancer and the suntan products in question are designed to enhance tanning. More rapid tanning leads to a reduced cumulative dose of UVA and UVB radiation whilst attaining a tan that is protective against subsequent exposure.

There is also the additional problem as to how 'clearance' could be given to specific products under the Cosmetics Directive which is designed to consider either allowed or prohibited ingredients rather than products.

The available data are summarised below:-

Toxicity of 5-MOP itself

The data on 5-MOP itself are reviewed in Appendix II and are summarised below, for background information.

5-MOP itself is well absorbed orally and may be absorbed through the skin. It is widely distributed in tissue, extensively metabolised, and rapidly excreted in the urine. It has very low oral toxicity, but is strongly phototoxic, producing marked skin reactions after exposure to UV radiation. Only limited data are available on sub-acute toxicity. The combination of 5-MOP plus UV radiation is clearly mutagenic, much information being available on the interaction of psoralens with DNA in the presence of UVA. Linear psoralens such as 5-MOP are bifunctional, forming DNA adducts that react further producing DNA cross-linking. 5-MOP plus UVA has been shown to produce positive results in mutagenicity studies in bacteria, yeasts and mammalian cells, and this combination is also clastogenic. Several studies have been carried out to investigate the skin carcinogenicity of 5-MOP plus UVA in animal bioassays. Such treatment has clearly been shown to produce skin cancer in mice. There are no adequate data available to assess the carcinogenicity of 5-MOP in the absence of UV radiation, nor are any adequate data available in man. However the combination of 8-MOP plus UVA (PUVA treatment) is clearly carcinogenic in man, and is included in the IARC Group I Carcinogens (ie known human carcinogens).

BERGASOL PRODUCTS; data submitted by Bergasol

Comprehensive data on both the toxicity of these products and the health benefits are given in Appendix I. These are summarised below.

Studies in animals and human volunteers showed no significant skin irritation or sensitization, nor any significant photo-irritancy or photo-toxicity. Furthermore the data provided by Professor Agache indicates that the absorption of 5-MOP from such formulations is minimal (less than 1%), although greater amounts of 5-MOP were present in the stratum corneum itself.

The crux of the problem is however the potential photomutagenicity and photocarcinogenicity with respect to the induction of skin cancer and this is considered in some detail below:-

Several studies have been carried out to investigate the mutagenicity of Bergasol preparations using Salmonella typhimurium TA102, but interpretation of these are difficult since somewhat conflicting results were obtained, possibly due to differences in the amount of material diffusing from the oil emulsion formulations through the agar medium. Some studies indicated a reduction in the number of revertants but positive results were obtained with commercial preparations and mutagenic potential cannot be ruled out from these studies. In addition it was pertinent that data from limited studies to investigate the mutagenicity of 'blister fluid' obtained from skin preparations of human volunteers treated dermally with a Bergasol preparation suggested that this had some photomutagenic potential when tested using the yeast strain Saccharomyces cerevisiae D7 (studies by Professor Dubertret). However, in view of the very limited nature of these studies (no dose response, no repeat of the results in an independent experiment) no firm conclusions can be drawn.

Long term bioassays for skin cancer in mice have given conflicting results. Limited data from an early study showed no protection using a UV filter (ethyl hexyl paramethoxy cinnamate). More extensive studies using a Bergasol formulation containing 5-MOP plus 2 UV filters (2 ethylhexyl, 4 methoxy cinnamate and 1,7, 7-trimethyl, 3-benzylidene-bicyclo (2,2,1) - 2 heptanone) revealed substantial protection against skin tumour induction in albino mice, but only during the treatment period (45 weeks). Animals exposed to SSR after treatment with 5-MOP plus sunscreens then went on to develop tumours sooner than those treated with sunscreen alone. In another study in pigmented mice of the same strain, much less protection was seen, but it was argued that these mice were not an appropriate model for human pigmentation due to the absence of epidermal melanin. It is thus difficult to assess the significance of these studies or to identify the mechanisms involved. It is not possible to assess the increased risk of skin cancer in those using formulations containing 5-MOP plus UV filter. The data do not allow concerns regarding this aspect to be discounted.

More recent studies have been reported using an 'accelerated' <u>in vivo</u> model for skin tumourigenicity based on hairless mice treated with UV radiation (with or without bergapten containing 50 ppm 5-MOP) as 'initiator' for 4 months together with croton oil, as promoter, for the first 2 months. At the end of the 4 month period a similar incidence of tumours was seen following treatment with UV radiation and bergapten (50 ppm 5-MOP) as UV radiation alone, <u>ca</u> 60% incidence of skin papillomas. A marked reduction of the incidence of skin tumours was seen when sunscreens were incorporated in the treatment formulation (20% incidence) but the most impressive results were obtained by the additional incorporation of antioxidants such as BHT or β carotene or by the addition of an inhibitor of polyamine (putrescene) synthesis.

<u>Benefits of incorporation of bergapten (5-MOP)</u>

Considering the protective effect of 5-MOP in sunscreens, data has clearly established that the inclusion of 5-MOP enhances pigmentation. In addition the series of studies by Dr Young showed that the 5-MOP induced tan protected against UV radiation induced DNA damage in humans as measured by unscheduled DNA synthesis (UDS) in individuals of skin type I to V. It was particularly noteworthy that the most sensitive population for UV induced skin cancer (ie types I and II) became like types III and IV in this regard.

However the mechanisms underlining the effects seen are not known and it is important to note that UDS is an indirect measure of damage, and a decrease may reflect a reduction in repair. The situation is complex and it is clear that there is no direct proportionality between UDS and SSR induced DNA damage. The extent to which 5-MOP induced cross-linking is detected is unknown. It is premature to draw any conclusions from this work. No meaningful assessment can be made of the relative risk of induction of a tan by 5-MOP plus SSR compared to SSR alone.

Conclusions

In summary, consideration of the health risks of products containing 5-MOP gives rise to concern. 5-MOP itself plus solar simulated radiation (SSR) is clearly mutagenic, and produces an increased incidence of skin cancer compared by SSR alone. Data provided on the Bergasol products containing 5-MOP plus UV filters gives a less clear-cut picture, but does not allow concerns about this aspect to be completely discarded.

Concerning the benefits, 5-MOP is clearly efficacious at inducing a tan. The data provided on the protective effect of this tan on subsequent DNA damage induced by SSR, as measured by Unscheduled DNA Synthesis, are difficult to interpret, in view of the lack of knowledge regarding the mechanisms involved. It is not possible to draw any meaningful conclusions regarding the relative risk of inducing a tan by SSR plus 5-MOP plus sunscreens, as compared to SSR sunscreens alone.

Much of the information provided is of research interest, but was not believed to be directly relevant to the Bergasol products on the market. For example the studies of Khettab et al relating to the incorporation of antioxidants with sunscreens. Such combinations were shown to offer complete protection against the development of skin tumours in the 4 month initiation/promotion model in hairless mice.

The Committee were concerned about procedures whereby specific cosmetic products could be considered acceptable. In this instance it was argued that certain products were less harmful than would be expected from the 5-MOP content alone, because of the protective effect of other substances present eg sunscreens, but this would entail a need to consider not only the concentration of the component of concern (5-MOP) but also the presence of other specific compounds at specified levels. [This would, in effect, result in the need for clearance of specific products]. The question of balancing health risks versus health benefits had been raised. The Committee felt that it was questionable whether it was appropriate to consider such arguments for cosmetics; it is a requirement that cosmetic products are not harmful.

On the available data the Committee felt that it was not possible to conclude that sunscreen products containing 15-60 ppm or more 5-MOP were without risk of producing harm due to their potential for photomutagenic and photocarcinogenic effects. It was the opinion of the Committee that the maximum amount of 5-MOP in all sun tan products should be 1 ppm.

European Communities — Commission

EUR 14208 — Opinion of the Scientific Committee on Cosmetology (11/86 — 10/90) — Commission Decision 78/45/EEC of 19 December 1977 concerning the institution of a Scientific Committee on Cosmetology (OJ L 13, 17.1.1978, p. 24)

Luxembourg: Office for Official Publications of the European Communities

1993 — VI, 540 pp., num. tab., fig. — 21.0 × 29.7 cm

Toxicology series

ISBN 92-826-5394-3

Price (excluding VAT) in Luxembourg: ECU 56

The Scientific Committee on Cosmetology was set up by Commission Decision 78/45/EEC on 19 December 1977 (OJ L 13, 17.1.1978, p. 24) in order to provide the Commission with informed opinions on any scientific and technical problems arising in connection with cosmetic products, and in particular on the substances used in their manufacture, on their composition and on the conditions for their use.

The members of the Committee are independent scientists, highly qualified in the fields of medicine, toxicology, biology, chemistry or other similar disciplines.